Connection Games

Connection Games

Variations on a Theme

Cameron Browne

A K Peters
Wellesley, Massachusetts

Editorial, Sales, and Customer Service Office

A K Peters, Ltd.
888 Worcester Street, Suite 230
Wellesley, MA 02482
www.akpeters.com

Library of Congress Cataloging-in-Publication Data

Browne, Cameron, 1966–.
Connection games : variations on a theme / Cameron Browne.
p. cm.
Includes bibliographical references and index.
ISBN 1-56881-224-8
1. Connection games. I. Title

GV1469.C66B76 2004
794–dc22

2004057137

Printed in the United States of America
10 09 08 07 06 05 10 9 8 7 6 5 4 3 2 1

Dedicated to Merlin

For always being there… even if he
did keep dribbling on the manuscript.

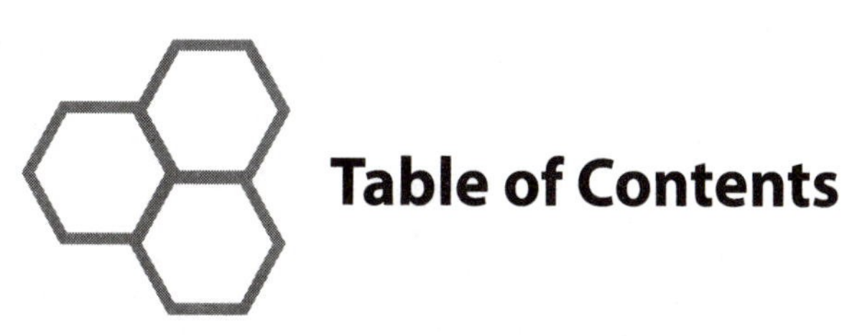

Table of Contents

Preface ix

Part One Defining Connection 13

Chapter 1 What Is a Connection Game? 15

Chapter 2 The Nature of Connection 19

Chapter 3 Games as Graphs 25

Chapter 4 Properties of Connection Games 37

Chapter 5 Common Plays 55

Chapter 6 Classification 71

Part Two The Games 77

Chapter 7 Pure Connection Games 79

Chapter 8 Connective Goal 263

Chapter 9 Connective Play 295

Chapter 10 Connection-Related Games 331

Part Three End Play 341

Chapter 11 Rolling Your Own 343

Chapter 12 The Psychology of Connection 353

Chapter 13 Afterword 363

Part Four Appendices 365

A Basic Graph Theory 367

B Solving the Shannon Game 371

C Hex, Ties, and Trivalency 375

D Strategy-Stealing 379

E Point-Pairing Strategies 381

F Y Reduction 385

G Sperner's Lemma 389

H Tessellations 393

References 397

List of Games 407

Index 411

Preface

Connection games are a recent genre of board games that take the basic idea of connection in a number of interesting directions. This aim of this book is to impose some structure on this increasingly large family of games, and to define exactly what constitutes a connection game.

This book contains many new and previously unpublished games; however, these have been screened for a reasonable level of quality. Those games catalogued in Part II have been found to play well or at least introduce some notable idea that may lead to greater things.

While it is obvious that some inclusions succeed more as games than others, value judgments of games have been avoided beyond pointing out those of exceptional quality. The amount of time spent on a particular game does not necessarily imply anything about its quality, but is more a function of its relevance to the concept of connection. Special emphasis is placed on new or previously unpublished games.

This is not an academic text. Although every effort has been made to ensure that the material is correct and complete, discussions are deliberately kept light and readable. Nor is this book a tutorial on how to play any particular game. Many games are examined in detail, but the emphasis is on exploring their fundamental nature, what distinguishes them from other games, and their place in the connection game family.

The book deals exclusively with board games. Computer games are not included unless they are manifestations of existing (or potential) physical board games. Boardless games may be included if they involve an implied board-like grid and can be played in a reasonable area, say a table top.

In many ways this is a collaborative work. It has been a joy to work with several people who share a passion for connection games, and who generously contributed ideas and material.

Note: The respective authors and publishers retain copyright to all of their games described in this book.

Overview

This book is organized into three main parts, followed by a set of appendices.

Part I covers introductory material including a general definition of connection. The groundwork is laid for a general study of the field, and a classification system is proposed.

Part II provides a catalogue of currently known connection games. Each branch of the classification is examined and key examples are explored in detail.

Part III summarizes the book's findings and touches on some of the more speculative aspects of connection games, including the art of game design and the psychology of connection.

The appendices include supplementary and technical material removed from the main text to improve readability.

Acknowledgments

Special thanks to the following people: Phil Bordelon for some excellent idea bouncing, editorial suggestions, and generous help and support; Richard Reilly for helping shape the classification framework; Frederic Maire for emphasizing the importance of graphs; Dan Troyka for suggesting patent searches; and my wife, Helen Gilbert, for encouragement. Also thanks to Michail Antonow, Jim Thompson, Scott Huddleston, Herbert Acree, Taral Guldahl Seierstad, Bill Taylor, João Neto, Jack van Rijswijck, Steven Meyers, Stephen Tavener, and Malcolm Hall from Blue Opal.

Richard Rognlie's play-by-email games server http://www.gamerz.net continues to provide a forum for testing and discussing new ideas, and many of the games described in this book can be played there. Thanks to Gamerz.net regulars Paul van Wamelen, Yves Rütschlé, and Karen Ridenour (the first to test any new game).

The BoardGameGeek web site http://www.boardgamegeek.com was an invaluable resource in researching obscure games.

Thanks to all proofreaders, especially Phil Bordelon and Helen Gilbert, for suggesting valuable improvements and corrections, and to those who generously contributed games, appendices, and other material.

Defining Connection

Connection games come in many different varieties, but all share a common theme: to develop or complete connections of some sort.

Part I looks at the history of connection games and introduces some concepts that help define exactly what constitutes a connection game. A basic graph analysis is introduced to help explain these games, and some common features, strategies, and tactics are discussed. Part I concludes by outlining a framework for classifying connection games, allowing each game to be placed within the context of the overall connection family.

What Is a Connection Game?

A connection game is a board game in which players vie to develop or complete a specific type of connection with their pieces. This might involve forming a path between two or more goals, completing a closed loop, or gathering all pieces together into a single connected group. In all cases, the size and shape of the connection do not matter; it is the fact of connection that counts.

Connection games are not necessarily abstract in nature; neither are they necessarily deterministic games of pure strategy. They may involve random elements such as dice, cards, or random tile draws, although the majority do not.

There has been to this point no universal consensus on what constitutes a connection game. It is something of a nebulous term used by players to describe the feel of a game, and is not always used consistently. The problem is that *all* board games (excluding perhaps some pure dice games) involve at least some degree of connection, be it as fundamental as the adjacency of cells on the board. One of the aims of this book is to provide a consistent definition.

There has been a recent surge of interest in the study and development of connection games. A classification scheme will be presented in an effort to impose some structure on this blossoming family of games, and develop a fundamental understanding of them.

1.1 History

The birth of connection games is generally equated with the invention of Hex in the early 1940s by Danish mathematician Piet Hein, and its subsequent reinvention by Nobel laureate John Nash several years later. Hex has attracted a dedicated following over the years and is still regarded as *the* classic connection game.

However, Jim Polczynski [2001] describes a connection game called Lightning, a historical curiosity dating back to the 1890s. This makes Lightning the earliest connection game by half a century, though it seems to have all but vanished from public knowledge and only recently emerged after more than one hundred years. Lightning belongs to a different connective category than Hex, and it is unlikely that it would have influenced the invention of Hex (or most other connection games) in any way.

Next came Zig-Zag in 1932, which introduced the concept of players establishing chains of pieces between opposite sides of the board. This game, too, seems to have been forgotten, and there is no evidence of its having inspired the invention of Hex a decade later.

The connective principle of Hex was further developed in the 1950s by Claude Shannon, Craige Schensted, and Charles Titus with the invention of Y. Y is arguably the most fundamental of all connection games, combining the clarity of Hex with the simplicity of identical goals for both players. Most subsequent connection games are indebted to one of these two great games.

The genre of connection games was first introduced to the general public through Martin Gardner's *Scientific American* article "Concerning the Game of Hex, which May be Played on the Tiles of the Bathroom Floor" published in 1957. Although Hex never became a commercial hit, a number of popular connection games were released shortly afterwards. These include Bridg-It in the 1960s, and Twixt and Thoughtwave in the 1970s.

During this time Schensted continued his investigations into connection and produced two notable games: Poly-Y and Star. These are both serious games for the dedicated player, which demonstrate the depth that connection games allow. More recent titles to have achieved some commercial fame include Trax and Havannah in the 1980s. Most current abstract board game players would be familiar with at least some of these games.

Over the years connective elements have seeped into many types of games: war games, rail network games, empire building games, and so on. The connective principle has outgrown the abstract strategy mould to be incorporated into many themed games. In fact, the concept of connection-based battle and capture has been around for thousands of years since the invention of Go.

Connection games appear to have finally earned some recognition as a distinct genre. Schmittberger ran an article titled "Making Connections" in *GAMES* magazine [2000], and Parlett's recent *Oxford History of Board Games* [1999] dedicated a section to them.

Schmittberger describes connection games as "the 20th century's greatest contribution to strategy games" [2000]. This claim is supported by the recent proliferation of connection games, and the increasing inclusion of connective elements in many board games of different varieties.

Figure 1.1 shows a timeline of the invention of Pure Connection games. It is evident that interest in connection games has developed some momentum in recent years, and that this group now represents a significant body of games. Many of these are high-quality games, but only time will tell which will become classics.

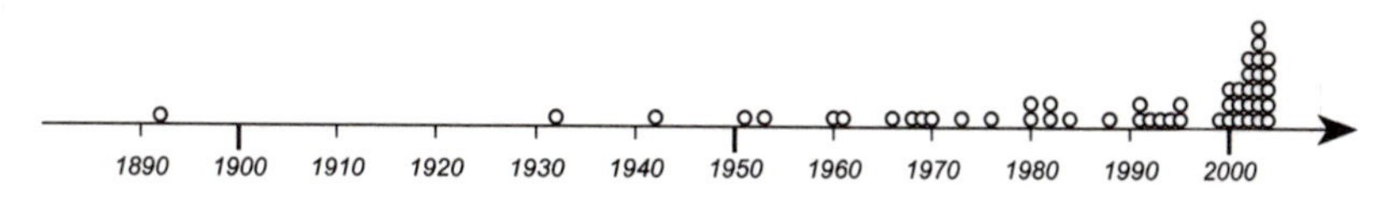

Figure 1.1. Timeline of Pure Connection games.

The Nature of Connection

Most board games feature at least some aspect of connection. This could be as fundamental as the adjacency of squares on a Chess board, or the fact that a winning pattern in Tic-Tac-Toe forms a connected line. To distinguish those games that are truly connection games, it is necessary to define precisely what "connection" means in this context.

2.1 Adjacency

The cells of a game board are generally *adjacent* to those cells with which they share an edge.

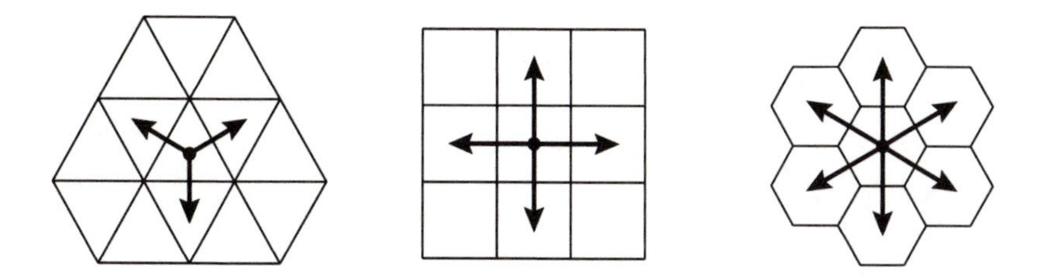

Figure 2.1. Adjacencies on the triangular, square, and hexagonal grids.

Notice that cells on the triangular and square grids have some neighbors that share an edge (*direct* neighbors) and some neighbors that share a corner only (*indirect* neighbors). Indirect neighbors can be bad news for connection games, as will be demonstrated.

Some games are played on grid intersections rather than cell interiors. In addition, the rules of some games specify connective adjacencies between cells or board points that are not physical neighbors; for instance, points in Twixt are not connected to their immediate neighbors but to those a knight's move away. However, it will be assumed that cells are adjacent to neighboring cells with which they share an edge, unless otherwise stated.

2.2 Connection

A *chain* is a set of same-colored pieces such that each member of the set can be reached from every other by a sequence of adjacent steps through the set. A single piece is the simplest possible chain. The pieces comprising a chain are *connected* and paths formed by connected sequences are *connections*. These terms are defined more rigorously in Appendix A, Basic Graph Theory.

Winning conditions in connection games typically include forming either

- a path between two or more goals,
- a cycle (a closed path that loops back on itself), or
- a single chain containing all pieces.

Barring a minimum length required to span the board, the size and shape of the connection does not matter; it is the mere fact of connection that counts. Kerry Handscomb, editor of *Abstract Games* magazine, describes this property as a form of *topological invariance* between such arrangements of pieces [2000].

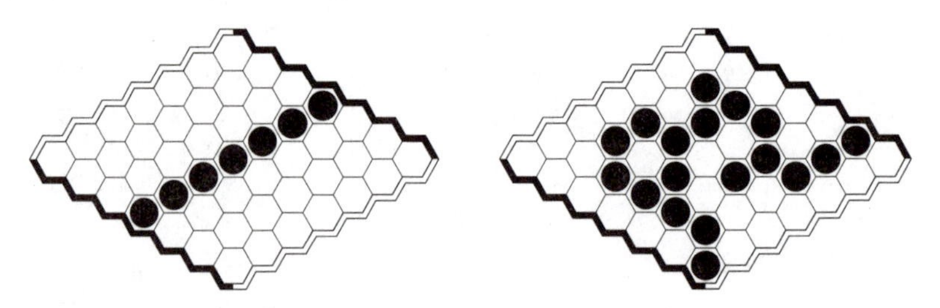

Figure 2.2. Two games of Hex with equally good connections for Black.

For instance, Figure 2.2 shows two games of Hex won by Black, who has completed a chain of black pieces between the black sides of the board. The two connections are equivalent in the sense that they both win for Black, even though they are of different size and shape.

2.3 Pattern

Consider the game of Tic-Tac-Toe. The winning white pattern shown in Figure 2.3 (left) connects two opposite sides of the board. In fact, if diagonal cells are considered adjacent then all possible winning patterns connect opposite sides of the board (Figure 2.3, center).

However, this is little more than coincidence, and it is possible to set up a board position that connects opposite sides of the board without either player winning (Figure 2.3, right). It is the pattern's size and shape that is important here (three pieces in a line), and the fact of connection is an irrelevant by-product of its formation.

Tic-Tac-Toe and any other game involving patterns of specific size and/or shape are not strictly connection games. It is worth noting that connected rings of pieces, although satisfying some geometric constraints, can come in many sizes and shapes and do not count as patterns in this sense.

2.4 Connection Quality

Some games are decided by the metric properties of connection or some by-product of connection. For instance, the winner of a game of Go is determined by the amount of territory enclosed by connected sets of pieces.

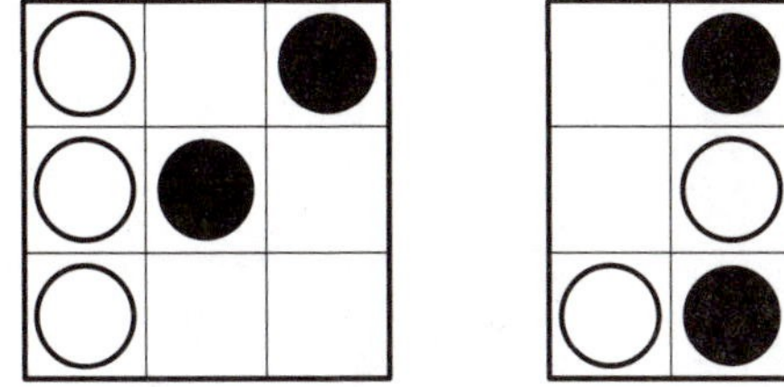
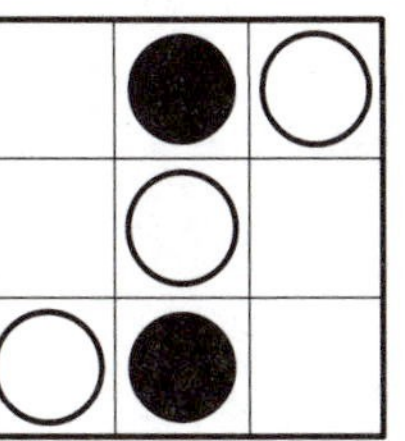
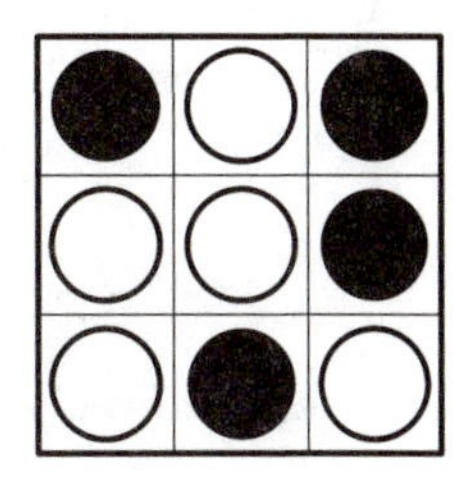

Figure 2.3. Tic-Tac-Toe is not a connection game.

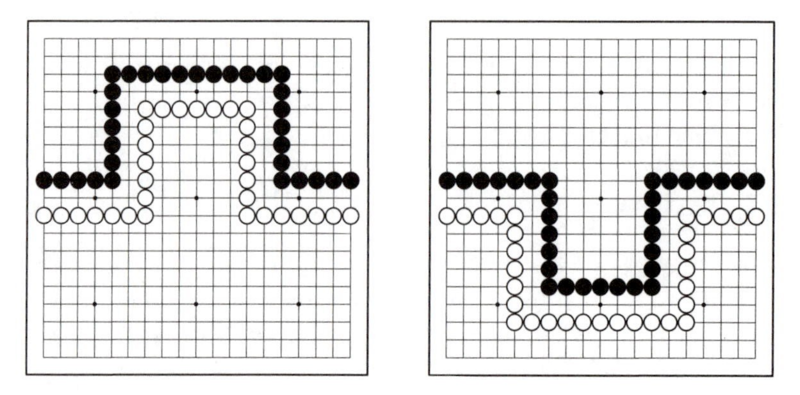

Figure 2.4. Similar paths can yield significantly different territories in Go.

Figure 2.4 shows two games of Go in progress. Each player's pieces form similar chains in both games: they are of similar length and have the same start and end points on the left and right sides. However, the shapes of the connections are critical here, giving a territorial advantage for White on the left board and a territorial advantage for Black on the right board.

Target connections in such territorial games do not have any specific size or shape, but their results are based on a measurable by-product of the connection shape. The goals of these games are therefore not strictly connection-based.

Note, however, that even in strictly connective games there is usually a minimum path length implied by board geometry. That is, a winning path must be at least as long as the board is wide, a ring on a hexagonal board requires at least six stones, and so on.

There is a subtle difference between games decided by connection length and games such as Trax, in which players win by creating a path that spans at least eight cells in either direction. This does not impose undue constraints on the connection, as shown in Figure 2.5, where Black's two paths are of significantly different size and shape yet both win the game.

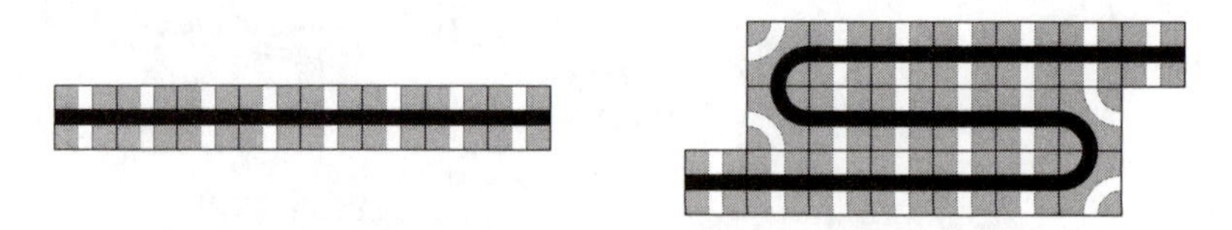

Figure 2.5. Dissimilar Trax paths that both win for Black.

Another way to explain this difference is to picture an imaginary board eight cells wide and eight cells high imposed on the Trax game in progress. The winning condition is then similar to that of many other Pure Connection games, such as Hex, where the aim is to connect opposite sides of the board. The difference here is that Trax is a boardless, tile-based game, hence the boundaries are not fixed in any particular place.

Games as Graphs

Connection games are tightly coupled with the topology of the board surface on which they are played. Much can be learned about a game by stripping it down to its underlying graph.

Figure 3.1 shows the symbols used for this analysis. Points at which pieces can be placed are described by *vertices*, which are generally neutral until owned by either player. Connections between adjacent vertices are described by *edges*. Similarly, edges are neutral until claimed by either player. More detailed definitions can be found in Appendix A, Basic Graph Theory.

3.1 Duals

A game board can be redefined as a graph using its *dual*. This is done by creating a neutral vertex for each board point, then connecting adjacent points with neutral edges.

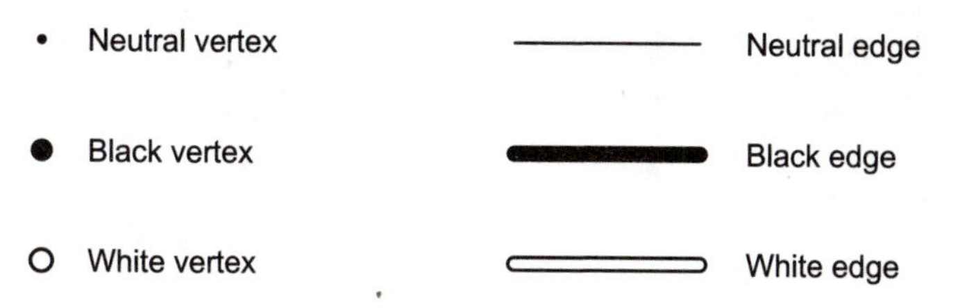

Figure 3.1. Key to the symbols used for graph analysis of games.

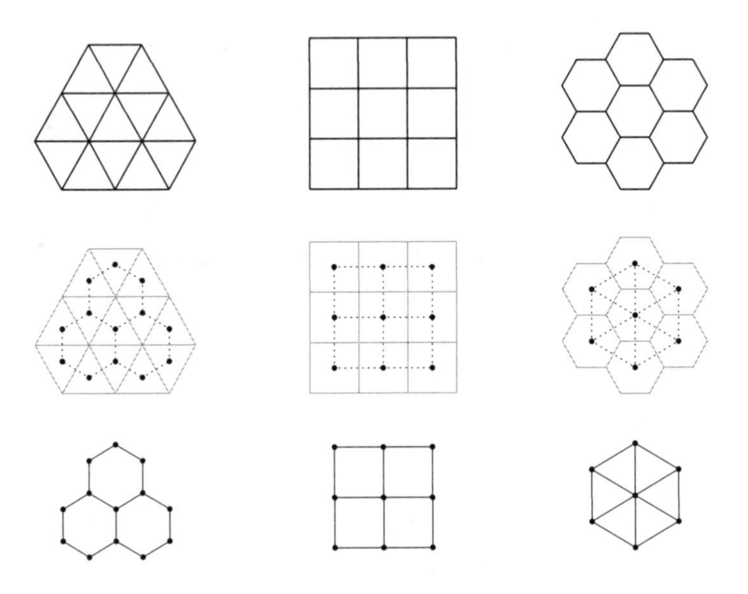

Figure 3.2. Some common grids and the construction of their duals.

This process is illustrated in Figure 3.2, which shows the duals of the three regular tessellations. The dual of the square grid is itself square, while the hexagonal and triangular grids are duals of each other. Notice that each edge of the dual graph crosses exactly one edge in the original graph. The dual representation of a game board showing the adjacency of board points is described as that game's *adjacency graph*.

For instance, Figure 3.3 shows the adjacency graph of a small Twixt board. Note that in this game bridges connect pieces a knight's move apart; physical neighbors are not necessarily adjacent in a connective sense.

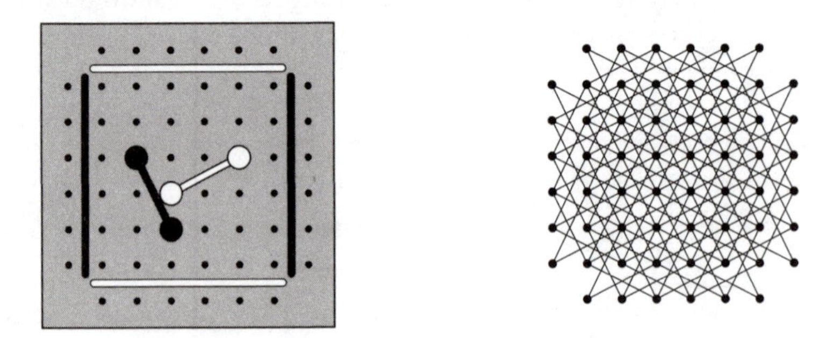

Figure 3.3. The adjacency graph of a small Twixt board.

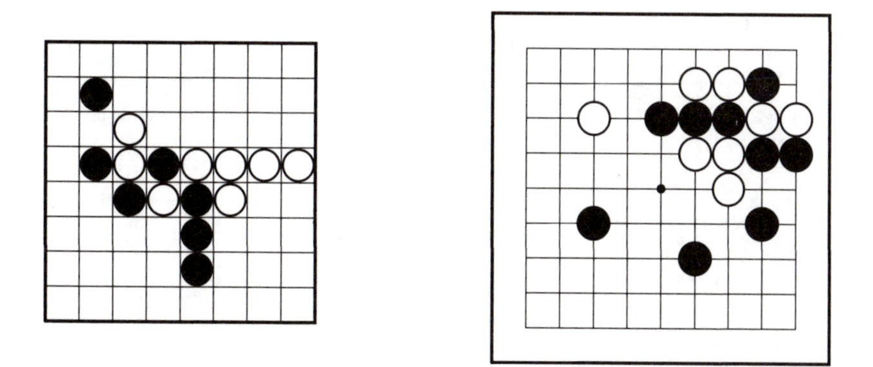

Figure 3.4. Lines of Action is played on the cells of a square board, while Go is played on the intersections (its dual).

Some games are already played on their adjacency graph. For instance, Go is played on points of intersection (Figure 3.4, right) that correspond directly to vertices in the adjacency graph. In such cases, the adjacency graph is taken directly from the board intersections, not the board cells.

3.2 The Shannon Switching Game

The *Shannon switching game*, or simply *Shannon game*, is an abstract two-person game played on a graph [Shannon 1955]. The graph is initially connected and all edges are in a neutral state. One player, *Join*, aims to permanently connect two distinguished vertices with a path of colored edges, while the other player, *Cut*, aims to permanently disconnect these vertices by cutting the graph.

Bridg-It, a board game from the 1960s, is the classic example of a Shannon game. Figure 3.5 (left) shows the interspersed grids of black and white pegs that make up a Bridg-It board. Players take turns placing a bridge between two orthogonally adjacent pegs of their color, in an effort to connect their two sides with a path of bridges.

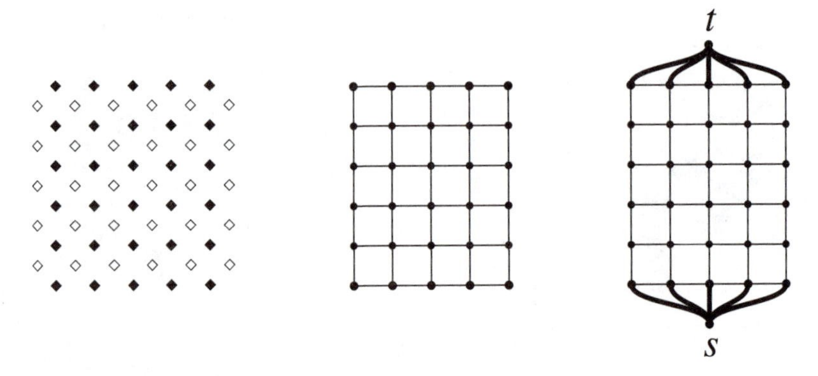

Figure 3.5. The Bridg-It board, its adjacency graph, and its game graph.

Figure 3.5 (middle) shows the adjacency graph corresponding to this board. Note that only black adjacencies are shown, because a white move implies the removal of the black connection that it crosses. Black adopts the role of Join and White adopts the role of Cut in this example.

Figure 3.5 (right) shows the adjacency graph with two distinguished terminal vertices s and t added. Each terminal is fully connected to those vertices along a Join side. The adjacency graph with terminal vertices added will be called the *game graph* throughout this book. This term is also used in some literature to describe a positional decomposition of a game; such structures will be called *game trees* to avoid confusion.

There exist two forms of the Shannon game, which are described as the *Shannon game on the edges* and the *Shannon game on the vertices*. Bridg-It is a Shannon game on the edges.

3.2.1 On the Edges

In the Shannon game on the edges, players take turns making the following moves:

- Join colors one neutral edge, and
- Cut deletes one neutral edge.

Figure 3.6 shows a game won by Black (Join) who has completed a path of bridges connecting the top and bottom board sides. It can be seen in the game graph that terminal vertices s and t have been connected.

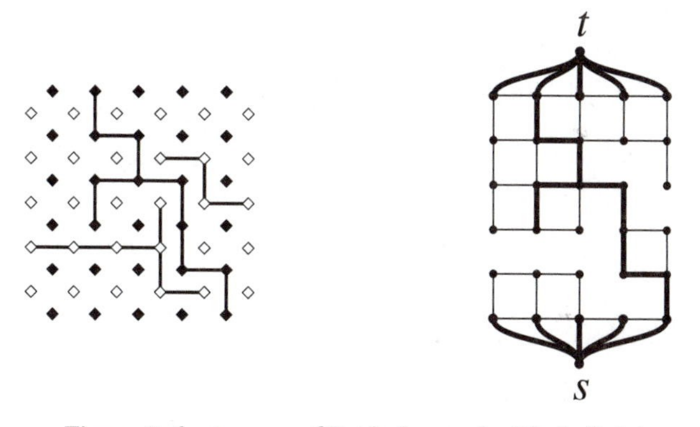

Figure 3.6. A game of Bridg-It won by Black (Join).

Figure 3.7 shows a game won by White (Cut) who has completed a path of white bridges connecting the left and right board sides. It can be seen in the game graph that this path of white bridges has cut the graph in two, permanently disconnecting terminal vertices *s* and *t*.

The Shannon game on the edges has a known solution (see Appendix B). The presence of a solution will turn some players off a game, but will intrigue another class of player who sees each game as a puzzle to be solved. In fact, Elwyn Berlekamp considers the Shannon game on the edges to be the most interesting game covered in this book, because it has an elegant solution yet still provides a worthwhile contest between players who do not know this solution.

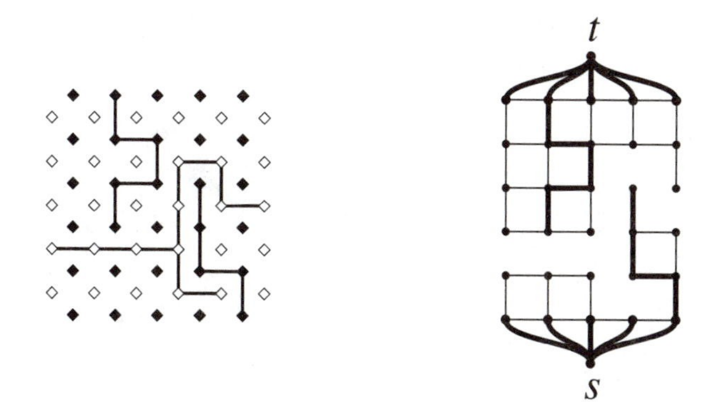

Figure 3.7. A game of Bridg-It won by White (Cut).

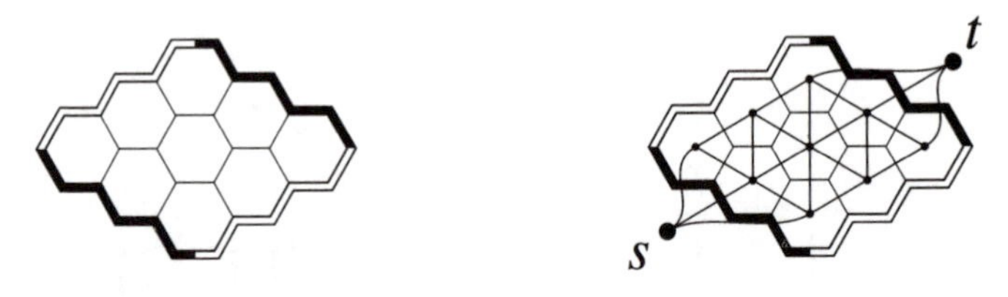

Figure 3.8. A 3 x 3 Hex board and its game graph.

3.2.2 On the Vertices

The Shannon game on the vertices is identical to the Shannon game on the edges except that at each turn,

- Join colors one neutral vertex and all incident neutral edges leading to another colored vertex, and
- Cut deletes one neutral vertex and all incident edges.

Hex is a Shannon game on the vertices. Figure 3.8 shows a 3 x 3 Hex board (left) overlaid with its game graph (right).

Again, Black adopts the role of Join and aims to permanently connect *s* and *t* with a chain of colored edges, while White adopts the role of Cut and aims to disconnect *s* from *t* by cutting the graph. Figure 3.9 shows the sequence of moves in a game won by Black (Join) and the modified game graph after each move.

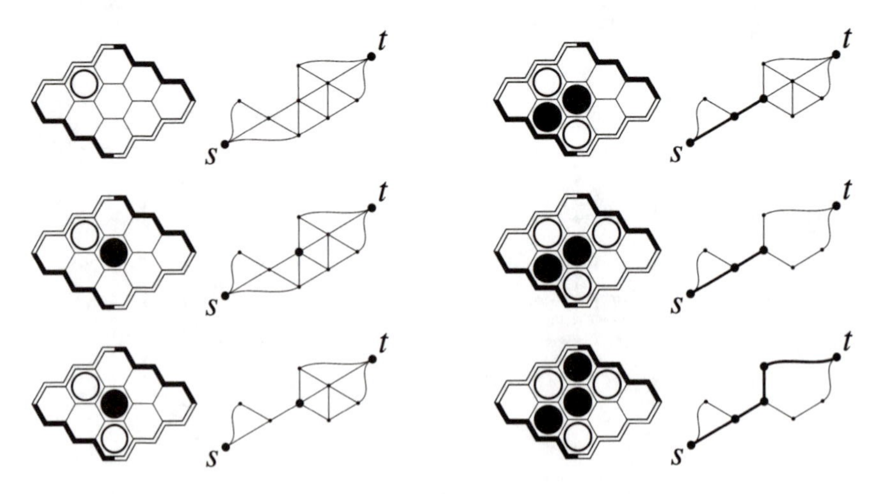

Figure 3.9. A 3 x 3 game of Hex won by Black (Join).

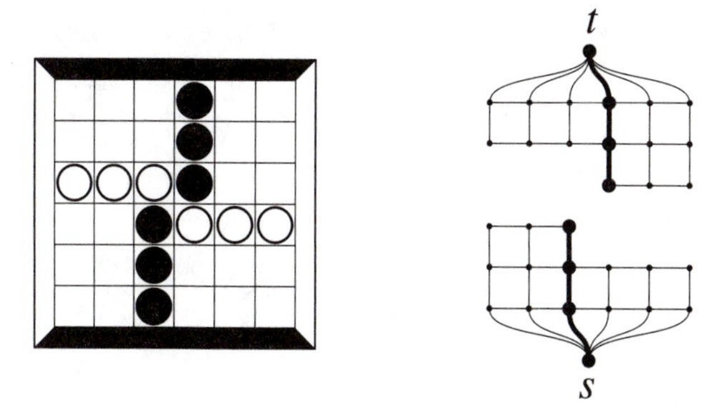

Figure 3.10. A game of Square at equilibrium ... or is it?

3.3 Deadlocks

It turns out that most connection games can be naturally described as a battle between Cut and Join. However, the serious problem of *deadlocks* must first be addressed. A deadlock occurs when opposing chains meet at a point but neither player can connect across that point.

Figure 3.10 shows a hypothetical game called Square, played by the same rules as Hex but on a square grid. The board position shown is deadlocked and can no longer be won by either player, however, its game graph (right) incorrectly suggests that White has achieved a cut and therefore won the game.

The problem can be traced to the connectivity of the underlying square grid. Consider the white piece ***a*** on the hexagonal grid shown in Figure 3.11 (left). The two black neighbors ***b*** and ***c*** are in consecutive clockwise order around this piece and are themselves adjacent, hence Black has a connection through ***b*** and ***c***.

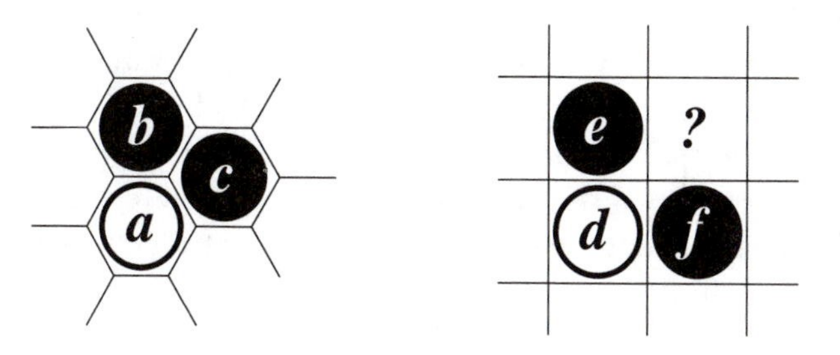

Figure 3.11. A white piece with two consecutive black neighbors on the hexagonal and square grids.

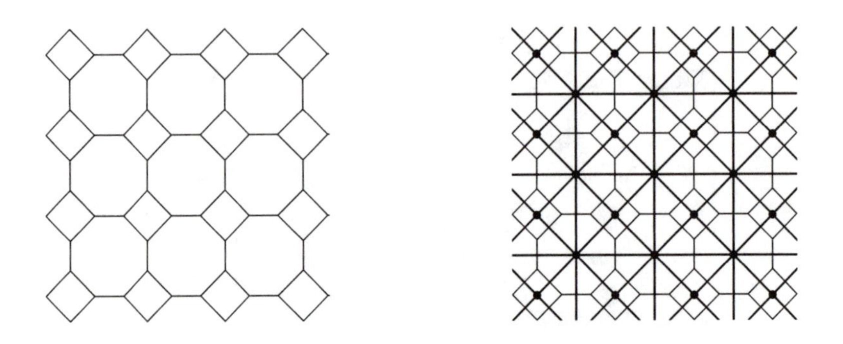

Figure 3.12. The 4.8.8 grid is not subject to deadlocks, as shown by its game graph.

On the square grid, however, consecutive neighbors around the white piece ***d*** are not necessarily themselves adjacent (Figure 3.11, right). If White plays at the critical point marked *?*, then a local deadlock is reached.

Games played on a planar graph can be deadlocked if the consecutive neighbors around each chain are not themselves adjacent. Luckily such potential deadlocks are easily spotted; any region of the game graph that is not *trivalent* (three-sided) is subject to deadlock. The reasons for this are explained in Appendix C, Hex, Ties, and Trivalency.

Looking back at Figure 3.2, it can be seen that the hexagonal grid is the only regular tiling guaranteed to avoid deadlocks, hence it is a popular choice for connection games.

The 4.8.8 tiling shown in Figure 3.12 is another tessellation that avoids the problem of deadlock. This tiling is used to good effect in games such as Quax and Stymie. See Appendix H for a further discussion of tessellations.

Figure 3.13 shows how the game graph can be modified to accommodate games that may be deadlocked. Note that players now have their own pair of terminals, and that White now claims edges and vertices rather than cutting them. The elegant Cut/Join simplicity of deadlock-free games has been lost; games of this type are Join/Join in nature instead.

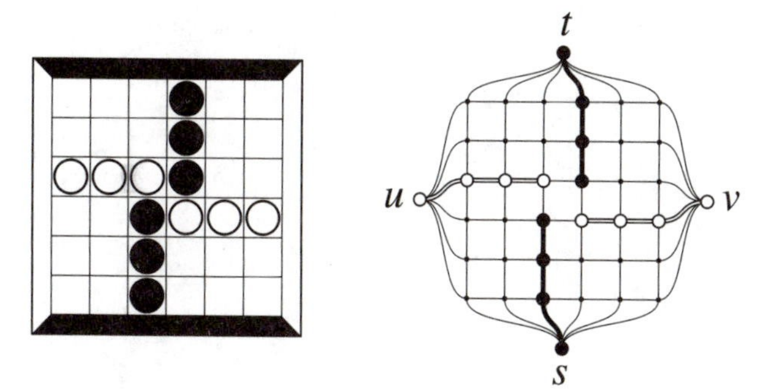

Figure 3.13. The game of Square correctly at equilibrium.

Connection games played on nontrivalent grids must involve special mechanisms to stop deadlocks spoiling the game. For instance, Gonnect allows capture to avoid global deadlocks, Trellis provides directional diagonal connections, Akron allows pieces to stack up to climb over blocking connections, and so on. Lynx provides an especially interesting solution to deadlocks on the square grid.

3.4 Example Graph Analysis

The following example shows how graph analysis can provide some insight into a given connection game. Figure 3.14 shows a game of Hex played on a map of the contiguous United States, as suggested by David Book [1998]. This idea is also included in an educational game suggested by Scott [1938].

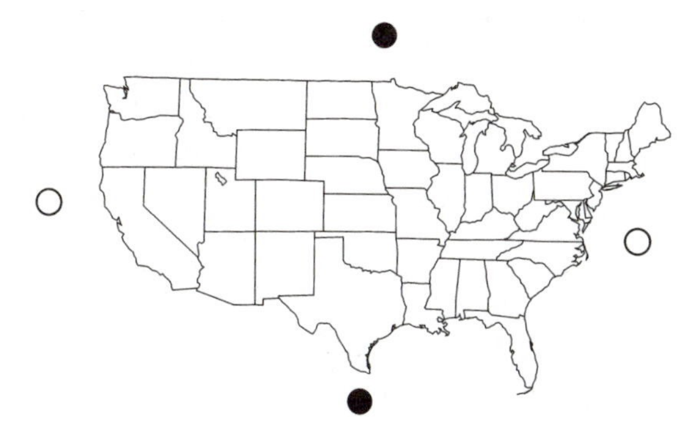

Figure 3.14. Hex played on the map of the contiguous United States.

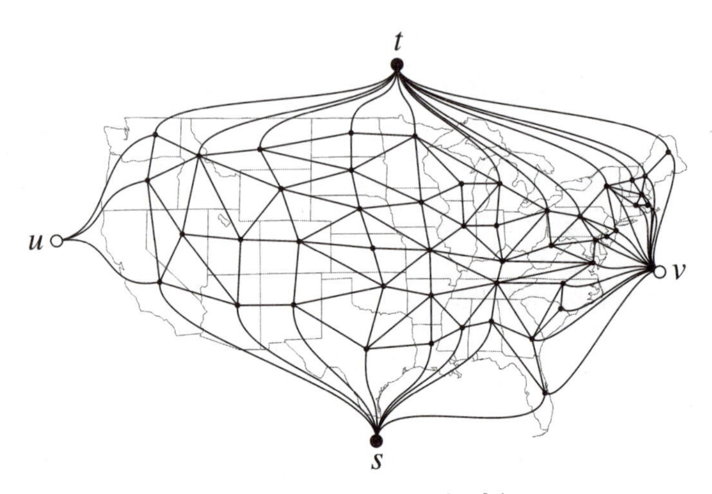

Figure 3.15. The game graph of the map.

Players take turns coloring a neutral state of their choice. Black aims to complete a black path between Canada and Mexico (including those states along the Gulf of Mexico), while White aims to complete a white path between the Pacific and Atlantic Oceans. As usual, adjacency is defined by borders shared between regions. The game graph of this map is shown in Figure 3.15.

Figure 3.16 shows the same game graph in a neater format. Terminals are shown for both players in the Join/Join style as potential deadlocks

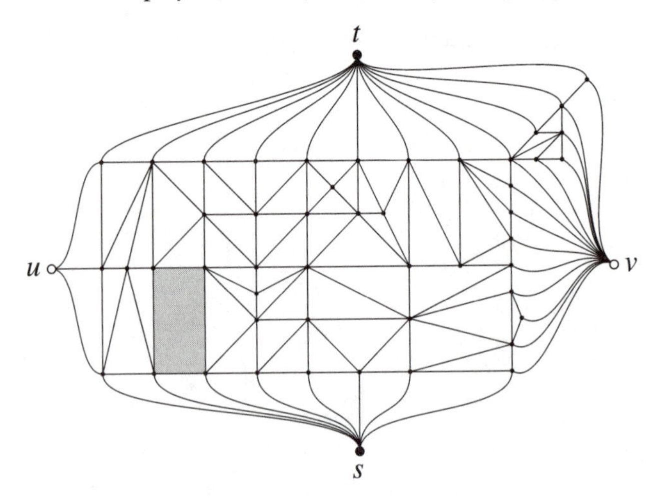

Figure 3.16. The game graph tidied up, with the potential deadlock region shaded.

may occur due to a nontrivalent region (shaded) that corresponds to the point known as the Four Corners where four states meet.

Despite the possibility of deadlock, Black can always win on this map no matter who moves first. The shortest path between Black's terminals requires only three moves, whereas the shortest path between White's terminals requires seven moves, putting White at an obvious disadvantage.

In addition, note that the six states in the top right corner are connected to the rest of the map through a single state (New York) that is adjacent to both t and v. This cluster is superfluous to the game and playing in any of these six states would be a wasted move.

This example demonstrates how the game graph can yield some useful insight into the underlying board design: whether the game can be tied, whether one player has a significant advantage, weaker regions to be avoided, and so on.

4 Properties of Connection Games

Connection games tend to share some general properties, which typically relate to equipment, rules of play, or winning conditions.

4.1 Elements of a Game

While all connection games are by definition variations on the basic connection theme, most can be traced back to a close relative from which they were derived.

The conceptual elements that make up a game may be described as *ludemes*. This term, attributed to Pierre Berloquin and used in passing by Parlett [1999], refers to game *memes*, or units of information that replicate from one person to another [Dawkins 1976]. Each individual game may be described as a unique combination of ludemes.

Equipment-related ludemes include the board's size, shape, type of tiling used, and special types of adjacency. For example, Figure 4.1 shows three board shapes commonly used with hexagonal tilings, each with approximately the same number of cells but different properties. The right-most design, a hexagon tiled with hexagons called the *hex hex board*, is of particular interest. The absence of acute corners means that edge cells are more uniformly distant from the center of the board, resulting in a more even distribution of power over all cells.

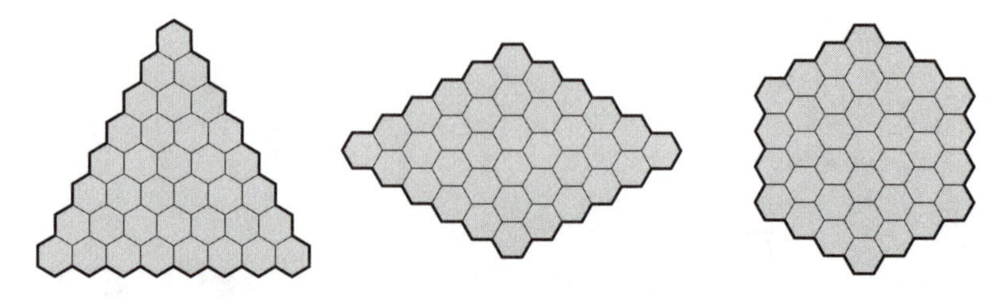

Figure 4.1. Triangular, rhomboid, and hex hex boards.

The game Trinidad (described in Section 8.1) introduces the idea of dividing the perimeter of the hex hex board into four equal regions, a ludeme that facilitates the migration of some games to this aesthetically appealing board shape. Equipment-related ludemes also include the properties of a game's pieces, whether plain stones or directional tiles are used, and so on.

In relation to the rules that make up a game, a *metarule* is defined as a rule or procedure that modifies an existing game in some fundamental and consistent way. Metarules generally apply to a game's winning condition or rules governing play and are similar to the *mutators* described by João Neto [2000].

Metarules governing play are many and varied, and can be as simple as whether players may pass on their turn or not. The no-pass metarule may sound trivial, but in fact can have a profound effect on a game, as demonstrated in the Gonnect example shown in Figure 4.2.

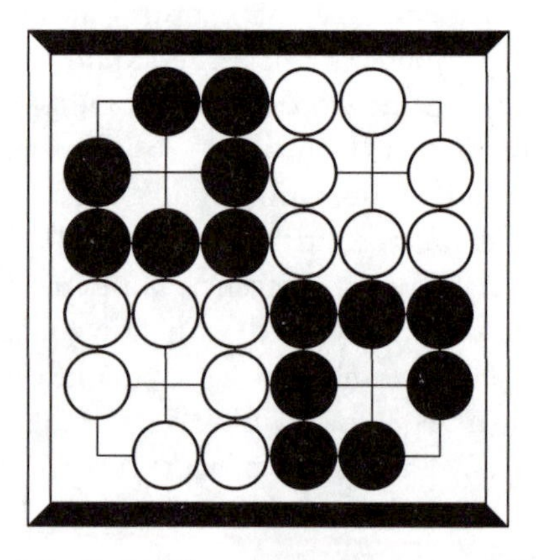

Figure 4.2. Forbidding players to pass saves Gonnect.

Gonnect combines the rules of Go with a connective goal metarule; rather than surrounding territory as in Go, players must connect opposite sides of the board. Figure 4.2 shows a temporary deadlock in which both players would rather pass than fill in one of their life-giving eyes. Adding the no-pass metarule, however, means that the next player to move *must* fill in an eye, losing that group and losing the game. This simple metarule saves Gonnect.

Games usually involve metarules of different types. For example, the connect-opposite-sides goal of Gonnect is an example of a winning condition metarule. A more fundamental example applicable to almost any board game is the *misère* metarule, which inverts a game's winning condition. For example, in Misère Hex the player to connect his sides of the board with a chain of his pieces *loses*.

Metarules may themselves be composed of compound sets of subrules, such as the Caeth metarule which may be applied to almost any connection game with a planar adjacency graph. Caeth players take turns claiming an edge of the graph; vertices are claimed by the player who claims at least half of the incident edges.

Caeth adopts its winning condition directly from a parent game or *overgame*. For instance, Figure 4.3 shows a game of Caeth Y won by Black and its standard Y equivalent. Similarly, Conhex and the related Noc are metarule systems that subsume parent games.

One of the more interesting metarules to emerge recently has been the either-color metarule, as found in Jade and Chameleon, in which players choose which color to play each turn and may win with either color. This

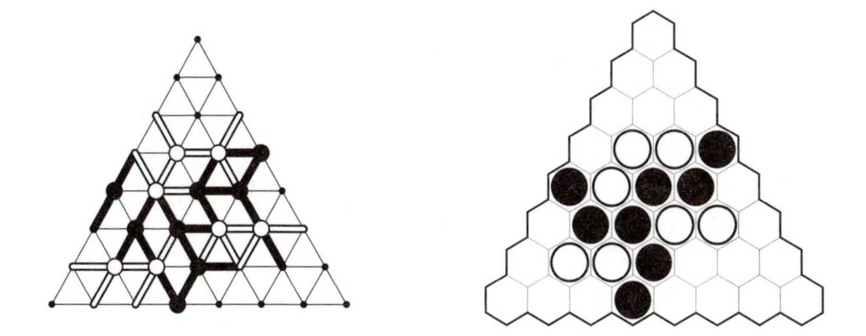

Figure 4.3. A completed game of Caeth Y and its standard Y equivalent.

introduces subtleties into play, as any piece may be used in the opponent's connection.

The game of Nex introduces an interesting neutral-piece metarule that may have broad applicability to abstract games of all types. Each turn, the current player may either place a piece of his color plus a neutral piece, or claim two neutral pieces and revert one of his colored pieces to neutral.

The Quax metarule, in which players may place bridges between diagonal pieces, saves games on some grids from deadlock, and therefore allows some hexagonally based connection games to be successfully ported to the square grid.

Metarules may be mixed and matched to produce some novel and wonderful new games. For instance, Phil Bordelon suggests the disturbing possibility of Misère Chameleon Nex Caeth Projex Det.

4.2 Board Design

As board geometry is a central element in most connection games, any singularities in board design can have profound effects upon the game. Players should take advantage of powerful cells and stay away from weaker cells. Powerful cells are those with greater connective potential or at which connective flow converges. The central point is usually the strongest point on the board, as shown in Figure 4.4, where the black pieces are optimally placed between each player's goals.

Points along diagonals equidistant from two edges of different color (indicated by white pieces) are usually stronger than the nondiagonal points nearby. However, diagonals get weaker the farther they are from the

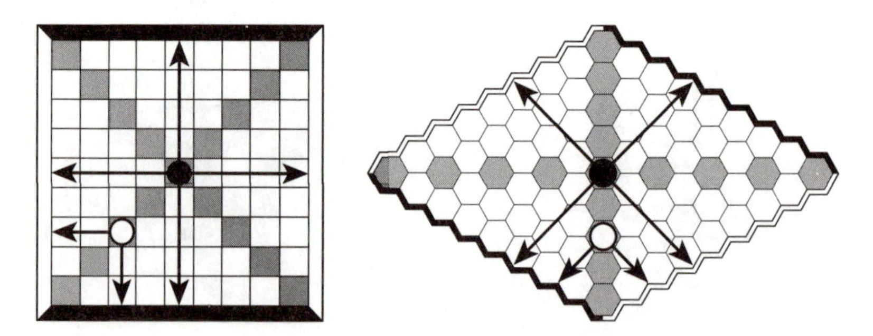

Figure 4.4. The board center is a strong point.

center, like any other board point. Acute corners are the weakest points on most boards, being the farthest from the center and tending to have the fewest number of neighbors.

4.2.1 Scale

Most connection games tend to scale up well. For instance, Hex can be played on a 20 x 20 board just as well as on a 10 x 10 board because the connections involved are independent of size.

However, connection games do not tend to scale down so well. Larger boards lead to more involved games with deeper strategy, while smaller boards degenerate into limited tactical battles. Smaller boards offer fewer lines of play, are more readily analyzed, and tend to be overshadowed by combinatorial edge tactics that reward rote learning of book positions rather than quick thinking over the board.

Larger games are generally more satisfying, but can take much longer to complete. Players will generally find a board size that balances their skill and depth of interest in a game with their patience to play it.

4.2.2 Corner Effects

One problem with the hexagonal grid is the presence of acute corners on some game boards, creating weak points far removed from the center. The hex hex board shown in Figure 4.1 is an aesthetically pleasing design that reduces this problem.

A more creative solution is to use a nonregular playing grid to achieve a regular board shape, although this usually introduces different flaws into the board geometry. Figure 4.5 shows a standard Y board (left) and Schensted and Titus's modified Y board (right).

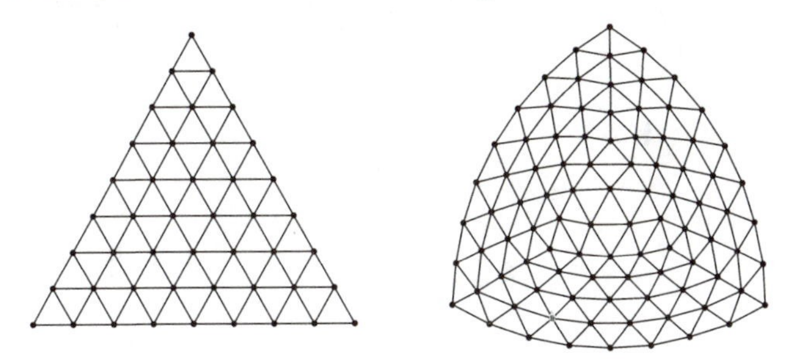

Figure 4.5. Standard and modified Y boards.

Note that three of the vertices towards the center of the modified board are 5-connected rather than 6-connected and the surrounding graph reshaped to fit. Not only is the modified board visually rounder, but the path distance from the center to the farthest edge point (acute corner) is shorter than on the standard Y board. In addition, the modified acute corners are 3-connected rather than 2-connected as per the standard board, making connectivity even more homogenous across the board.

The price for this elegant design is that weak points have been introduced. When playing on the modified board it is generally wise to maximize connective potential by avoiding the 5-connected points if possible.

4.2.3 Surface Types

A planar graph is one in which no edges cross (see Appendix A, Basic Graph Theory). A connection game with a planar game graph can be drawn as a single map on which all connected components are adjacent.

Figure 4.6 shows a cylinder (left) pinched at one end to form a punctured hemisphere (middle), and pinched at both ends to form a punctured sphere (right). These surfaces are described as punctured because there is still a hole at each end, even though those ends are constricted to a point. Games played on all of these shapes can be represented by planar game graphs.

A *projection* is the transformation of points and lines in one plane onto another plane by connecting corresponding points on the two planes with parallel lines [Weisstein 1999]. *Projective connection games* can be defined based on the principles of projective geometry as those games whose surfaces contain at least one connection between nonadjacent cells. The only known projective connection games are ZeN and Projex.

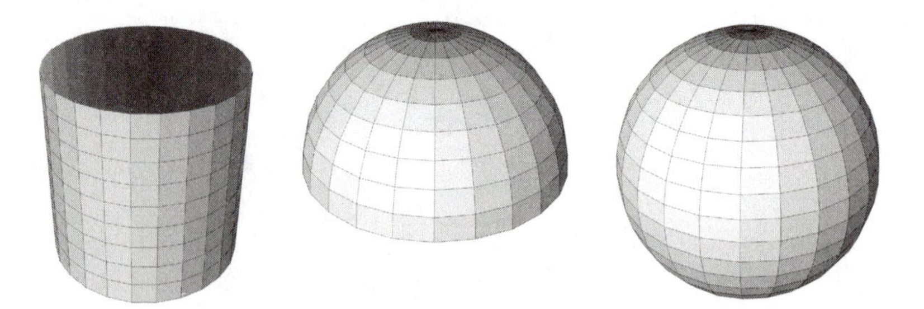

Figure 4.6. A cylinder pinched at one end forms a punctured hemisphere, and pinched at both ends forms a punctured sphere.

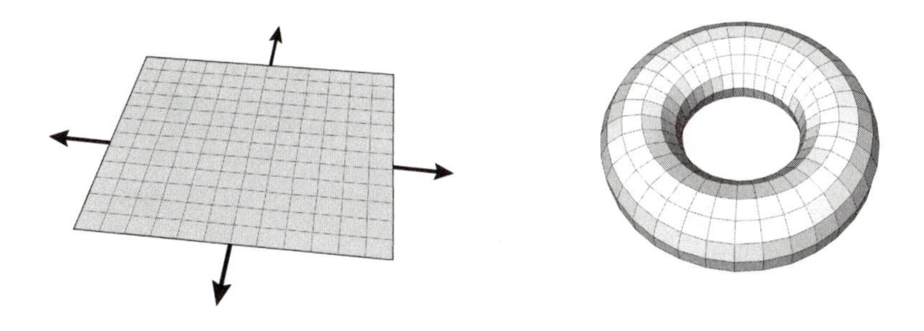

Figure 4.7. The projective plane and a torus.

ZeN is played on a torus formed by wrapping the edges of a projective plane in two directions (Figure 4.7). Unlike the cylindrical board mapping, the toroidal board surface cannot be represented by a planar game graph. Projex takes place on a hexagonally tiled projective plane which cross-wraps in three directions, on which players strive to form a global loop of their color.

Table 4.1 summarizes the various surface types upon which connection games are played, and some of the games played on them. Almost all connection games belong to the Plane category; each of the other categories have only a single known member.

A more complicated surface called the *Klein bottle* (Figure 4.8, left) might make a promising playing board except that it cannot be physically realized in three-dimensional space [Weisstein 1999]. The Klein bottle can be entirely tiled by hexagons, a very attractive property for a connection game. It can be formed from two *Mobius strips* (Figure 4.8, right) twisted in opposite directions and joined at their edge. Similarly, a projective plane can be made by joining a Mobius strip to a disk.

Planar Game Graphs	Nonplanar Game Graphs
Plane (Hex, Y, Bridg-It, etc.)	Torus (ZeN)
Cylinder (Cylindrical Hex)	Projective plane (Projex)
Hemisphere (Escape Hex)	
Sphere (Antipod)	

Table 4.1. Summary of board surface types and the games played on them.

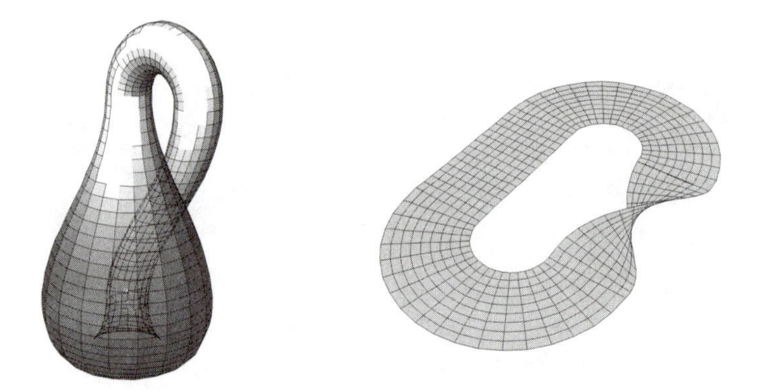

Figure 4.8. The Klein bottle and the Mobius strip.

A mathematician named Klein
Said "The Mobius strip is divine."
He observed, "If you glue
Up the edges of two,
Then you get a strange bottle like mine."
—Unknown author

The Klein bottle and Mobius strip have no inside or outside, hence choosing a cell on either of these surfaces would color that cell on both apparent sides, raising some interesting possibilities for play.

Many boardless tile-laying games such as Trax and Andantino are effectively played on an infinite or unbounded surface. Artificial bounds can be imposed, such as the 8 x 8 floating window used in Trax to detect lines of length eight. This constitutes a floating-window metarule that may be applied to many connection games. Such games would be played on an infinite board, and won by the player to achieve the required condition on any subsection of the board. Misère versions of such games are not practical as players could exploit the unbounded nature of the board to play infinitely far away from other pieces each turn; hence the losing (winning) condition will never be met.

4.3 Rules of Play

Rules of play govern the interaction between pieces on the board, and shape the stages of play as the game progresses.

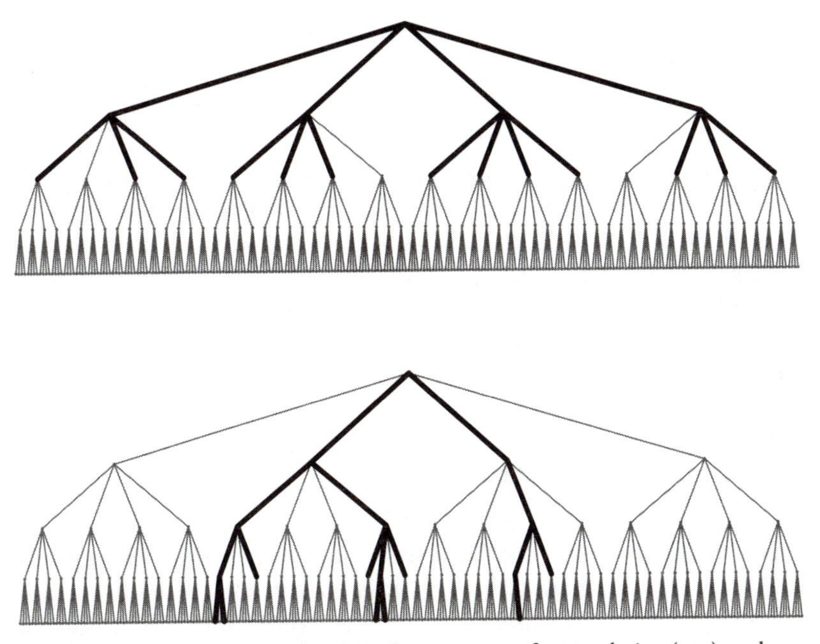

Figure 4.9. Planning ahead in the presence of poor clarity (top) and good clarity (bottom).

4.3.1 Clarity

Clarity, in the context of abstract board games, is the ease with which a player can understand what is going on [Abbott 1975]. Games like Hex with transparently simple rules and goals, and no special conditions or hidden complexities to distract the mind, have excellent clarity.

Abbott explains that the clarity of a game determines how far you can see down the game's strategy tree, and hence how deep the game is. Games with overly complex move mechanics or excessive piece movement tend to have poor clarity and hence limited depth.

Figure 4.9 illustrates the concept of clarity in terms of the *game tree*, which shows all possible choices from a given position, then all possible choices from those positions, and so on. The gray lines indicate possible lines of play, and the black lines indicate those moves that the player has deemed worth investigating this turn.

The game described by the top-most tree has poor clarity. The movement rules or goals are not clear in the mind of the player, who must evaluate almost every choice to decide which to play. The game described by the bottom tree has much better clarity. The player can readily identify and concentrate on key moves, and anticipate the opponent's most likely replies.

In both cases the player has investigated the same number of possible moves. However, poor clarity leads to a shallow exhaustive search while good clarity allows a deeper and narrower search; the player is more likely to formulate worthwhile strategies in this second game.

Game-playing computer programs tend to be more successful for games with better clarity, as they can search deeper along more promising lines of play and ignore the remainder. In this context, the clarity of a game is defined by how easily a player can identify good moves.

Note that the game trees shown in Figure 4.9 are representative only. They have a branching factor of exactly four each turn, whereas many connection games start off with a branching factor in excess of one hundred. Good clarity is therefore critical in these games.

An important feature of many connection games is the complementary nature of their goals: a player's connection cuts all possible opponent's connections, hence there can be no draws and exactly one player must win. This inherent property of absolute connection allows complex games to be created from surprisingly simple and elegant sets of rules, giving these games excellent clarity in most cases. The tired old phrase applied to many abstract strategy games, *a minute to learn, a lifetime to master*, is especially apt for almost all connection games.

Moreover, such games tend to be finite in nature and guaranteed to end before a certain number of moves are made. This combination of features means that most connection games are guaranteed to converge to a solution within a specified time, often starting with a dazzling choice of potential connections which converge to smaller sets of increasingly well-defined choices as the game progresses. Connection games without complementary goals tend to be race games, which may be less satisfying for some purists.

It appears that clarity of *purpose*, not *method*, is most important in connection games. In other words, if the player can clearly understand what must be achieved with each move, then the complexity of making the move does not matter so much. This point is discussed further in the Havannah and Knots sections in Part II.

A small note on etiquette: given the excellent clarity of most connection games, it is often clear who will win a game well before the winning connection is actually met. It is generally good form for the losing player to resign in such circumstances rather than play dead games out to their pedantic end.

4.3.2 First-Move Advantage

Most Pure Connection games suffer from a severe first-move advantage as the opening player can win with perfect play. This is due to a well-known strategy-stealing argument (see Appendix D).

Move transformers are a balancing mechanism used in some games to counteract the first-move advantage. For instance, the move transformer 12333 means that the first move is a single move, the second move is a double move, and all subsequent moves are triple moves. This mechanism is used in games such as Scorpio, Master Y, Trellis, Stymie, and Orion. It can speed up play but can also introduce unnecessary complexity, reducing clarity and making it extremely difficult to anticipate future moves and formulate strategies.

Multiple moves should be used judiciously in connection games. The best games tend to balance on a knife's edge, and this subtlety is lost if the advantage swings wildly each turn. Some players adopt local rule restrictions to facilitate multiple moves, such as specifying that moves made in the same turn may not be adjacent or belong to the same group.

4.3.3 The Swap Option

The *swap option* is a more elegant way to address any first-move advantage. The opening player makes a move, then the opponent has the option of either making a move in reply or swapping colors. If the swap is made, then the second player effectively steals the opening move and it becomes the first player's turn to move again. That is, if a white piece is played first then a black piece will be played next, whether the swap took place or not.

The swap option discourages the first player from making an overly strong opening move. It should generally be used for any game in which playing on a particular cell results in a stronger position than playing on another cell, such as Hex.

Sometimes a single-move swap option is not sufficient, and a more subtle *three-move swap* or *three-move equalizer* is preferred. In this case, the first player places three pieces on the board, usually two of his own and one of his opponent's, then the second player chooses whether to move or swap colors. This technique is especially applicable to boards on which all cells have equal or nearly equal status. Two-move swaps are also possible but not very common.

The swap option is colloquially known as the *pie rule* in reference to a method for fairly sharing a pie: I cut, you choose the piece.

4.4 Winning Conditions

The goal or objective of a game is usually the single most important factor in defining the nature of that game.

4.4.1 Types of Goals

Goals in connection games tend to fall into one of the following categories:

- symmetric,
- equivalent, or
- unequal.

Symmetric goals are similar for each player but distinguished in some way, usually by differentiated goal areas. Most connection games feature symmetric goals. For instance, Hex has symmetric goals as each player aims to connect their differentiated sides of the board (Figure 4.10).

Symmetric goal areas are generally placed so that any winning connection will cut across every possible opponent's winning connection. Such games with mutually exclusive goals are called *win-loss complementary* or just *complementary* and do not allow ties. These games are also described as *categorical* [Weisstein 1999].

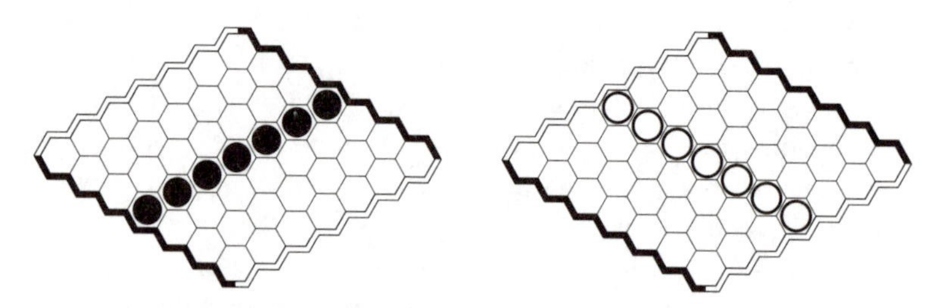

Figure 4.10. Hex has symmetric complementary goals.

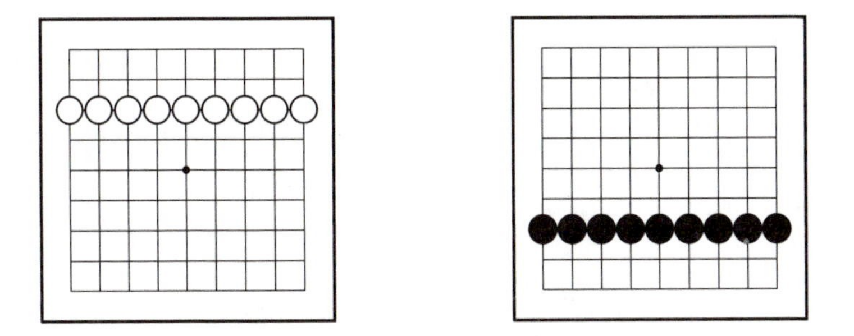

Figure 4.11. Gonnect has equivalent goals.

Equivalent goals are those that are identical for all players, and in which no distinctions are made regarding ownership of board edges or goal areas. For instance, players have equivalent goals in Gonnect: to connect either pair of opposite sides (Figure 4.11).

Connection games with equivalent parallel goals tend to be race games, as there is little incentive to cut across opponent's paths other than to slow them down. However, some games, such as Projex and Y, have both equivalent and complementary goals (Figure 4.12).

Lastly, a small number of connection games, such as Cylindrical Hex, Jade, Unlur, and Antipod, have *unequal* goals. For instance, one player in Jade attempts to form a parallel chain of each color connecting any two opposite sides, while the other player attempts to connect all four sides with a same-colored cross (Figure 4.13).

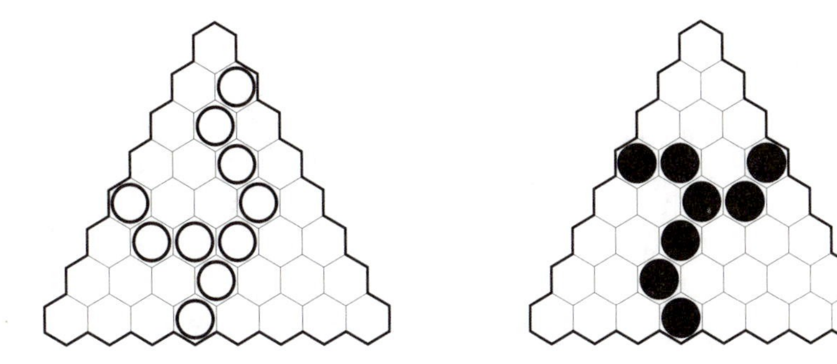

Figure 4.12. Y has equivalent complementary goals.

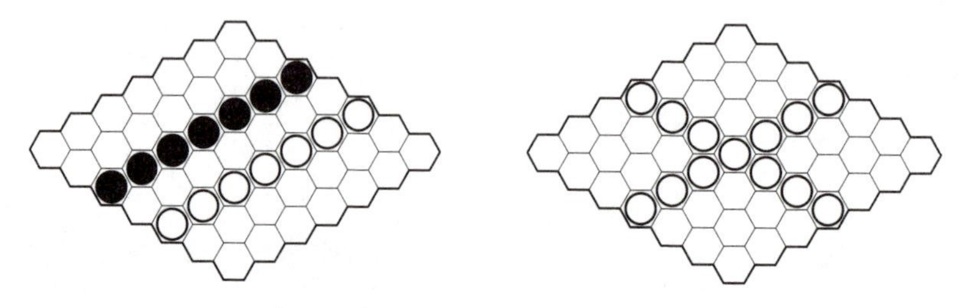

Figure 4.13. Jade has unequal goals.

There are a small number of *anticonnection* games, such as Black Path and Puddles, in which players strive to avoid making connections. The misère metarule can also be used to turn a connection game into an anticonnection game.

4.4.2 Multiple Terminals

Some games involve the connection of multiple terminals. One such case is the game of Y, in which both players try to connect all three edges with a chain of their pieces. Figure 4.14 (left) shows a game of Y won by White.

It is possible to phrase Y as a game graph with two terminals by reflecting the board and its pieces along one of the goal edges, as shown in Figure 4.14 (right). The game of Y can now be treated as a game of Hex, which is won by connecting either the left and right edges, or the top and bottom edges. This technique can be useful for converting games with multiple terminals into Shannon game representation with two terminals.

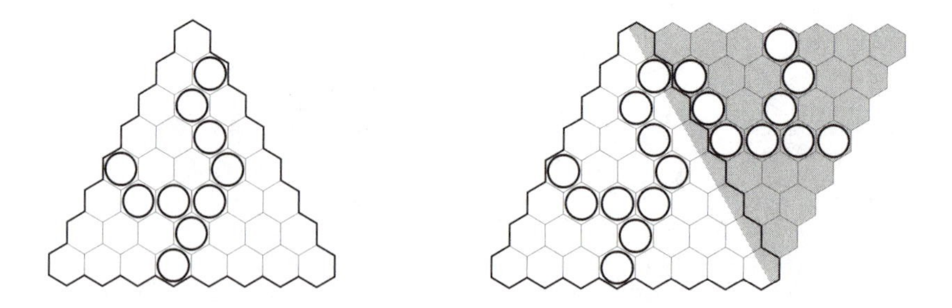

Figure 4.14. A winning white chain in a game of Y and its reflection.

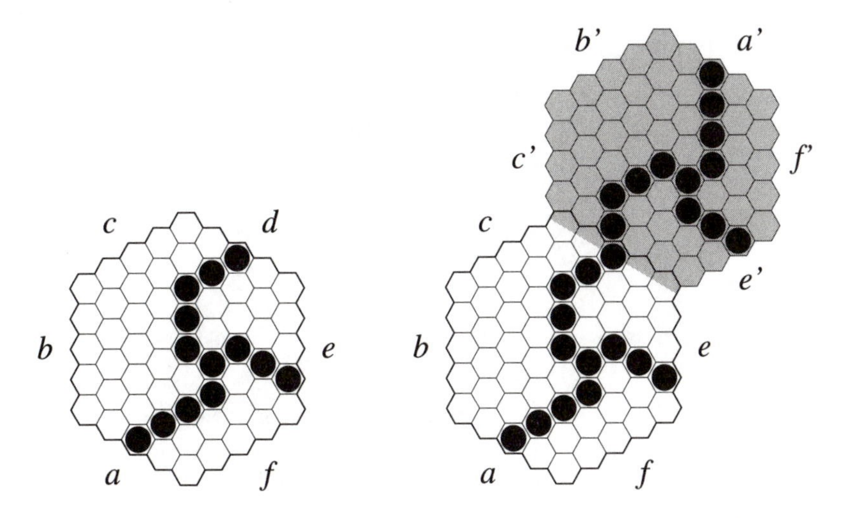

Figure 4.15. A winning black fork in a game of Havannah and its reflection.

Figure 4.15 (left) shows a game of Havannah with a winning formation by Black, who has connected three board edges with a chain of black pieces. A similar technique can be used to reduce this problem to one of connecting two sides across a reflected edge; Black has won this game by connecting terminals ***a*** and ***e'*** via ***d***, as shown in Figure 4.15 (right).

Note that this is only one of ten possible reflections that must be tested for the triple connection, and that the triple connection is only one of the three possible ways to win the game; Havannah has an unusually complex set of winning conditions for a connection game. The complexity of a game's graph generally indicates the clarity of the game's rules and winning conditions.

4.4.3 Tie Breakers

In addition to deadlocks (discussed in Section 3.3), a game may be tied if winning conditions can simultaneously be met for more than one player. This can generally occur only in path race games in which players share pieces or paths. Such cases are usually resolved in one of two ways: *Mover Wins* or *Mover Loses*.

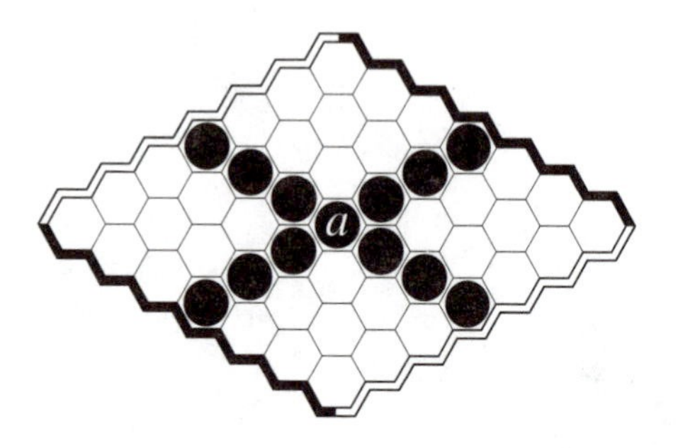

Figure 4.16. Who has won this game of Chameleon?

For instance, Figure 4.16 shows a game of Chameleon, in which a player wins by completing a chain of either color connecting his sides. Chameleon is played with the Mover Wins rules, so the player who made move ***a*** wins the game (this must have been the last move played).

Mover Wins is a simple, direct solution that leads to elegant and clear resolutions to games. Mover Loses is less direct and can reduce clarity, but can also add another level of complexity and intrigue to the end game.

Rule sets requiring such tie breakers are bad for *hot* games (games in which it is better for players to move than not to move) as they add an unnatural element of coldness. Players are obliged to make negative time-wasting moves rather than move somewhere that will harm them. Cooling is described in detail in Conway [1976] and Berlekamp et al. [1982].

4.4.4 Multiple Players

Even though the Cut/Join nature of connection games is ideally suited to two-player games, games involving three or more players can be implemented successfully. The hex hex board naturally accommodates three players.

However, care must be taken as three players may create deadlocks even on the hexagonal grid. For instance, Figure 4.17 shows a game of Three-Player Hex, in which players aim to connect their sides of the board with a chain of their pieces. A deadlock between the three players is shown on the left. Fortunately, the rules of Three-Player Hex allow for this contingency: *as soon as it is no longer possible for a player to connect his sides, that player is eliminated and may not place any more pieces.*

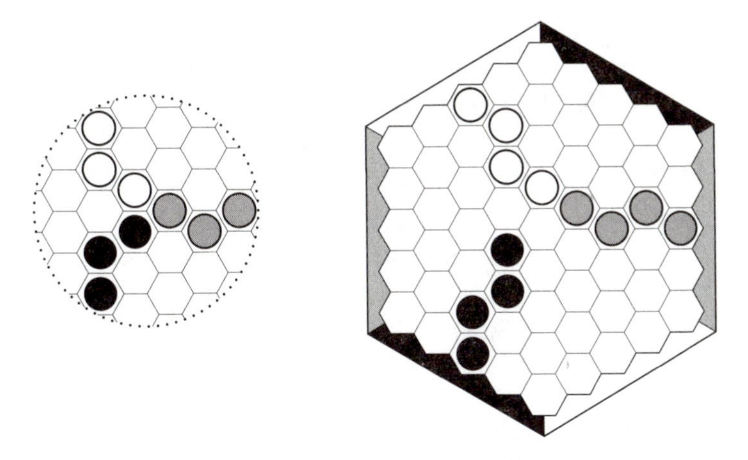

Figure 4.17. Deadlocks (left) are not a problem in Three-Player Hex (right).

Referring to Figure 4.17 (right), this means that Black, whose turn it is to play, has been eliminated from the game and the immediate deadlock has been averted. Any further deadlocks will eliminate another player and leave a single winner.

Multiplayer connection games are also subject to the *petty diplomacy* problem that plagues most three-player board games: temporary coalitions may form between the two weaker players at any given point to thwart the leading player. This can lead to interesting and often hilarious situations, although it can affect the strategic purity of a game by encouraging players to lie low [Schmittberger 1992].

Another danger of multiplayer games is that a losing player may attempt to also bring down a particular opponent. The ability for a losing player to dictate the outcome of a game is called the *kingmaker* effect [Kramer 2000] and may be detrimental to a game.

Philip Straffin [1985] recommends the use of McCarthy's Revenge Rule, which can be paraphrased as follows: *I will hurt the player who has hurt me the most.* McCarthy's Revenge Rule makes excellent sense but can be difficult to apply in practice. The losing player may not be able to decide which opponent has contributed most to his downfall, or may choose to forego this revenge in order to penalize a particular player.

Bill Taylor suggests the more direct and enforceable Stop-Next Rule: *a player cannot make a move that will give the next player an immediate win, unless there is no other choice.* However, this rule tends to impair the clarity of the game, as players must then look two moves ahead in order to determine the immediately available legal moves.

4.4.5 Cold Wars

Games in which it is not always to the player's advantage to move are described as *cold* games, and are subject to the problem of *cold wars*. Paul van Wamelen coined this term to describe situations in which a game devolves into a battle of attrition as players struggle to avoid playing a cold (losing) move.

For instance, Figure 4.18 shows another game of Chameleon in which it would be suicidal for either player to play a piece of either color at any of the cells marked x. Players are obliged to burn moves elsewhere on the board until they have no other choice.

The winner in this simple example can be determined by counting the number of empty noncritical cells. Since this number is even, then the next player to move will eventually be forced to play at one of the danger cells first. The game should be awarded to the opponent immediately.

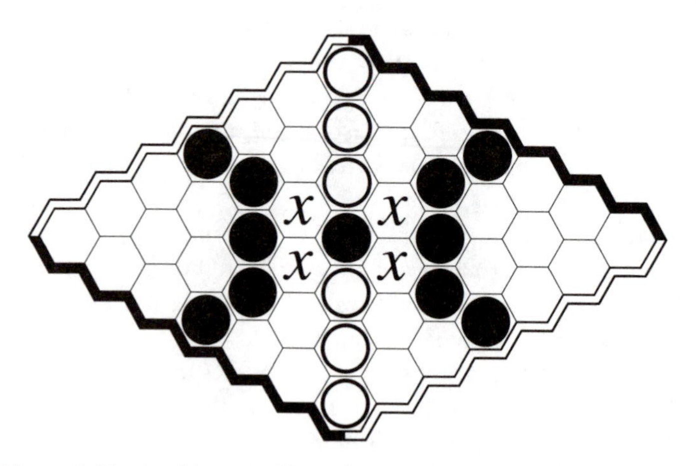

Figure 4.18. A cold war in Chameleon. Neither player wants to occupy any point marked x.

5 Common Plays

This chapter describes some general points of strategic and tactical play common to many connection games. A player proficient in one connection game can usually transfer at least some of his skills to related connection games with some degree of success. Although some fundamental concepts are endemic to the genre of connection games, it is the manner in which these are applied to individual cases and the resulting tactical nuances that make each game interesting.

5.1 Strategic and Tactical Play

Throughout this book, a move, sequence of moves, or even an entire game may be described as either strategic or tactical. It is worth clarifying exactly what these terms mean in the context of connection games.

Strategy is the art of war and concerns the overall management of forces in conflict with an enemy. The ability to conceptually decompose play into worthwhile subgoals, such as in Chess, is an indicator of strategy.

Tactics involve the low-level deployment of those forces to achieve specific goals. Larry Levy describes strategy as long-term thought and tactics as those decisions that change with every game turn [2002]. Levy personally prefers the challenges thrown up by each turn of a tactical game, rather than having to study a game as a whole to come up with a perfect strategy.

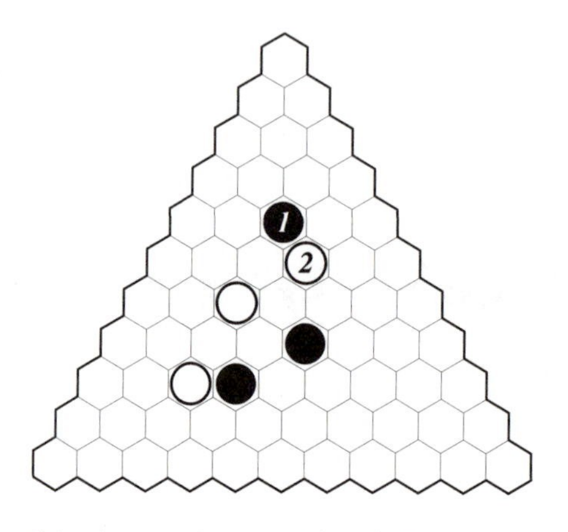

Figure 5.1. A strategic move *1* by Black in a game of Y.

As a practical example, Figure 5.1 shows a rather strategic Y move by Black. Move *1* poses little immediate threat and White is able to isolate it with cutting move *2*. However the true worth of Black's move soon becomes apparent.

Figure 5.2 shows an extension of play in which Black's move *3* threatens to connect to the left edge, triggering a sequence of forced tactical moves that eventually allow Black to connect with piece *1* for an unbeatable connection to all three sides. The strategic move *1* embodied an overall plan, and the tactical moves *3* to *9* realized that plan.

Table 5.1 presents a list of evocative comparisons that explore the relationship between strategy and tactics.

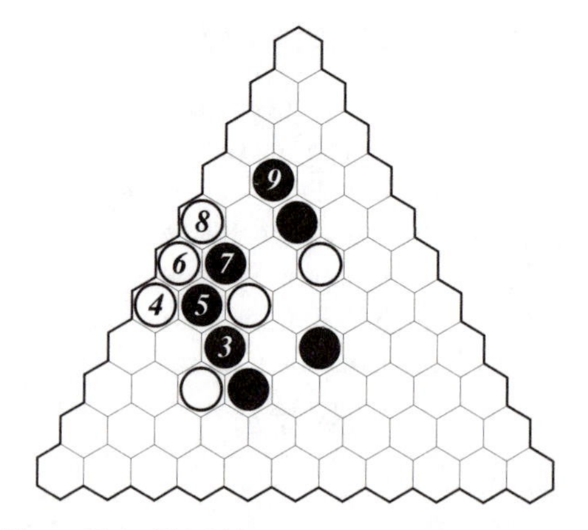

Figure 5.2. Tactical sequence *3* to *9* by Black.

Strategy	Tactics
Global	Local
Long-term	Short-term
Planning	Execution
Ideas	Action
Communication	Language
Structural	Dynamic
Intuition	Routine
Estimation	Calculation
Imagination	Method
Geometry	Arithmetic
Creativity	Algorithm
Wisdom	Cleverness

Table 5.1. Comparisons between strategy and tactics.

This list was compiled by Bill Taylor [2002], who also offers the following aphorisms:

Strategy is what you want … tactics is what you get.

Tactics is about doing things right. Strategy is about doing the right thing.

Tactics is knowing what to do when there's something to do,
Strategy is knowing what to do when there's nothing to do.

As a group, connection games tend more towards tactical combinatorial play than most other abstract games, and the distinction between strategy and tactics is not as clear. Within this scope, however, most connection games achieve a reasonable balance between the two.

Ideally these two aspects will complement each other in a given game. The player should be able to develop creative, deep strategies and be able to lay subtle traps for opponents within the practical choices available to them each turn. Conversely, strong combinatorial play should generally improve a player's position and open up new avenues to explore.

Overly tactical games can get bogged down in the minutiae of low-level combinatorial play that rewards rote learning rather than over-the-board thinking; overly strategic games can feel intangible and boring to all but expert players.

Nudging a game towards either end of the strategic-tactical spectrum can help improve its balance. For instance, the games of Go and Y can feel overly strategic and intangible in the opening moves as players maneuver to stake their territory, whereas two variants based on these games, Gonnect and Caeth Y, introduce tactical play right from the start, encouraging direct conflict much earlier.

When scaling a game's size, smaller boards tend to emphasize tactical play while larger boards allow greater strategic freedom.

5.2 Forks

The *fork* is a fundamental play in most board games, but connection games tend to make them quite explicit. A fork in this context refers to two or more simultaneous threats, such as alternative routes along a potential path.

For instance, White cannot stop the black pieces in Figure 5.3 from eventually connecting. If White takes one of the intermediate points then Black can take the other to complete the connection, as can be readily seen in the graph representation (right). A safely forked path such as this is called a *virtual connection* and corresponds to the OR deduction rule described by Anshelevich [2002].

Figure 5.4 shows a slightly more complex virtual connection between black pieces *s* and *t*, though the same principle applies; for every white move there is a corresponding black move that keeps the connection alive. It is often useful to think of such virtual connection systems as a single unit, to simplify higher-level strategy.

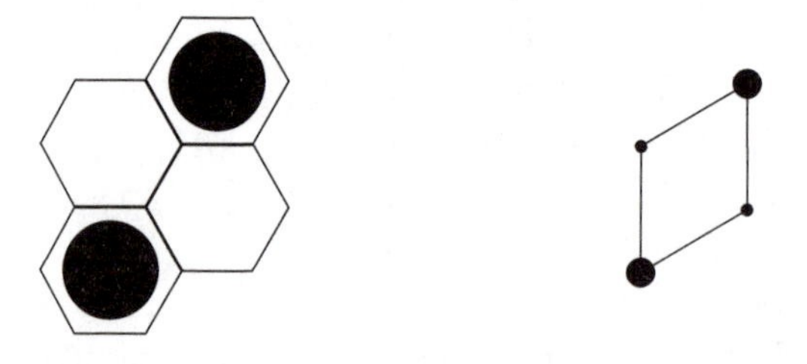

Figure 5.3. Alternative routes between targets form a fork.

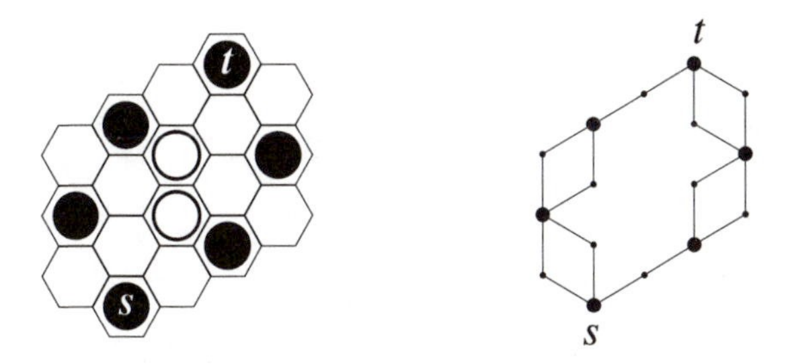

Figure 5.4. A more complex forking system.

This system works for games with two players, as each player is able to immediately respond to an opponent's move. For a fork to succeed in multiplayer games, there must be as many alternative routes as there are players (Figure 5.5).

Forking moves may take different forms, but all share the common principle of posing two or more independent threats that the opponent is unable to counter next turn. For instance, Figure 5.6 shows forking move ***a*** in a game of Trax. Black threatens to make two loops (dotted). White can block one of these next turn, but not both. Note that tile ***b*** is an automatic placement caused by the placement of ***a***.

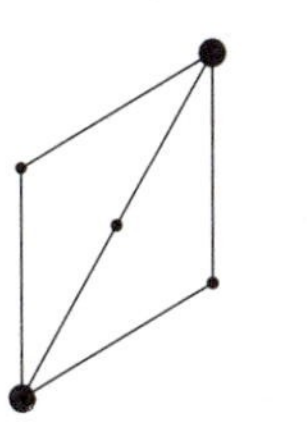
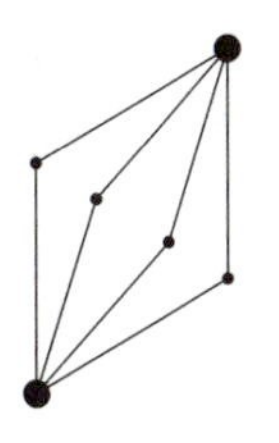

Figure 5.5. Three-player games require three alternative routes, and four-player games require four alternative routes.

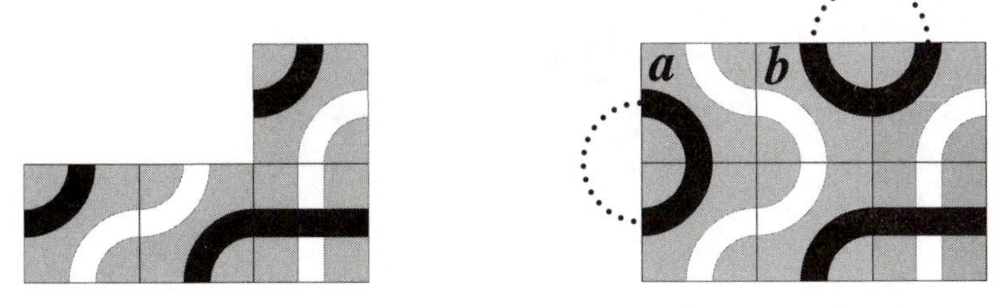

Figure 5.6. Forking move ***a*** in Trax.

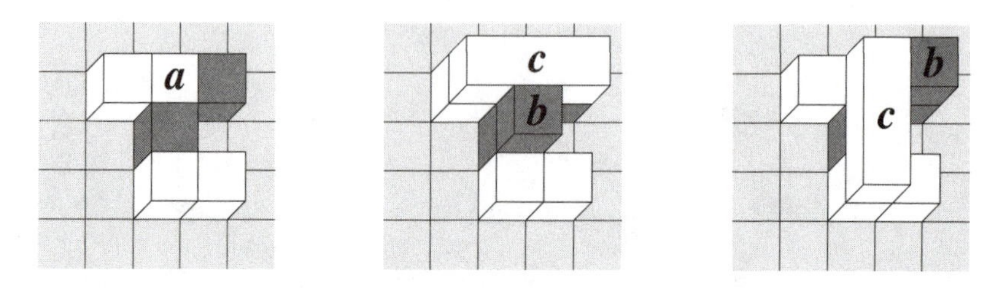

Figure 5.7. Forking move *a* in Druid.

Similarly, Figure 5.7 (left) shows a forking move *a* in a game of Druid, in which White aims to push his connection horizontally across the board. Black has two obvious blocking moves (marked *b*), but in each case White has an alternative reply *c* that extends his connection horizontally.

5.3 Convergence of Flow

The concept of *flow* is central to most connection games, especially those with complementary goals. Flow in this context refers to the number of potential connections flowing through empty board points.

For example, two critical cells are shaded in the game of Hex shown in Figure 5.8. These are the points of greatest convergence of flow, as the most likely winning paths for both players pass through them. In fact, the next player to move can establish an unbeatable connection by playing at either point.

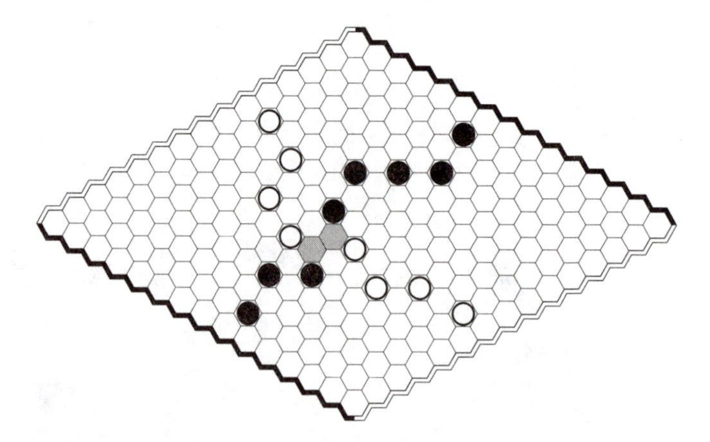

Figure 5.8. The points of convergence are obvious in this game of Hex.

The weakest points in each player's strongest connection are generally the key points in any given connection game. A good rule of thumb is: *improve your flow, impede your opponent's*. Any move that does both is usually a good move.

One of the features of complementary-goal connection games such as Hex and Y is that the opponent's best move is often the player's best move as well. If in doubt, players can just move where they think the opponent would most like to move next turn, and more often than not this will be a good play. In fact, if a player gets confused and mistakenly plays as if the opponent, there will often be little harm done.

Most players can visualize flow and intuitively see what would be a reasonable move even in the absence of any strategic or tactical knowledge of a given game. This is a very attractive feature of connection games. Of course, good players can see beyond merely reasonable moves to spot the truly brilliant ones. Often the best move is simply the most obvious. It can be tempting to play an unexpected move to throw the opponent into confusion, but this can be a dangerous ploy.

5.4 Forcing Moves

A *forcing move* is one that forces an immediate reply from the opponent that puts them at a disadvantage. Forcing moves can be used to great effect, as the example shown in Figure 5.2 demonstrated. While the presence of additional enemy pieces will generally not help a player, it can be good to make the opponent waste a move if some positional advantage is gained.

For instance, Figure 5.9 shows a killer forcing move *1* by Black in a game of Hex. White is forced to block the immediate threat with *2*, however Black can push through with move *3* to eventually connect with *1* and

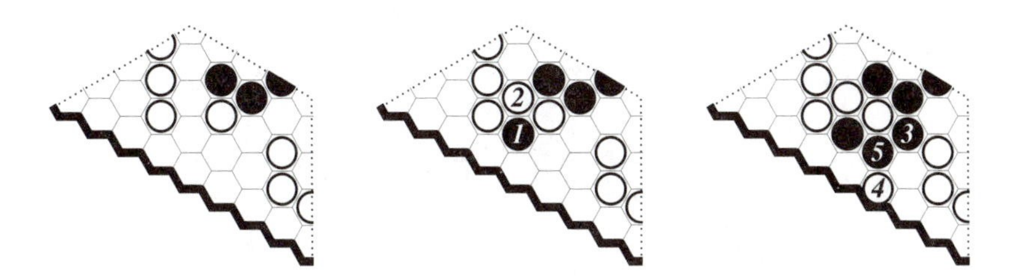

Figure 5.9. Forcing move *1* wins the connection for Black.

therefore the edge. Like most good moves, move *1* served two purposes at once. White's ineffectual forced replies *2* and *4* were wasted moves while Black developed his connection.

5.4.1 Tempo

A player is said to be forcing the *tempo* of a game if the momentum or initiative is in his favor. Forcing moves are especially good for setting the tempo, as the defender has little choice but to struggle for survival while the attacker is free to follow a strategy.

Most connection games are so well balanced that the tempo between evenly matched players swings with each move. In fact, the player with the first move should generally win with perfect play, hence the importance of first-move equalizers such as the swap option (see Section 4.3.3).

5.5 Ladders

Ladders are consecutive sequences of forcing moves in a particular direction, which tend to form a solid line of defense or attack. For example, Figure 5.10 shows a game of Y with White to move and win.

White is forced to play move *1* as shown in Figure 5.11 to keep the game alive. Black is forced to block with move *2*, then White is forced to play move *3*, then Black is forced to block with move *4*, and so on.

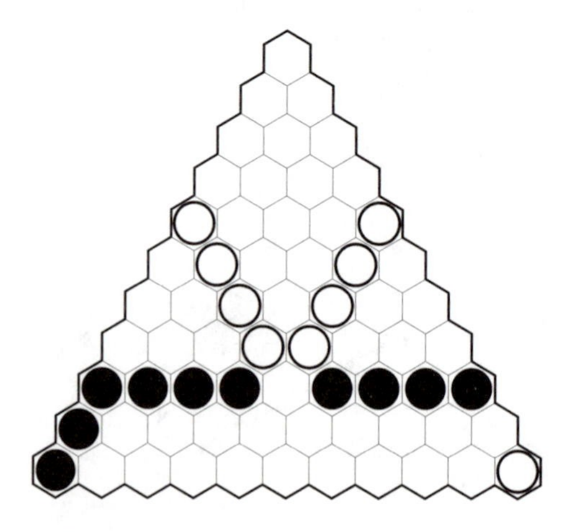

Figure 5.10. White to move and win.

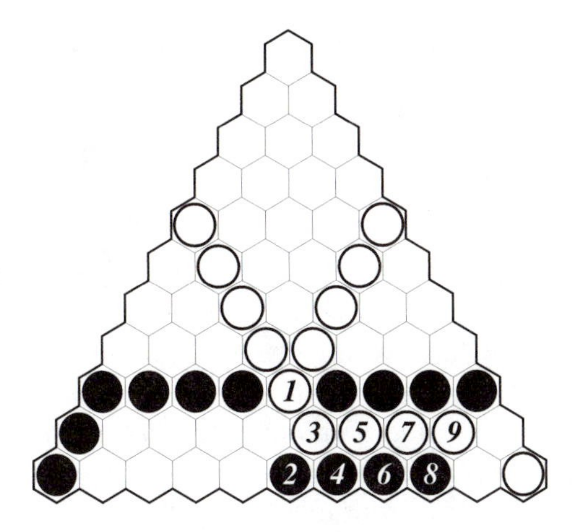

Figure 5.11. Forcing move *1* creates a ladder to victory.

The ladder ends with move *9* when White creates a virtual connection with his piece in the bottom right to create an unbeatable position. This life-saving piece is called the *ladder escape* or *ladder breaker*.

Ladders tend to occur when one player is trying to push the envelope of his connection whilst the opponent is trying to form a solid wall against it. They often develop parallel to a board edge, but may be found anywhere that a player is squeezed through a gauntlet of enemy pieces.

A slightly different form of a ladder plays an important role in Go, as shown in Figure 5.12, where white piece ***a*** finds itself in difficulty. White and Black exchange a series of forcing moves until White is able to escape via the ladder breaker ***b***.

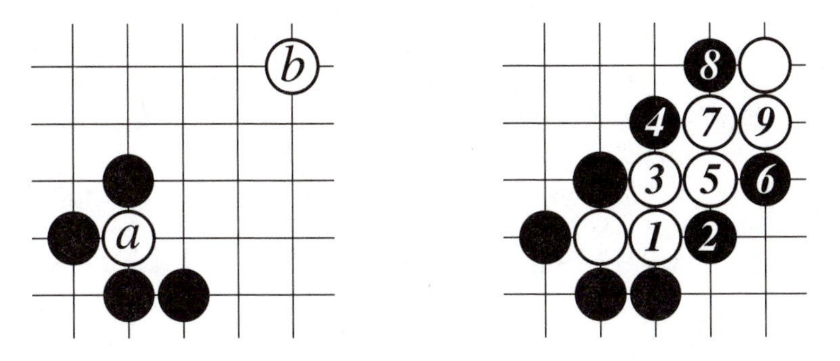

Figure 5.12. White piece ***a*** escapes this Go ladder.

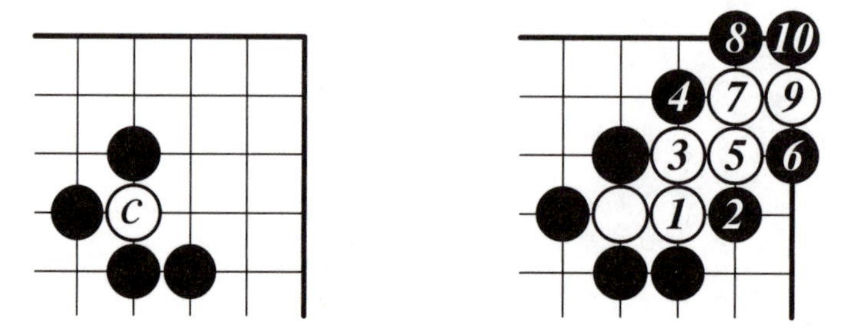

Figure 5.13. White piece *c* is not so fortunate.

Figure 5.13 shows the disastrous outcome if there is no escape piece in the ladder's line of play. The white group is eventually run into the edge of the board, surrounded, and captured.

Ladders are common in most piece-based connection games, but tend to be less common in games played with directional tiles.

5.6 Defensive Play

A good rule of thumb for complementary-task connection games is that *defense equals attack*. Blocking the opponent's connection by definition implies a win for the player.

The value of defense is quite often underestimated. It is common for players to complain about how bad their position is, only to hear the opponent reply with the same complaint. A strongly defended position is a good one, and players should resist the urge to play excessively aggressive moves that overreach and lead to disaster.

This is not to say that connection games need be overly defensive or passive. Gonnect, a purely connective variant of Go, is a prime example. While opening moves in Go tend to be positional and spaced well apart, opposing forces in Gonnect may engage in conflict from the very first moves. Here the distinction between offense and defense blurs somewhat, and the key is to be aggressive but not overly so.

An active form of defense is to exploit weak points of overlap in the opponent's potential connections. For instance, White can exploit a point of overlap in Black's potential connection between *s* and *t* in Figure 5.14. White piece *u* interferes with two black forks simultaneously (as can be seen in the graph analysis on the right) making a black connection between *s* and *t* unlikely.

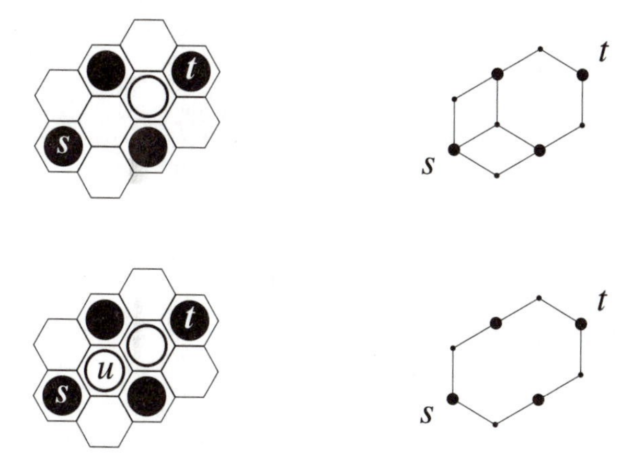

Figure 5.14. Points of overlap can be exploited.

As well as playing close-in moves to interfere with specific connections, it is often important to block a group's progress across the board. Such positional defense is usually best done from a distance.

For example, Thoughtwave involves two players, Light and Dark, striving to form a path between their edges of the board. Players share the black path, which may be used in either winning connection. Imagine that Light has the upper hand in a game and that Dark must block the path's horizontal progress across the board (Figure 5.15, left).

Blocking the path directly (Figure 5.15, middle) is a bad play for Dark. The Dark piece caps the path's immediate horizontal growth but Light is able to bypass this attempted block and continue unimpeded (Figure 5.15, right). Dark has wasted a move and a valuable Terminator piece.

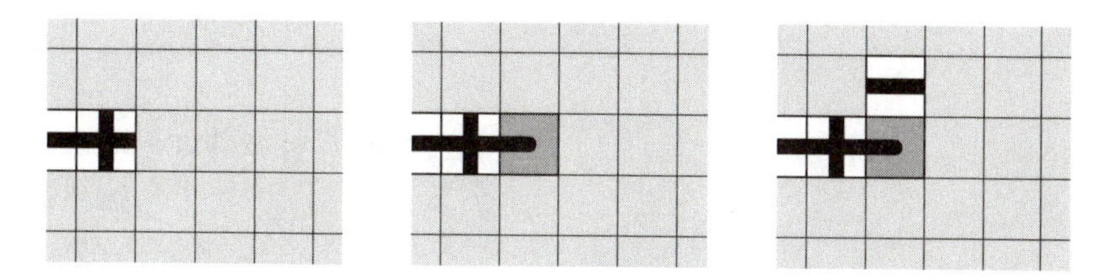

Figure 5.15. A poor close block by Dark in Thoughtwave.

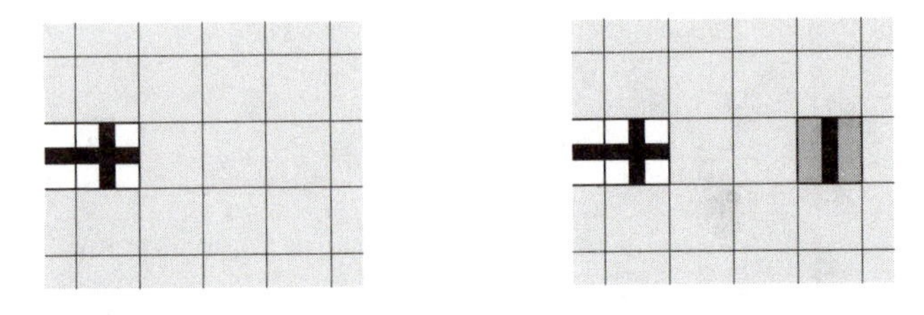

Figure 5.16. A distant block is more effective for positional defense.

Figure 5.16 shows a more effective defense by Dark. This more distant move impedes Light's horizontal progress and gives Dark room to maneuver in response to the inevitable counter-plays.

Close blocking can be nothing short of disastrous in some cases. Consider the Druid board position shown in Figure 5.17. If Black tries to block the white pieces directly, then White can just play a higher-level piece to cover it. Black has not only wasted a move but has given White an easy step up to the next level.

5.7 Staking Your Claim

It is generally good to spread pieces out in order to influence a wider area of the board. Most connection games tend to begin with widely spaced pieces that form a loose connective framework, which fills in and becomes better defined as the game progresses. Players tend to map out large areas quickly, then focus on the points of friction where those areas collide.

Consider the two formations shown in Figure 5.18. The formation on the left is solidly connected but does not cover much distance. By comparison, the loosely connected formation on the right has a greater reach, making it more threatening as a whole. There is a fine balance between stretching a connection to cover as much territory as possible, and keeping it sufficiently solid to withstand attack.

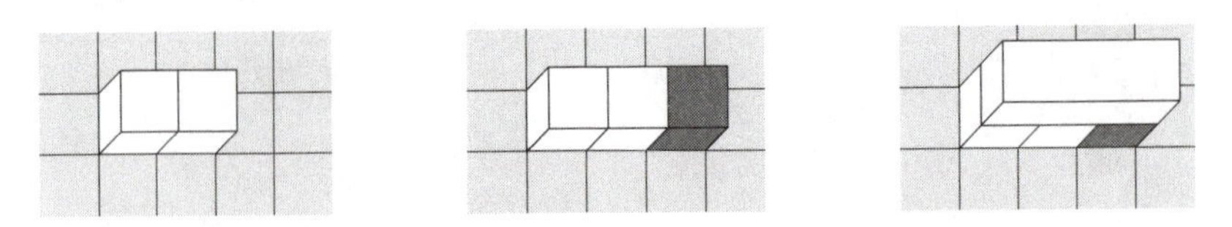

Figure 5.17. Adjacent blocks can be disastrous in Druid.

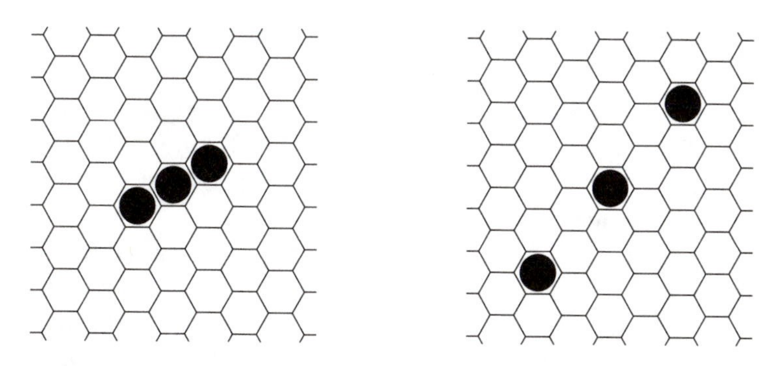

Figure 5.18. A solid chain and a more threatening spread of pieces.

It should be pointed out that this observation does not apply to all games. For instance, it is good to keep all pieces connected in Andantino if possible, as the opponent can threaten to surround stray pieces and force the play. Indeed, the entire aim of the Convergent class of games is to connect all pieces. Akron involves a fine balance of tension between these two approaches, as it is just as important to establish a spread of pieces across the board as it is to clump pieces together to improve mobility.

5.8 Efficiency

An *efficient* move is one that achieves multiple purposes at once. Schmittberger [2000] describes the importance of efficiency in connection games; since most games confer a theoretical winning advantage to the first player, every move counts.

For instance, Figure 5.19 shows a situation proposed by Schmittberger in which White must move to connect the two white pieces. The three moves ***a***, ***b***, and ***c*** all achieve this goal, but which is best?

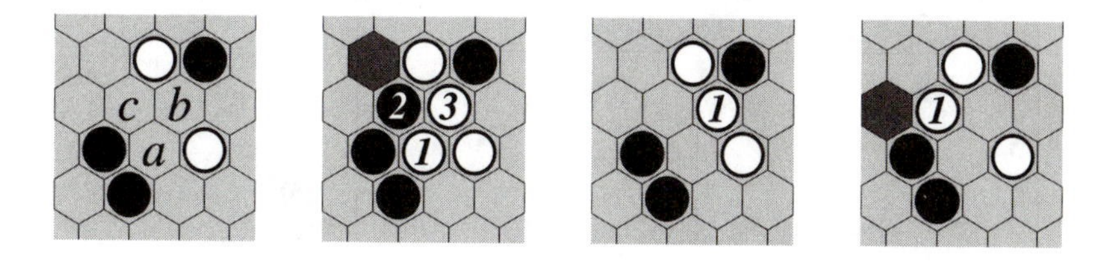

Figure 5.19. White to move and connect. Which move is best?

Figure 5.19 (second figure) shows the result after moving at cell ***a***. Black can intrude with move ***2***, forcing White to play a wasteful move to save the connection. White move ***3*** covers no new territory while Black has extended his adjacent territory by one cell (shaded).

Figure 5.19 (third figure) shows the result after moving at cell ***b***. The White pieces are now solidly connected but no territory has been gained.

Figure 5.19 (fourth figure) shows the result after moving at cell ***c***. This is the best move for White; not only does it guarantee a virtual connection, but it extends White's influence by one cell (shaded). This move is the most efficient as it performs two tasks at once. The killer move shown in Figure 5.9 is another example of an efficient move.

Keeping pieces spread out rather than bunched up can help improve efficiency [Schmittberger 1992]. Pieces too close together may end up doing the same job, and any piece that proves superfluous represents a wasted move.

5.9 Sacrifice

It is good practice to make any move serve at least two purposes (usually two threats). This can be an especially good play if one of the threats is obvious and the other more subtle and dangerous. The more obvious threat is a *sacrifice*; it is hoped that the opponent will take it and fail to see its subtle partner.

For instance, Figure 5.20 shows White in a dire (losing) Hex position, but with a trick up his sleeve: the sacrificial move ***1***. If Black takes the bait and immediately responds to the more obvious threat with move ***2*** (Figure 5.21, left), then White can push through and initiate a short ladder to win the game as shown.

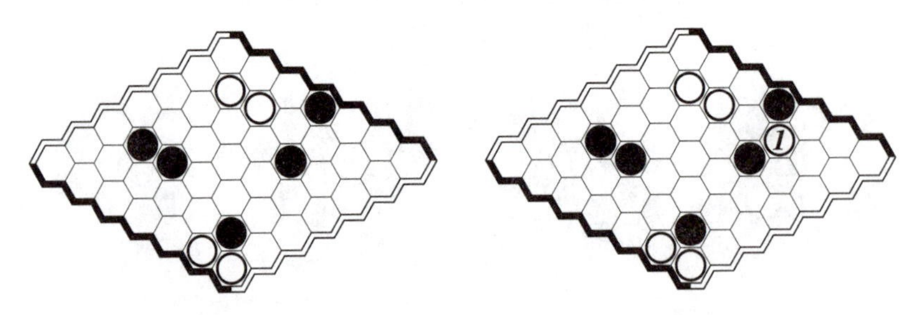

Figure 5.20. White in a losing position, to move.

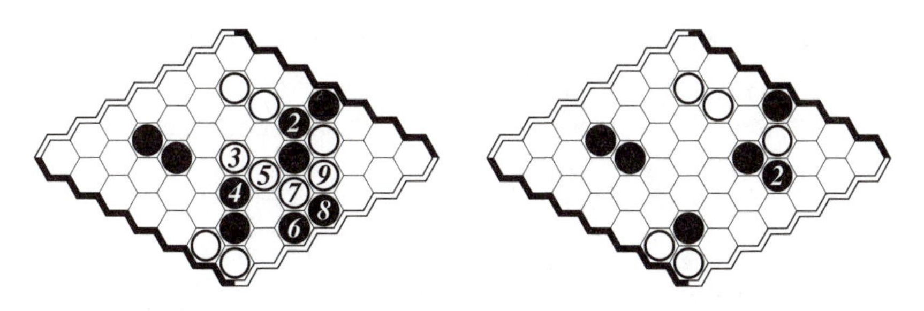

Figure 5.21. Black either falls for the trap and loses (left) or makes the correct play and wins (right).

A much better reply *2* by Black is shown on the right. This move intrudes into both the obvious and subtle threats simultaneously, and guarantees victory for Black.

5.10 Broadsides

While players usually strive to build their connection directly towards their goal, it can sometimes be a surprisingly good move to build perpendicular to this direction. Such moves are called *broadsides*.

Move *1* in Figure 5.22 illustrates a winning broadside by White in a 5 x 5 game of Hex. This move develops White's connection perpendicular to his direction of connective flow, yet provably wins the game; in fact, this is the only move that does so. Note that the winning connection flows around both sides of the Black pieces on both sides of the board.

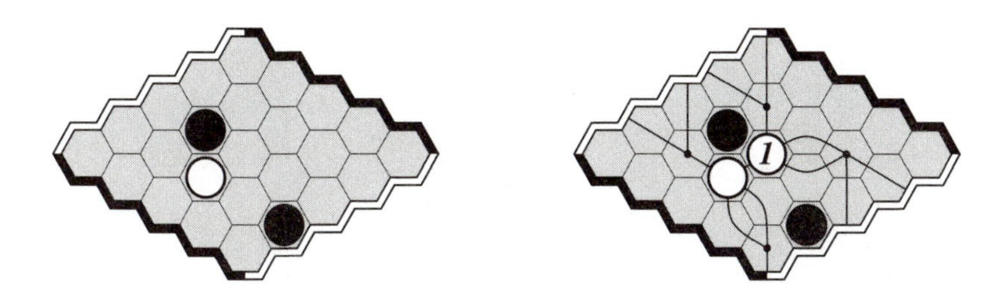

Figure 5.22. A winning broadside by White.

Broadsides can be strong moves as they usually

- attack both sides of the board at once,
- provide more attachment points for connection than a straight-on attack, and
- extend a solid central block to gain some territory.

In addition, they have the element of surprise; a perpendicular attack is often not what the opponent is expecting.

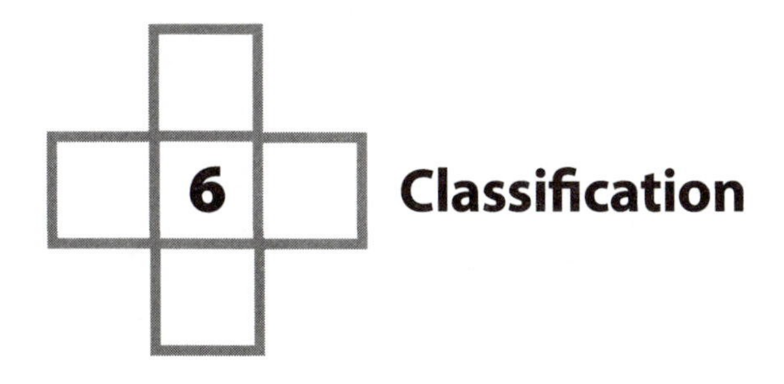

6 Classification

Out of the many connection-based board games it is obvious that some are more connective in nature than others, and that some games often described as connection games are only so in a very weak sense. A classification system is proposed below in an effort to impose some order, identify related games, and help define the boundaries of this group.

Several classification schemes were considered. It was not until an observation by Richard Reilly that connection can be a means or an end to a game (or both) that a truly useful system clicked into place. The following scheme is based upon this idea.

6.1 Strictly Connective Games

Three categories form the core of this classification scheme. The third, most important, category is an intersection of the first two.

Connection games are not necessarily pure abstract games; chance may legitimately play a part. For instance, rolling dice or drawing random tiles to play each turn may not impinge on the connective nature of a game, and is not necessarily a basis for disqualification from the strictly connective categories.

6.1.1 Connective Goal

Connective Goal games are those that end as soon as a specified connection, independent of size or shape, is achieved; connection is paramount in deciding the winner. Such games can be described as involving connection at a global or strategic level.

6.1.2 Connective Play

Connective Play games are those that feature at least some connective aspect and no nonconnective aspects during general play; connection between pieces is paramount during play. Such games can be described as involving connection at a local or tactical level.

Nonconnective aspects include unconstrained piece movements, jumps, or flips. Similarly, the movement, rotation, or removal of tiles to arbitrarily change connections disqualifies a game from this category. Moves must be primarily dictated by connection.

The placement of pieces on the board from an outside pile does not exclude games from this category. In some games, such as Antipalos, Visavis, September, and Network, players may pick up pieces from the board and play them elsewhere in the rare instance that their stockpile of pieces runs out before the game is over. This constitutes a form of unconstrained piece movement, but does not disqualify a game from the Pure Play category if this rule is deemed to be an after-the-fact fix to address the limited supply of pieces rather than a central part of the game.

Majority connection counting is allowed (as in Poly-Y, Conhex, and Caeth) as this is a categorical rather than a relative measure. Piece capture is allowed, as long as it is strictly connection-based (as in Go) and does not involve size or shape constraints.

Although adjacency and connection are inseparably related concepts, a distinction is made between adjacency-based moves and connection-based moves. For instance, Figure 6.1 (left) shows a game in which piece *a* may influence or be influenced by adjacent pieces. This alone is not sufficient to warrant inclusion in the Connective Play category; instead, *a* must influence or in some way be influenced by all pieces connected to it (right).

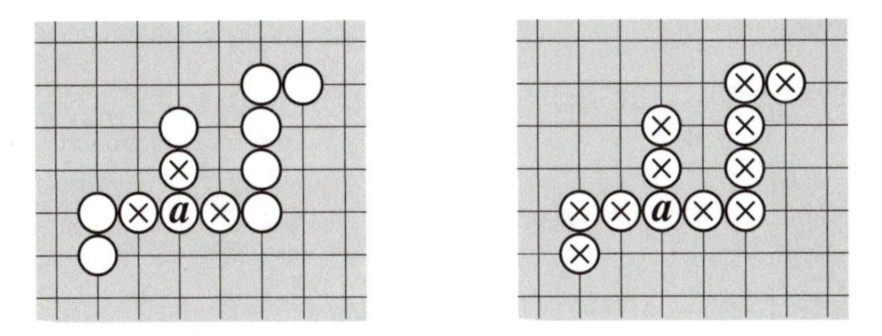

Figure 6.1. Pieces adjacent to ***a*** (left) and pieces connected to ***a*** (right).

6.1.3 Pure Connection

Pure Connection games are those with both strictly connective play and strictly connective goals. These can be described as games involving connection at both the local and global levels. Figure 6.2 illustrates the relationship between these three main categories.

A basic rule of thumb is: *as soon as size, shape, or pattern is introduced into a game, then it is likely to be disqualified from at least one of the strictly connective categories*. Note that describing a game as Pure Connection does not imply that it is necessarily better than a non-Pure Connection game.

6.2 Taxonomy

The three main categories are further illustrated with subcategories and lists of member games in Figure 6.3. This taxonomy fits within existing classifications of the overall family of board games suggested by previous authors, such as Parlett's *The Oxford History of Board Games* [1999] and Keller's brief "Taxonomy of Games" printed in issue number 1 of *World Game Review* [1983].

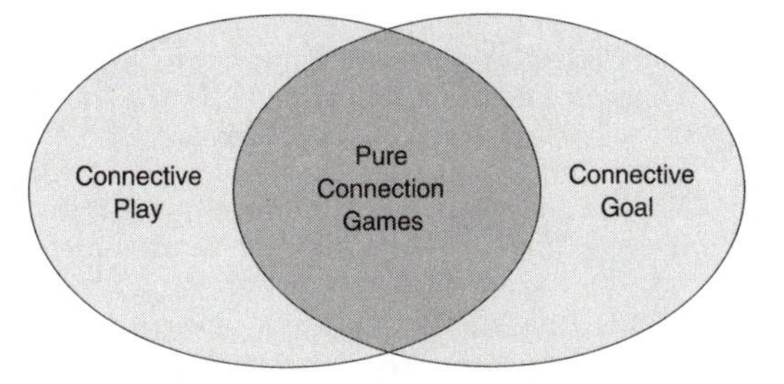

Figure 6.2. The relationship between the three main categories.

Connection Games

Pure Connection: Games with both strictly connective play and a strictly connective goal.

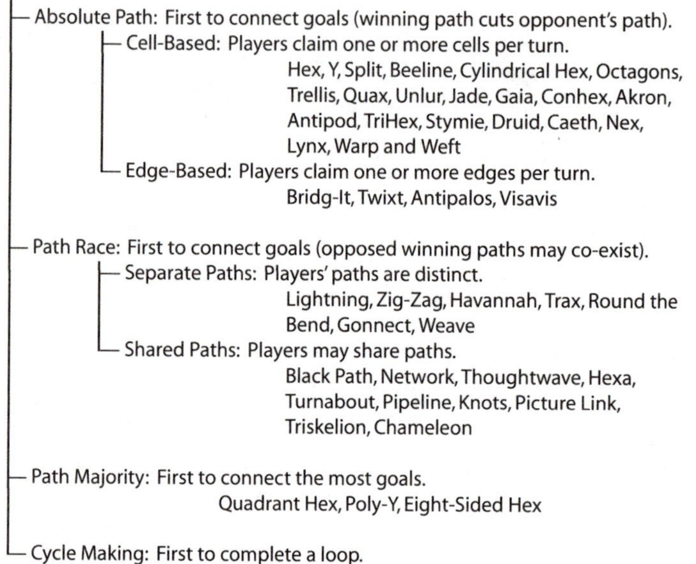

Connective Goal: Winner is decided strictly by connection.

- Path Making: First to complete a specified path.
  McGaughey's Game, Orion: Hydra, Crosstrack, Square Off, Turn, Morse's Game, Suspension Game, Meander, Onyx, Link, Tack, Fan, Eynsteyn, Trinidad, Tara, Bez's Game, Troll, Panda's Game, Sisimizi, Proton, Kage, Split, Apex, Square Board Connect, Notwos, Alta, Ecoute-Moi!, Quintus, Creeper
- Convergent: First to converge to a single group of pieces.
  Octiles: Team Up, Orion: Banana Boat, Lines of Action, Groups, Moloko

Connective Play: Moves are defined by connection.

- Path Making: Players make or extend paths with each move.
  Mongoose Den, Nile, Rivers, Roads & Rails, The Great Downhill Ski Game, Trails to Tremble By, Waterworks, Weber's Game, Le Camino, Galdal's Game, Deaton's Game, Star, Kaliko, Hextension, Yaeger's Game, Tantrix, Würmeln, Andantino, Schlangennest, Puddles, Metro, Ta Yü, The Very Clever Pipe Game, Meander, Nexus, TransAmerica, Speleo, Snap, The Legend of Landlock, Rumis
- Territorial: Players strive to enclose territory with each move.
  Go, Pathfinder, Worm, Steppe/Stak, Tanbo, Rum's Game, Through the Desert, Nasca, Blokus, Anchor, Symbio, Orbit, HexGo, Waroway's Game, Block, Occupier, Parcel, Great Walls, Feurio!

Figure 6.3. Taxonomy of connection games.

If a game belongs to multiple categories, then the most representative one is chosen.

There are other ways to classify these games. For instance, a distinction is often made between piece-based and tile-based connection games. However, Figure 6.4 shows that a piece has identical connectivity to a completely paved tile. Tiles act as filtered pieces, but piece-based games are in fact a subset of tile-based games.

The distinction between piece-based and tile-based games focuses on surface level mechanics, and it is hoped that the proposed classification scheme reveals a deeper understanding of connection games. It focuses on types of connections rather than the types of pieces that make up those connections.

6.3 Drawing the Line

This scheme provides practical guidelines for deciding whether a given game is a connection game or not. Many games fall just short of meeting these criteria, and it is useful to examine such games to define where the boundary is drawn.

Connection-Related games are those in which connection plays a significant but not exclusive role during play or in the winning condition. Some nonconnective element excludes them from the strictly connective classes and they are not considered connection games as such. For instance, some by-product of connection is used rather than the fact of connection itself.

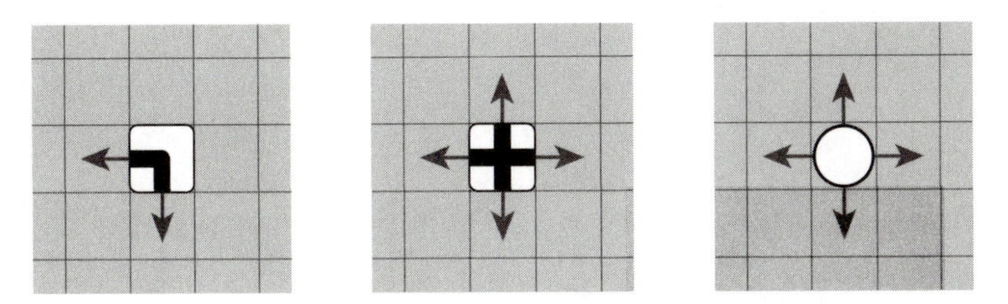

Figure 6.4. Tiles are filtered pieces.

Part II

The Games

Part II is a catalogue of connection games, with key games examined in detail. Games are arranged chronologically within each category, and those with identical names are distinguished by year. Each subcategory is represented by an icon for quick visual reference.

The Pure Connection category is the most important group. While every effort has been made to make the list of games in this category exhaustive, it should be pointed out that new (and sometimes old!) Pure Connection games continue to emerge.

The game lists for the Connective Goal and Connective Play categories are intended to be representative rather exhaustive, but should still be reasonably complete. Notable games from each category are discussed in detail.

A brief list of Connection-Related games is then presented, and a few notable examples shown. No attempt has been made to compile an exhaustive list of Connection-Related games, as this would potentially contain thousands of games.

Several game descriptions are drawn from US patents belonging to the category 273/275 (Amusement Devices: Games/Path Forming). Due to the nature of patents, these descriptions are general and not tied down to a specific embodiment of a game. Some patents describe devices on which players compete to close electrical circuits across an array (literally switching games) but these are not board games as such and hence not included in this catalogue.

Games described as "Joris games" are taken from Walter Joris's *100 Strategic Games for Pen and Paper* [2002]. Thus "Joris game #57" refers to the 57th game in this collection.

The term "Zillions of Games" refers to a computer program for playing a wide range of board games: http://www.zillions-of-games.com/. Zillions of Games features a script-based rule language that allows new games to be defined.

7 Pure Connection Games

Pure Connection games are those with both strictly connective play and strictly connective goals. Connection is paramount in these games, which may be described as being connection-based at both the tactical and strategic levels. This group forms the core of the connection game family.

7.1 Pure Connection > Absolute Path

Absolute Path games are those in which a player wins by completing a path that precludes any possible winning path by the opponent. These games generally have complementary goals (exactly one player must win) allowing simple rule sets but significant depth. Good defense is generally equivalent to good attack.

Most of the games in this category are played on the hexagonal grid to avoid deadlock issues. Those games played on other types of grids have some alternative mechanism to avoid this problem.

Absolute Path games fall into one of two categories when stripped down to their underlying graph:

- played on the cells (vertices), or
- played on the edges.

They are also sometimes called *border-to-border* games.

7.1.1 Pure Connection > Absolute Path > Cell-Based

Cell-Based games typically involve placing a colored piece on an empty board point each turn. These games are equivalent to Shannon games on the vertices.

The icon for this group shows a path of black pieces cutting a path of white pieces.

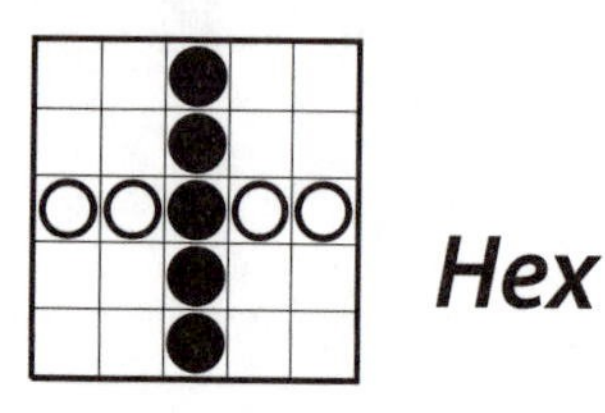

Hex

Hex is the game that kick-started the connection game genre in the middle of the twentieth century. It has extraordinarily simple rules yet remains one of the most difficult and interesting of all connection games.

Rules

Hex is played on a rhombus of hexagons, typically 11 x 11, which is initially empty. Two players, Black and White, own alternating sides of the board that bear their color. Players take turns placing a piece of their color on an empty cell.

The game is won by the player who connects his two sides with a chain of his pieces. Exactly one player must win (see Appendix C). Figure 7.1 shows a game won by White, who has completed a chain of white pieces between the white sides of the board.

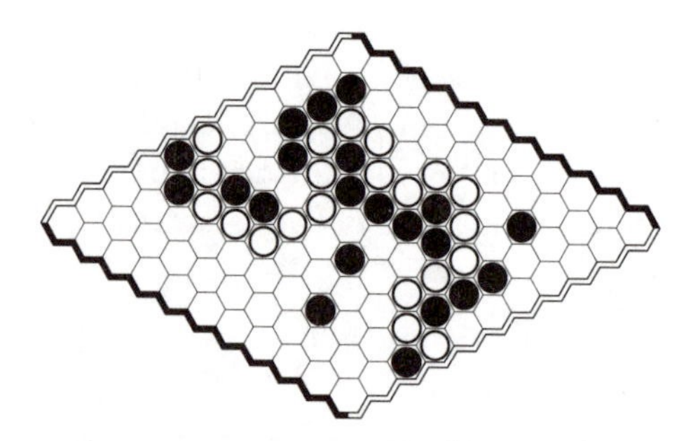

Figure 7.1. A game of Hex won by White.

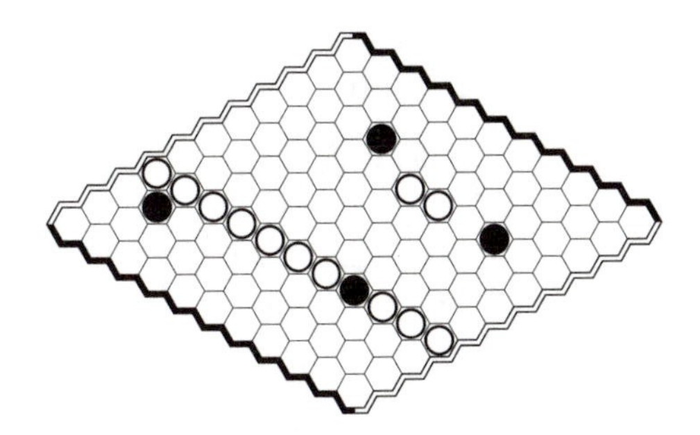

Figure 7.2. Puzzle by Piet Hein: Black to play and win.

The first player has a huge (winning) advantage, especially if allowed to open near the center of the board. It is recommended that a single-move swap option be used (described in Section 4.3.3).

Notes

The basic nature of Hex can be summarized by the following example. Figure 7.2 shows a simple puzzle devised by Hex's inventor, Piet Hein, over 50 years ago; Black to play and win.

Figure 7.3 shows the winning move ***1***. Black now has a *virtual connection* (a potential connection that the opponent cannot block in isolation) to both black sides.

The virtual connection to the top right side is guaranteed by the path indicated. If White intrudes into this path at any point then there is always a matching move that Black can play to keep the connection alive.

The virtual connection to the bottom left side is guaranteed by the ladder that will develop in the direction indicated by the arrow in Figure 7.3. This ladder has an escape piece and is guaranteed to succeed.

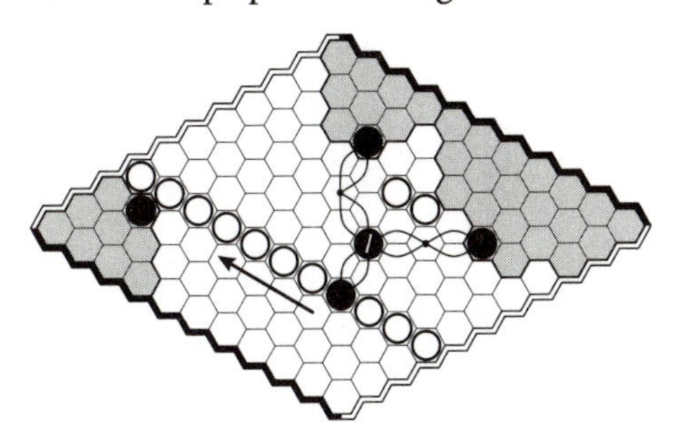

Figure 7.3. Move *1* wins for Black.

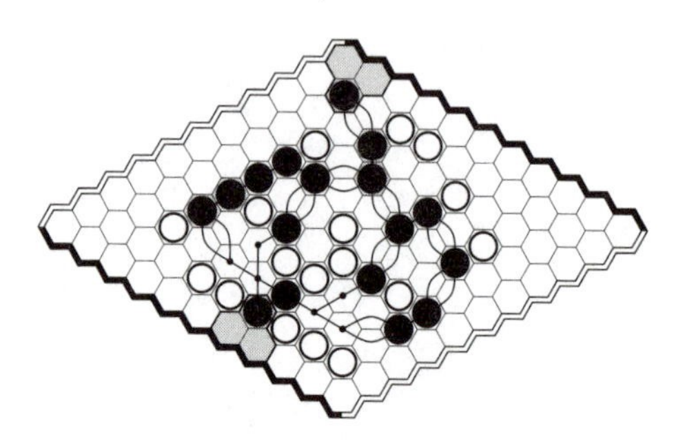

Figure 7.4. A provable win for Black.

Figure 7.4 shows a more complex example of board analysis. Black has already won this game, as every possible White intrusion has a matching reply that restores the connection.

Hex is a typical Shannon game on the vertices. See Section 3.2.2 for details.

Strategy and Tactics

Note the presence of bounded and shaded regions in Figures 7.3 and 7.4. These are *edge templates*, which represent known patterns that guarantee safe connection of a piece to the nearest edge. The shaded cells must be empty (or occupied by friendly pieces). Templates are a way of simplifying board analysis by parceling up known connections into units that can then be used in higher-level analysis.

Figure 7.5 shows further detail of the edge template in the left-most corner of Figure 7.3. It can be seen that this template is safe by revealing disjoint path pairs linking the black piece to the black edge. If White intrudes anywhere within this template, then Black has a reply that re-establishes the connection.

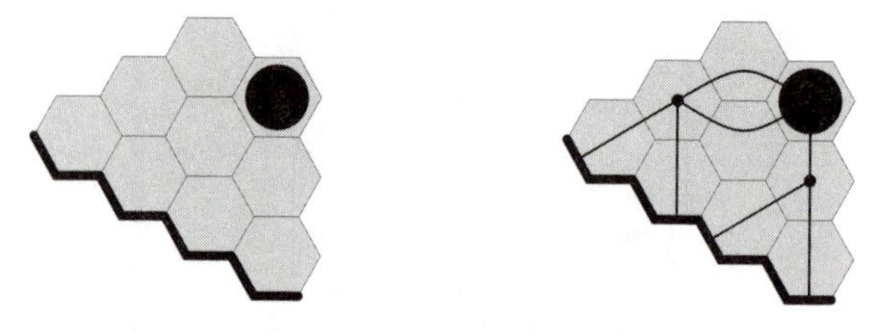

Figure 7.5. An edge template.

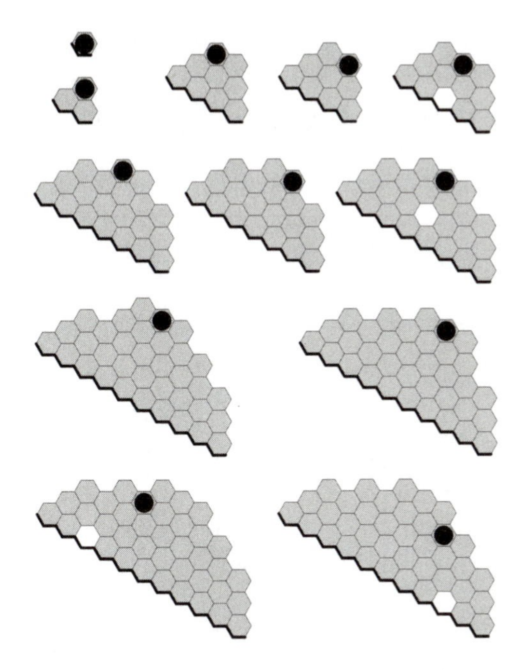

Figure 7.6. The 12 single-piece edge templates.

There are 12 known single-piece edge templates, as shown in Figure 7.6. Familiarity with these templates is a useful Hex skill that can help the player quickly evaluate board positions. Unshaded points do not belong to the template, and may be empty or occupied by either player without affecting the connection.

These templates decrease in utility as they increase in size, since it is more difficult to fit larger templates on the board. It is unlikely that the two largest templates would ever be used on an 11 x 11 board.

All of the common plays discussed in Chapter 5 apply to Hex. Forcing moves, as described in Section 5.4, are especially relevant and usually decide the outcome of most games of Hex.

Mathematical Background

Hex has probably the most famous heritage of any recent abstract board game. It was independently invented by two of the twentieth century's greatest mathematicians, Piet Hein and John Nash.

Hex first occurred to Piet Hein in 1942 while contemplating the famous four-color theory of topology [Gardner 1957]. This is evident from the basic map-coloring nature of Hex and many of the connection games derived from it.

Hex was independently reinvented by John Nash in 1948 while developing his pioneering work on game theory, for which he later received a Nobel prize.

Hex has appealed to mathematicians ever since its invention, largely due to the fact that the game itself embodies two elegant mathematical proofs:

- one player must win (see Appendix C), and
- the first player has a winning strategy (see Appendix D).

There has been significant research interest in Hex. Some important findings are that a generalization of Hex is PSPACE complete [Even and Tarjan 1976], Hex itself is PSPACE complete [Reisch 1981], and that Brouwer's fixed point theorem is related to the no-tie property of Hex [Gale 1979].

History

Hex was originally called Polygon by Piet Hein, and has also been called John, Nash, Join, Con-Tac-Tix, and other names over the years. It has been briefly marketed from time to time under various names, but did not achieve wide recognition until Gardner's *Scientific American* article [1957] propelled Hex to fame and kick-started the connection game genre. Further details regarding the history of Hex can be found in Browne [2000] and Milnor [2002].

Variants

Although Hex is the inspiration behind many subsequent connection games, some can be singled out as direct Hex variants. These are games that use the standard Hex rules with minor modifications.

Beck's Hex

Beck's Hex is played as per standard Hex, except that the second player dictates where the first player must open. This is a very harsh form of first-move equalizer; the second player can specify a losing move for the opener [Beck 1969].

Misère Hex

The first player to connect his sides of the board with a chain of his pieces loses. Misère Hex was first described as Reverse Hex by Ronald Evans [1974]. Lagarias and Sleator demonstrate that the first player has a winning strategy on an $n \times n$ board when n is even, and the second player has a winning strategy when n is odd; furthermore, the losing player has a strategy that guarantees that every cell of the board must be played before the game ends [1999].

Vex (1975)

The first player opens in an acute corner and wins by connecting this piece with either of the two edges opposite. The second player wins by cutting this connection off. Vex was proposed by Ronald Evans [1975] and named by David Silverman.

Vertical Vex

Vertical Vex is identical to Vex, except that the first player opens anywhere along an edge and wins by connecting this piece with the opposite edge. Vertical Vex was proposed by Ronald Evans [1975].

Tex

Tex is played on an infinite board, on which the second player wins by completing a cycle around the first player's opening move. This game was invented by Ronald Evans [1975] who named it Tex due to the size of the board. Tex is more a theoretical exercise than a suggestion for a real game; it is not known whether or not the first player can avoid defeat indefinitely.

Three-Player Hex

Three-Player Hex is played on the hex hex board, typically with five cells per side. As in standard Hex, players take turns placing a piece of their color on an empty cell, and the first player to connect the opposite sides of the board marked his color with a chain of his pieces wins.

As discussed in Section 4.4.4, deadlocks are elegantly avoided by the following additional rule: *as soon as it is no longer possible for a player to connect his edges, that player is eliminated from the game and may not place any more stones.*

David Straffin [1985] explains that this version of Three-Player Hex was invented by Thomas Sibley and shown to him in 1975.

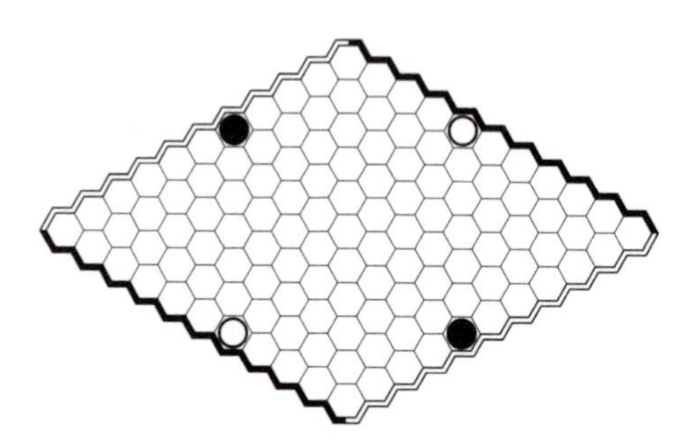

Figure 7.7. Head Start Hex.

Kriegspiel Hex

Each player moves on his own board and is unable to see the opponent's board. A referee with a master board has the following duties:

- to advise players when it is their turn,
- to declare an illegal move so that the offending player can try a different move, and
- to state when a player has won.

Kriegspiel Hex was proposed by William McWorter [1981] who states that there is no winning strategy for the first player on an n x n board if $n \geq 4$.

Head Start Hex

Head Start Hex is played as per standard Hex, but starts with the board set up as shown in Figure 7.7.

This starting position, suggested by Larry Back [2001a], reduces the strength of the board center compared to that of standard Hex, and brings the corner and edge regions more into the game. While standard Hex games generally develop around the center of the board before branching out to the edges, this trend is somewhat reversed in Head Start Hex, where players are just as likely to secure corner connections before developing them towards the center.

Square Hex

Square Hex, played using standard Hex rules on the board shown in Figure 7.8, was designed by Larry Back [2001a]. Of the many exotic tilings on which a game of Hex can be played, the Square Hex board has some properties that lend itself to the game especially well.

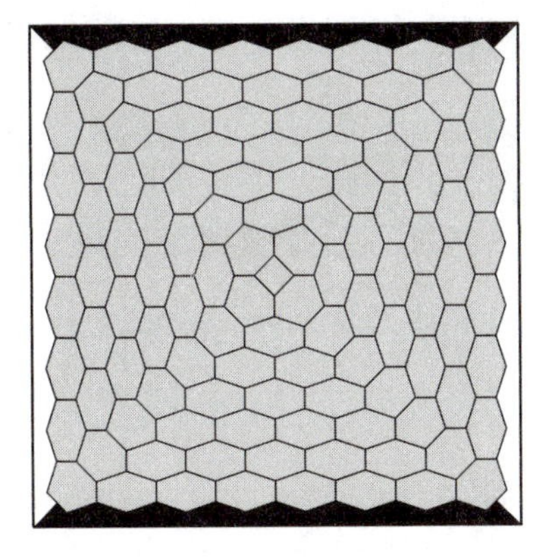

Figure 7.8. The Square Hex board.

Firstly note that players' shortest winning paths are along the sides of the board. This makes the corner and edge spaces more important than in standard Hex, yet a path along the edge is still the most easily blocked [Back 2001a]. The middle of the board is less important than in standard Hex. In stark contrast to Hex, the central point is a weaker four-sided square cell that makes it the least attractive move. These factors mean that each cell on the Square Hex board is of relatively equal importance; cell strengths are distributed more evenly across the board than in standard Hex.

The Square Hex board is the same design as one of the Poly-Y boards shown in *Mudcrack Y and Poly-Y* [Schensted and Titus 1975, 181] with modified edge markings.

Scorpio

Scorpio is played as per standard Hex, except that players place two nonadjacent stones of their color per turn. William Chang proposed the above rules under the name Scorpio in a 2001 posting to the newsgroup rec.games.abstract, although Double-Move Hex was first described by Martin Gardner [1959]. Double moves lead to a fast-paced and exciting game.

Pass Hex

Played as per standard Hex except that players may elect to pass on their turn. The eventual winner scores a number of points equal to 1 plus an additional point for each passing move. Pass Hex was suggested by Taral Guldahl Seierstad in July 2003.

Figure 7.9. The Bipod starting position.

Vex (2004)

This second game called Vex (obviously a popular name that rhymes with Hex) includes a neutral piece that the second player may elect to place on an empty cell at any point during the game, even before the opening move. This neutral stone does not count in either player's connection and means that every game should end in a draw with perfect play [Neto 2004].

Bipod

Bipod is played on the trapezoidal board shown in Figure 7.9. The game starts with a black piece and a white piece placed as shown.

In an initial contract phase, players take turns placing a gray piece on an empty cell. A player may choose to pass, at which point he becomes the *Runner* and the opponent becomes the *Blocker*. Thereafter, the Blocker continues to play gray pieces, but the Runner may play either a black or a white piece per turn. The Runner wins by connecting both

- the initial black piece to any of the three short sides with a chain of black pieces, and
- the initial white piece to any of the three short sides with a chain of white pieces.

The Blocker wins by preventing either or both connections.

Bipod was devised by Bill Taylor in 2004. It is essentially Vex (1975) played with two paths on a modified board, and with an Unlur-like contract phase.

Other Variants

A number of Hex variants constitute significant games in themselves and are discussed further in their own sections. These games include

- Cylindrical Hex,
- Chameleon,
- Eight-Sided Hex,
- Nex, and
- Diskelion (see Triskelion).

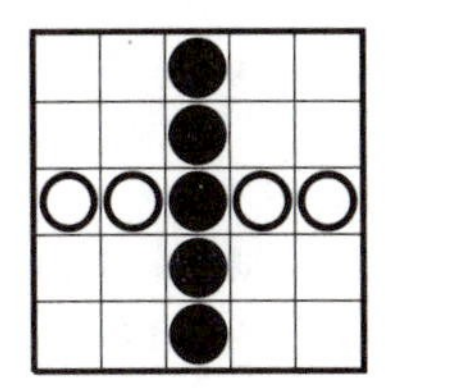

Y

Y was invented shortly after Hex and remains one of the giants in the connection game genre. These two games embody the principle of connection and influenced most connection games that followed.

Rules

Y is played on a triangular board tessellated by hexagons. The board is initially empty. Two players, Black and White, take turns placing a piece of their color on the board.

A player wins by connecting all three sides with a chain of his pieces. Exactly one player must win (see Appendices F and G).

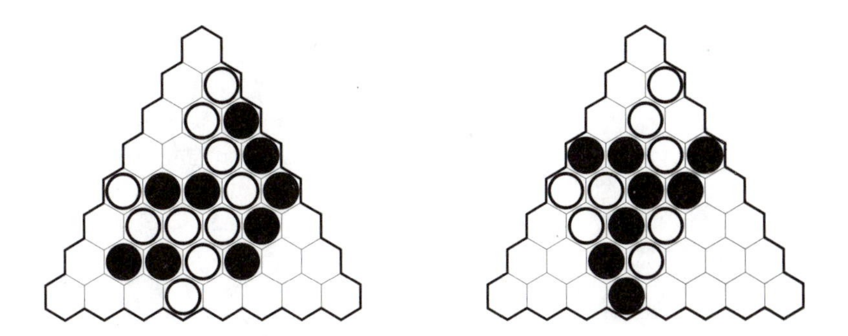

Figure 7.10. A game won by White (left) and a game won by Black (right).

As in Hex, the first player has a significant first-move advantage, especially when opening near the center of the board. It is recommended that Y be played with a single-move swap option.

Notes

The Y board is usually a regular hexagonal tessellation, as shown in dual form in Figure 7.11 (left). Schensted and Titus designed the elegant board (right) in which three of the interior cells are pentagonal rather than hexagonal. Although this introduces weaker cells onto the board, it has the benefit of distributing cell importance more evenly; central cells become slightly less powerful and the acute corners become slightly less weak. This rounding of shape means that center-to-edge distances are more uniform around the board, both conceptually and visually.

Schensted and Titus introduce the *mudcrack principle*, the fact that games of this type can be played on a wide variety of tilings. They provide hundreds of exotic board designs for precisely this purpose [Schensted and Titus 1975]. For simplicity, this discussion will focus on Y as played on the standard board.

It is sometimes argued that Y is a more fundamental game than Hex, due to two main reasons:

- the goals of Y are simpler than those of Hex, and
- Hex is a special case of Y.

The goals of Y are about as pure as can be. It is one of the very few board games of any sort with equal and complementary goals; both players aim for the same goal but exactly one must win. By comparison, the goals of

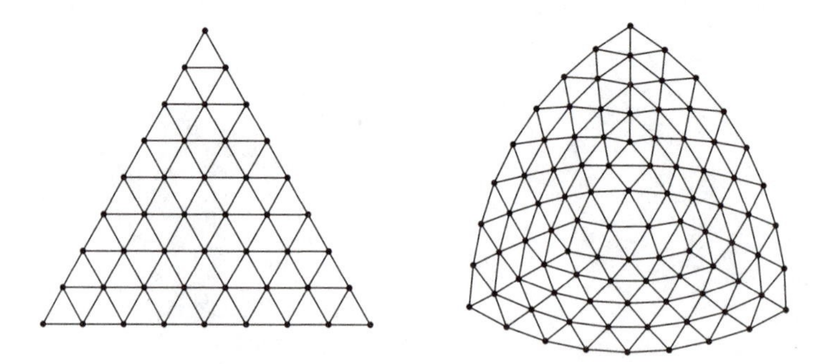

Figure 7.11. The standard Y board in dual form and the commercial Y board.

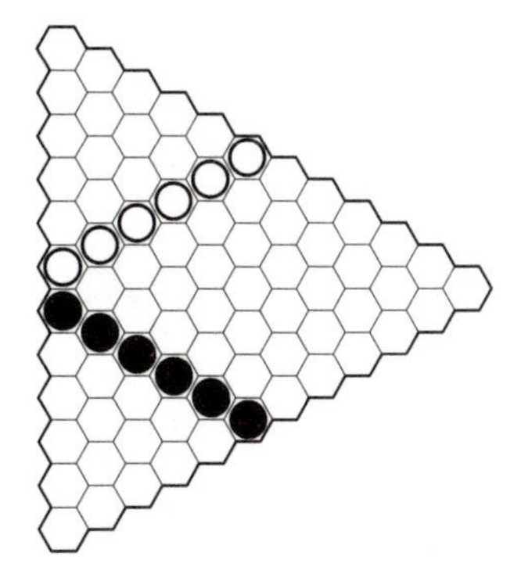

Figure 7.12. Hex is a special case of Y.

Hex are complementary but symmetric rather than equal. However, it can be argued that the goals of Hex are more intuitive, and that the concept of one path cutting across another will be clearer and more familiar to the average player than the goal of connecting all three sides.

Setting up a Y board as shown in Figure 7.12, it can be seen that completing a Y connection is then equivalent to completing a game of Hex. Black wins if and only if the black line of pieces connects to the parallel board edge opposite, and White wins if and only if the white line of pieces connects to the parallel board edge opposite it. Hex can be formulated as a special case of Y without modifying the rules.

Schensted and Titus [1975] argue that Y is a more interesting game than Hex for this reason. Yet no matter how valid this is as a theoretical result, it is extremely unlikely that this formation will ever occur in practice. Each game has its own strengths and distinct character.

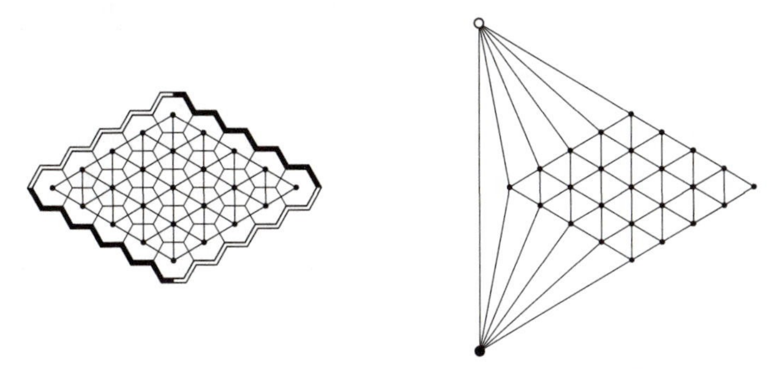

Figure 7.13. Hex can be phrased in terms of Y.

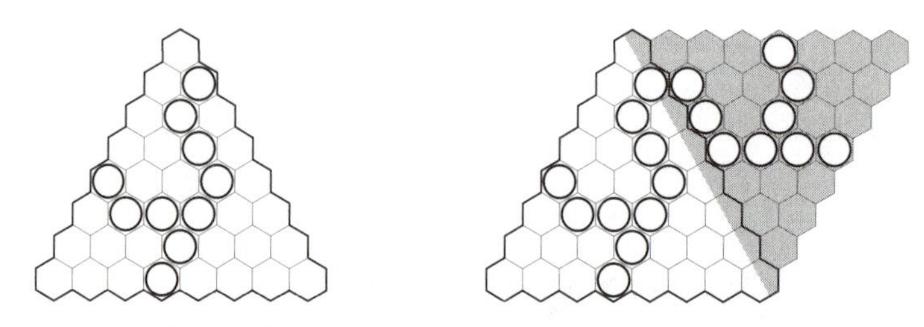

Figure 7.14. Y can be phrased in terms of Hex.

Figure 7.13 further shows the relationship between Hex and Y, as suggested by Chris Hartman. On the left is the adjacency graph of a 5 x 5 Hex board. On the right is the same graph with a white and black vertex added. By Sperner's Lemma (Appendix G) we know that a completely filled Y board will contain a chain of one color connecting all three sides. By a similar argument it can be shown that the graph on the right will define a win for either Black or White when completely filled.

Furthermore, it can be argued that in fact the opposite is true, and that Y can be formulated in terms of Hex (with some modification). Figure 7.14 shows how this can be achieved by reflecting the Y board and its pieces along one edge. The same Shannon game analyses that apply to Hex may then be applied to the reflected game graph.

By arguments analogous to those used in Hex, it can be seen that Y cannot end in a draw and that it must be a theoretical win for the first player [Van Rijswijck 2002].

Strategy and Tactics

Strategies and tactics used in Hex such as forks, ladders, forcing moves, and edge templates are also important in Y. However, one major difference is that a player may attack two adjacent target edges at once in Y.

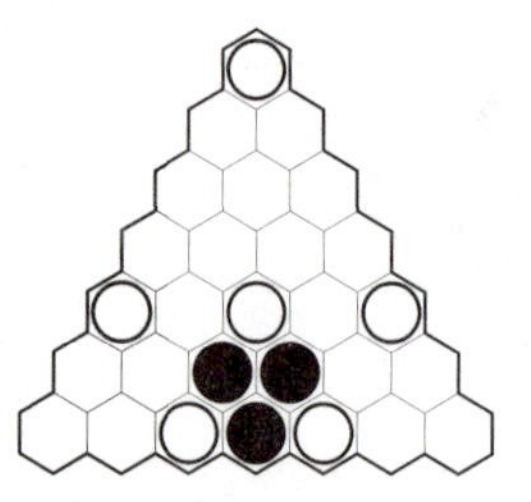

Figure 7.15. Puzzle: Black to play and win.

The simple puzzle shown in Figure 7.15 (Black to play and win) demonstrates most of these points. Black's winning play is a forking move that provides ladder escapes to two edges at once.

History

Claude Shannon first described Y in the early 1950s, not long after the invention of Hex [Gardner 1959]. Y was independently reinvented around the same time by Craige Schensted and Charles Titus, who studied the game in depth and produced the excellent book *Mudcrack Y and Poly-Y* [1975]. The modified Y board with pentagonal cells shown in Figure 7.11 was invented by Schensted and Titus in 1969.

Variants

Below is a brief list of games derived from Y.

Master Y

Master Y is played as per standard Y, except that players place two pieces per turn (except for the first turn, which is a single piece to address the first-move advantage). Described by Schensted and Titus [1975], Master Y is similar to the Hex variant Scorpio.

Fortune Y

Players shuffle a deck of cards and turn up one card per move: if red then White moves, if black then Black moves. Dice or coins can be used instead to determine how many pieces each player can place per turn. Schensted and Titus designed these randomized variants [1975].

Fifty-Fifty Y

Fifty-Fifty Y is similar to Fortune Y except that the order of play for each turn is determined by some random factor, such as the draw of a card. If the card is red then that turn White plays first, followed by Black; if the card is black then that turn Black plays first, followed by White.

This variant was suggested by Schensted and Titus [1975], who point out that it is not always desirable to move first in a round, as the only way to move twice in succession is to get the second move of one round and the first move of the next.

Tabu Y

Tabu Y is played as per standard Y, except that it is forbidden to play adjacent to the opponent's last move. This leads to some interesting and subtle developments of play. Draws in which a player has no legal move are possible, but not likely in practice. Tabu Y was devised by Schensted and Titus [1975].

Holey Y

Holey Y is a standard game of Y played on a board with holes or forbidden cells. Schensted and Titus [1975] propose a number of possible board designs.

Obtuse-Y

Obtuse-Y, suggested by Greg van Patten in 2003, is played on the hex hex board. A player wins by establishing a chain of his pieces connecting at least three sides, such that at least one side from each opposed pair is included in the connection. Obtuse Y is a race game but cannot be drawn.

Gem

In 2003, Bill Taylor proposed two games played on the board shown in Figure 7.16. In both games, players take turns placing a piece of their color on an empty cell. A single-move swap option is recommended.

> Gem-Y: A player wins by establishing a chain of his pieces connecting any four sides, or any side and its opposite two sides.
>
> Gem-EE: A player wins by establishing a chain of his pieces connecting any four sides, or any two chains of his pieces connecting three sides each.

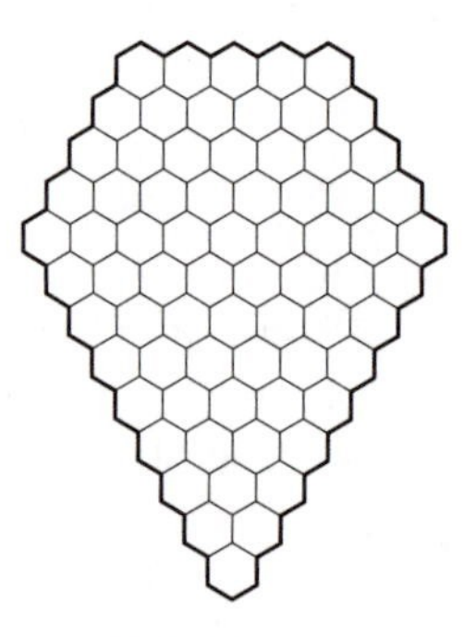

Figure 7.16. The Gem board.

Why Not (Misère Y)

The first player to connect all three sides loses. Suggested by David Molnar, who recommends using smaller board sizes as games generally fill all available cells unless a player makes a mistake.

Other Variants

The following variants are discussed in detail in their own sections:

- Poly-Y,
- TriHex, and
- Caeth Y (see Caeth).

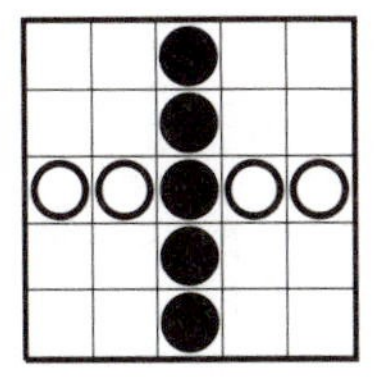

Split (1966)

Split is an early connection game that introduces a round board with player-defined goals.

Rules

Split is played on a small circular board consisting of 26 cells. Each cell is connected to four, five, or six neighbors, and the outer ring of edge cells are marked "S-P-L-I-T-S-P-L-I-T" (Figure 7.17). The board is initially empty.

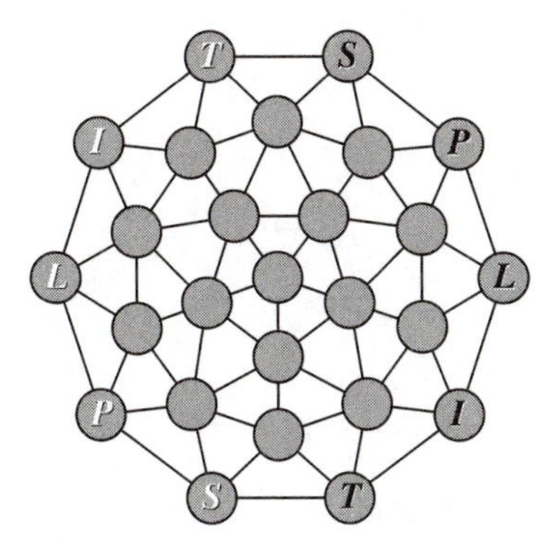

Figure 7.17. A Split board.

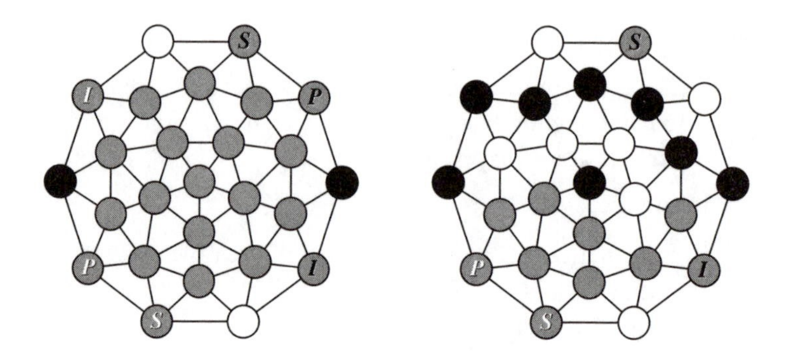

Figure 7.18. A game won by Black.

Two players, Black and White, each select a pair of opposed edge cells with the same letter to be their goals, and mark them with pieces of their color. Players then take turns placing a piece of their color on an empty cell. The first player to complete a chain of his color between his goals wins the game.

Notes

Figure 7.18 (left) shows the start of a game in which Black has chosen the letter *L* to be his goal pair, and White has chosen the letter *T*. Figure 7.18 (right) shows this game after a win by Black.

The Split board corresponds directly to its game graph. Games are short, straightforward, and tend towards deadlock; the game graph is trivalent, but it is easy to cut off an opponent's goal cell (Figure 7.19).

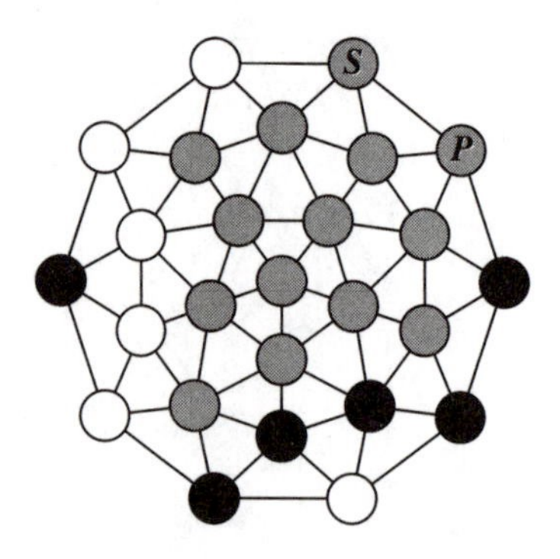

Figure 7.19. A deadlocked game.

Split's most notable contribution to the connection game genre is the introduction of user-defined goals.

History

Split was published in 1966 by the Western Publishing Company. Little is known about Split apart from the information provided by Mark Thompson: http://www.flash.net/~markthom/html/split.html.

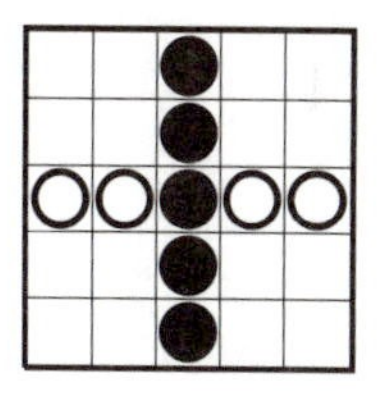

Beeline

Beeline introduces unusual playing pieces that incorporate both players' colors with each move.

Rules

Beeline is played on a hex hex board with nine cells per side. The board is initially empty. Two players, Black and White, share a common pool of 24 of each type of tile shown in Figure 7.20. Each tile, when placed on the board, covers exactly three empty hexagons.

Black starts by placing a tile of either type on the board. Players then alternate placing a tile of either type on the board such that it touches at least one edge of at least one existing tile.

A player wins by establishing a path of hexagons of his color

- between any two opposite sides of the board,
- between three alternating sides of the board, or
- surrounding at least two hexagons of the opponent's color.

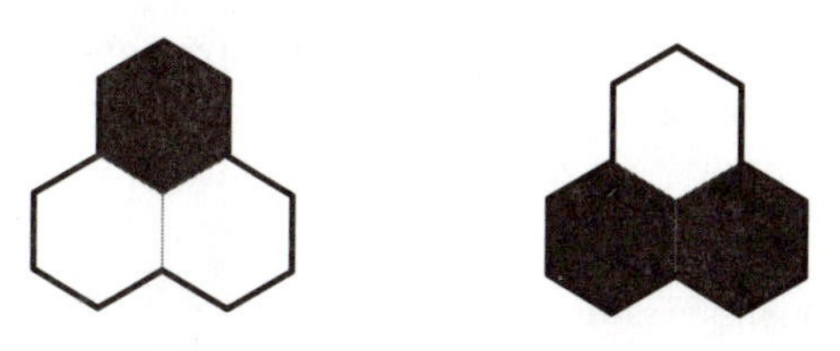

Figure 7.20. The Beeline tiles.

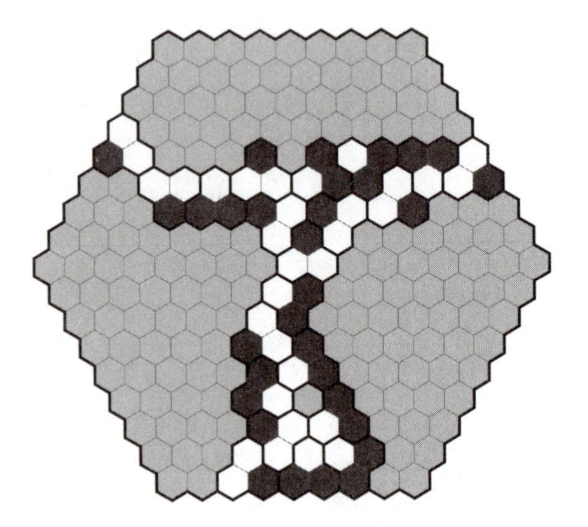

Figure 7.21. A game won by White.

Figure 7.21 shows a game won by White, who has completed a path of white hexagons between three alternating sides of the board.

The game is drawn if the tiles run out before either player wins. The rules do not state who wins if a tile placement achieves a winning connection for both players (presumably the mover wins). Corner cells belong to both adjacent sides.

Notes

The Beeline board and winning conditions are almost identical to those of Havannah. However, packaging three adjacent hexagons in a single tile is an interesting variation. The fact that every tile includes at least one hexagon of the opponent's color means that players must plan their moves carefully.

Strategy and Tactics

It goes without saying that players should always choose a tile with two hexagons of their color for every move.

History

This version of Beeline, as described by William Chang, is attributed to John Brassell (1984).

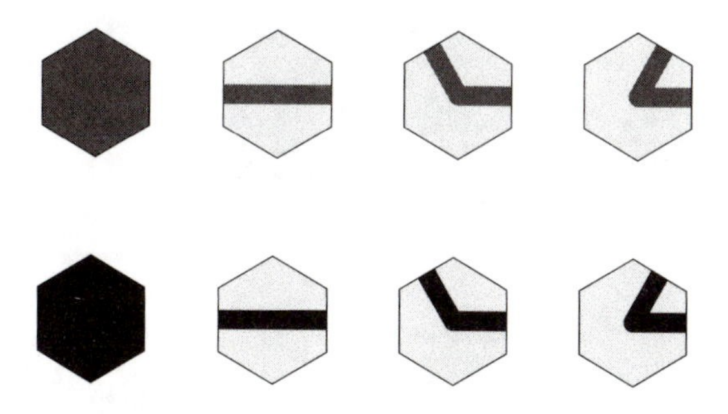

Figure 7.22. Tiles from an earlier version of Beeline.

Variants

There is an earlier version of Beeline published in 1968 by Good Games and in 1974 by Brassell Games. Little is known about this version, beyond that it is played on the same board using the tiles shown in Figure 7.22. Presumably each color (light and dark) belongs to a player, and the solidly colored tiles are goal pieces that players aim to connect.

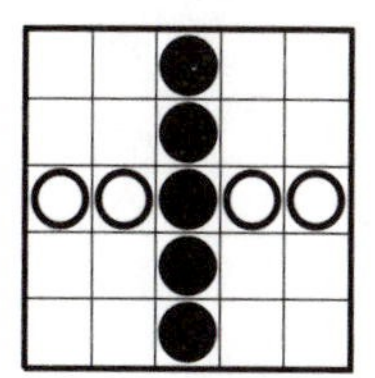

Cylindrical Hex

Cylindrical Hex is effectively Hex played on a board wrapped around a cylinder. A general solution is known.

Rules

The board is a standard Hex board except that cells connect across the left and right sides. It is initially empty. Two players, Black and White, take turns placing a piece of their color on an empty cell.

Black wins by establishing a black chain between the top and bottom sides, and White wins by establishing a white chain between connected cells on the left and right sides. Exactly one player must win.

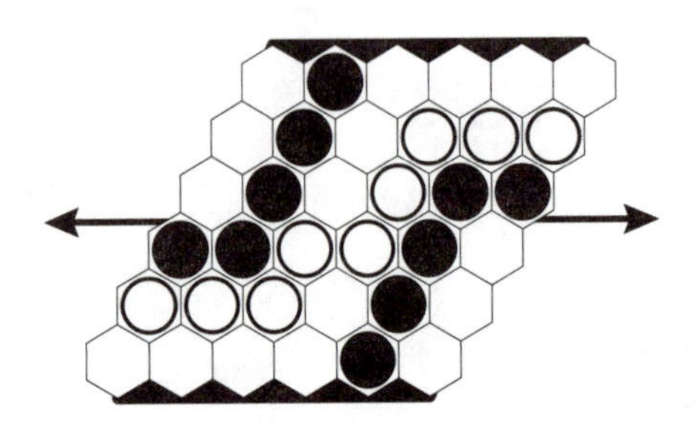

Figure 7.23. A game won by Black.

Figure 7.23 shows a game won by Black. Note that the winning black chain crosses between connected cells on the left and right sides.

Notes

The cylindrical board can be deformed into an annulus to show the wraparound connections explicitly. For instance, Figure 7.24 shows the same game won by Black in square offset grid format (left) and annular format (right). It can be seen that Black has won this game as there exists a chain of black pieces connecting the top (inner) side of the board with the bottom (outer) side.

Strategy and Tactics

Cylindrical Hex is one of the few connection games with unequal goals. Black obviously has the easier task, and Alpern and Beck [1991] have in fact proven that Black can win on any n x n board when n is even, even if moving second.

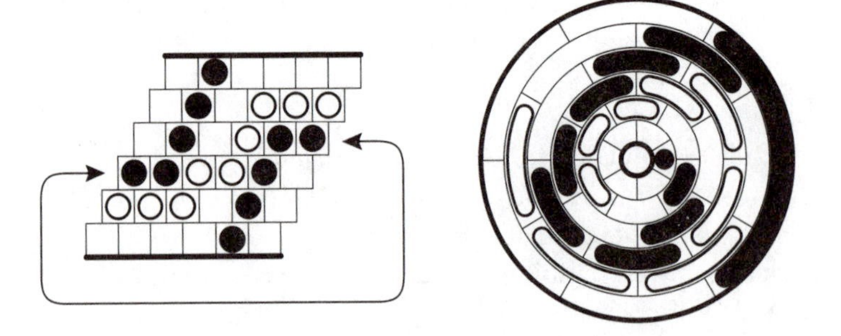

Figure 7.24. The same game in annular format.

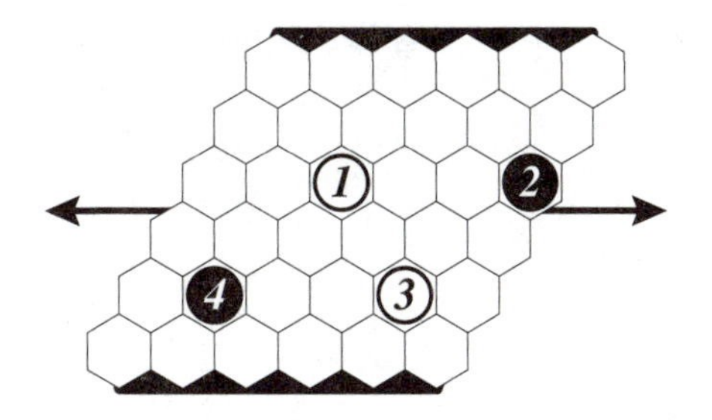

Figure 7.25. Black's winning point-pairing strategy.

The winning strategy for Black involves playing on the same row but half a board away from the last White move. In terms of the cylindrical nature of the game, this is the same as playing in the cell that is the same distance along the cylinder's axis but on the far side from White's last move. If the cell that Black wants to occupy is already taken (for instance by Black's first move) then Black can play at any empty cell without harming their position.

Figure 7.25 shows Black's winning *point-pairing* strategy in action. Black's winning replies to White moves *1* and *3* are *2* and *4*, respectively. Alpern and Beck believe that this strategy also applies when n is odd, but have not proven this. See Appendix E for a further discussion of point-pairing strategies in connection games.

History

Cylindrical Hex was proposed by Alpern and Beck [1991]. Their paper introduces the game, proves that exactly one player must win, and outlines the winning point-pairing strategy for Black.

Octagons

Octagons is an elegant connection game played on an unusual grid. Multiple moves per turn balance out inequities between cell sizes.

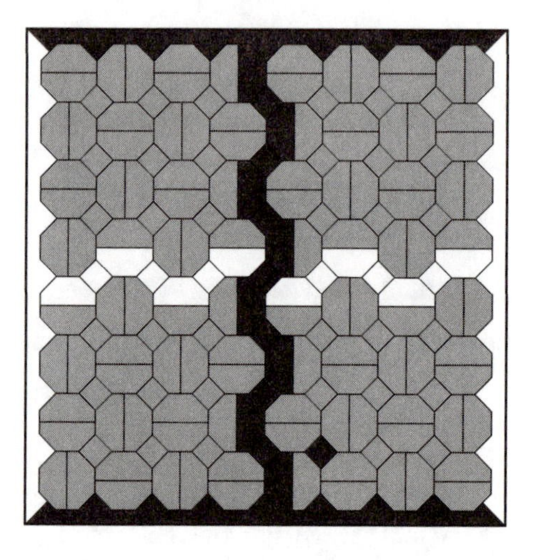

Figure 7.26. A game won by Black.

Rules

Octagons is played on the board shown in Figure 7.26, which is composed of half-octagons and squares. Two players, Black and White, own alternating sides of the board that bear their color. The board is initially empty.

Players alternate taking turns, in which the current player must claim either an unclaimed half-octagon, or two unclaimed squares. The first player to complete a chain of cells in his color between his edges of the board wins the game. Exactly one player must win. A single-move swap option is recommended.

Notes

The Octagons board looks similar to the 4.8.8 Archimedean grid used in Stymie but is fundamentally different due to the division of octagons into half-octagons. Interior half-octagon cells have seven neighbors.

Figure 7.27 shows the dual of the Octagons board. Its game graph is equivalent to that used for Onyx, as observed by Larry Back [Handscomb 2001b]. All regions of the game graph are three-sided, hence this game is free from deadlock.

The 4-connected square cells are substantially weaker than the 7-connected half octagons. As with Stymie, the ability to claim more than

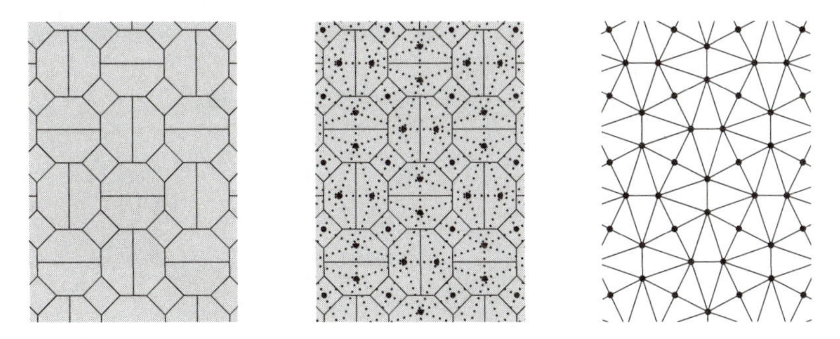

Figure 7.27. The Octagons board and its dual.

one square compensates for this weakness and encourages players to bring square cells into the game. It is possible for a player to achieve a winning connection using no square cells, but this is unlikely.

Strategy and Tactics

The two black pieces shown in Figure 7.28 are strongly connected. White is not able to stop Black from connecting these pieces (when this system is considered in isolation).

Note that three alternative and nonoverlapping routes between pieces are required for a safe connection (Figure 7.28, right), as the opponent may make double moves each turn.

Figure 7.29 shows a weakly connected pair of black pieces with only two direct routes between them. If White is able to occupy one of these links and threaten elsewhere at the same time, then this tenuous black connection can be broken. White pieces correspond to Cut moves in the game graph, as with all Absolute Path games played on the vertices of a planar deadlock-free graph.

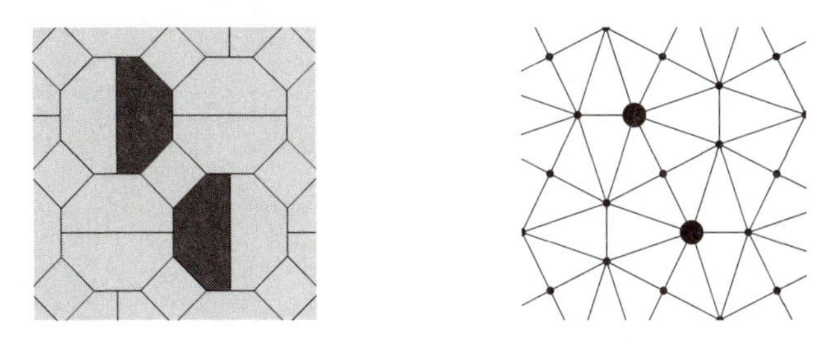

Figure 7.28. A strongly connected pair of black pieces.

Figure 7.29. A weakly connected pair of black pieces.

Figure 7.30 illustrates a successful double-move intrusion with an example similar to that given by Schmittberger [2000]; White to play and connect to the left edge.

White can break through Black's defense by playing the double square move shown in Figure 7.30 (right). Although Black can answer one threat, White can claim the other path next turn to complete his connection to the white edge. The double move must threaten two different connections at once for this attack to work; in other words, it must be a forking move. This is a standard Octagons tactic.

Kerry Handscomb [2001b] describes weakly connected pieces as *semiconnected* and strongly connected pieces as *fully connected.* Figure 7.31 shows a commonly occurring formation called the *dog's leg.*

The two dog's legs (Figure 7.31, left and center) are weakly connected and prone to disconnection. However their union (Figure 7.31, right) is strongly connected: White cannot stop Black connecting the top piece to the bottom two.

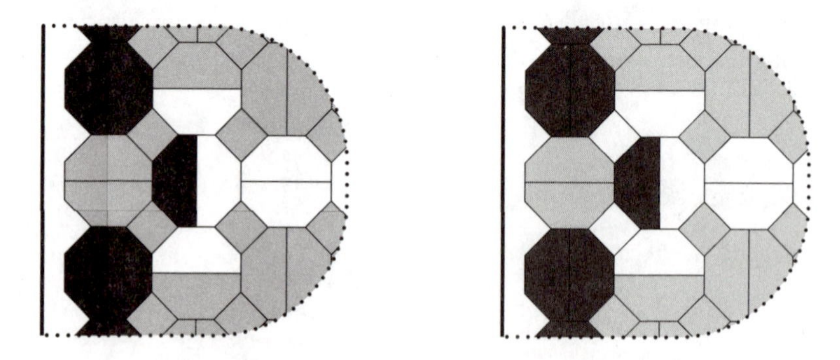

Figure 7.30. White can break through to connect.

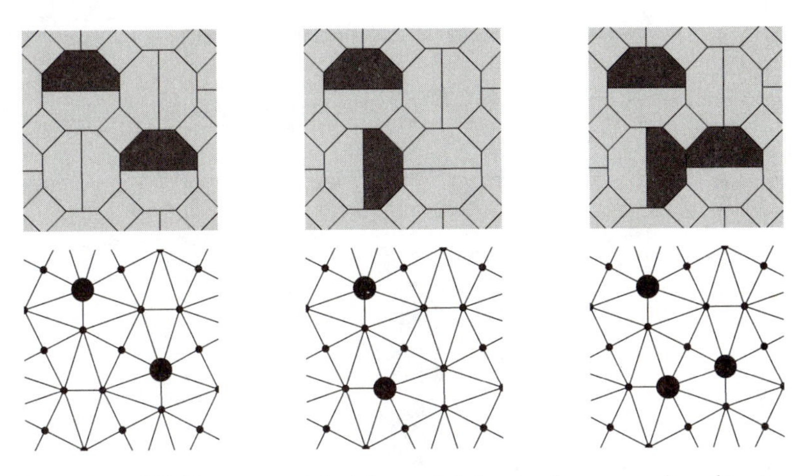

Figure 7.31. Weakly connected dog's legs and their strongly connected combination.

As with most connection games of this class, the best offense is a good defense. It is often easier for players to block the opponent than develop their own connection, but this is not a problem as a successful defense implies a win.

History

Octagons was invented by R. Wayne Schmittberger, and is described in *New Rules for Classic Games* [1992].

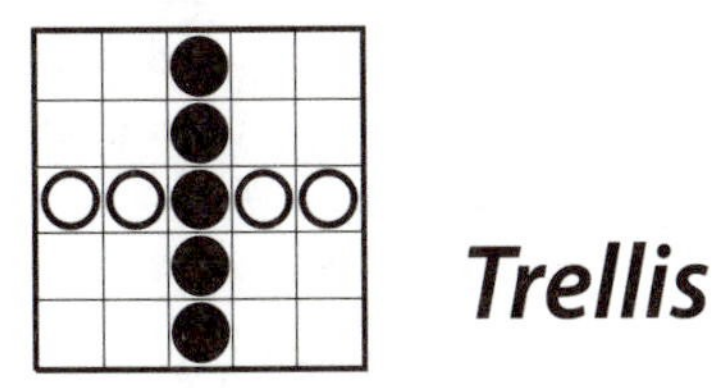

Trellis

Trellis is an intriguing game that uses conditional diagonal connections to avoid deadlocks on the square grid.

Rules

Trellis is played on the intersections of a checkered 14 x 14 square grid. Two players, Black and White, own alternating sides of the board that bear their color. The board is initially empty.

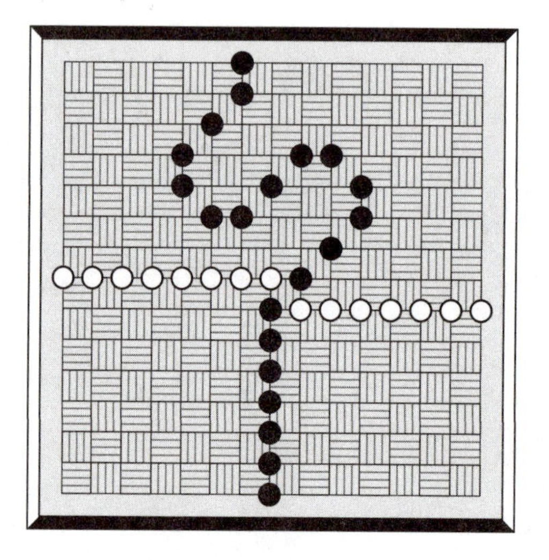

Figure 7.32. A game won by Black.

Players take turns placing either one or two pieces of their color on empty board points. If two pieces are placed then they must be three edges apart (explained below). Players may pass. Players may not mirror their opponent's moves for more than ten moves in succession. A three-move swap option is recommended.

Black (vertical) owns the board squares marked with vertical stripes, and White (horizontal) owns the board squares marked with horizontal stripes. Pieces connect with orthogonally adjacent neighbors, and also diagonally adjacent neighbors across squares belonging to the player. In other words, black pieces on opposite corners of a vertically striped square are connected, and white pieces on opposite corners of a horizontally striped square are connected.

Black wins by establishing a chain of black pieces between the black (top and bottom) sides of the board. White wins by establishing a chain of white pieces between the white (left and right) sides of the board.

Figure 7.32 shows a game won by Black, who has completed a chain of black pieces between the black sides of the board. White does not have a winning horizontal connection because the two chains of white pieces do not connect diagonally across the black (vertical) square near the middle.

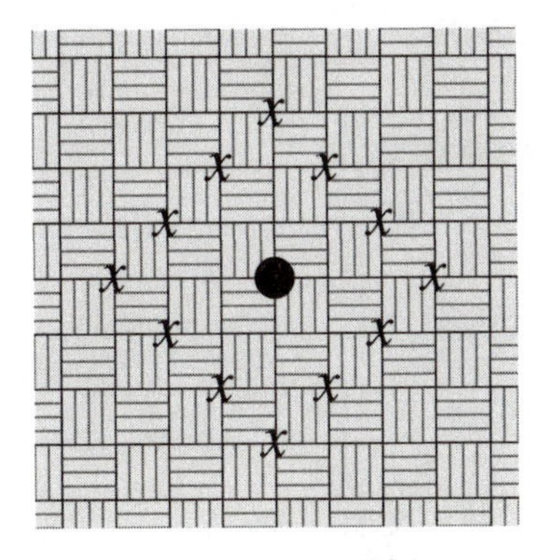

Figure 7.33. Possible move pairs for Black.

Notes

Figure 7.33 illustrates what is meant by placing pieces three edges apart. Given the black piece shown, Black is then entitled to place a second piece at any of the empty points marked *x*.

Figure 7.34 demonstrates diagonal connectivity in Trellis. Both pairs of pieces on the left are diagonally connected across squares belonging to their owners. However, neither pair of pieces on the right is connected, as they occupy the corners of enemy squares. This concept of diagonal ownership is crucial to the game's success, otherwise it would suffer from the usual deadlock problems on the square grid.

Figure 7.34. Pairs that are connected (left) and not connected (right).

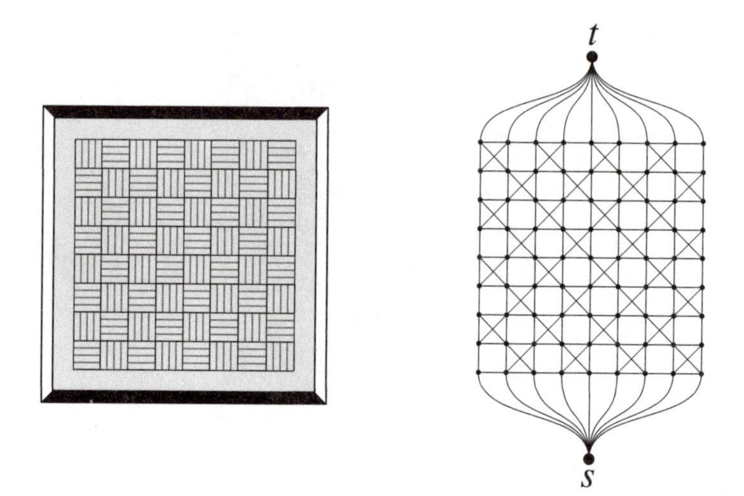

Figure 7.35. A 9 x 9 board and its game graph.

Figure 7.35 shows a small 9 x 9 board and its game graph. The conditional diagonals provide escape routes for potential deadlock formations.

History

Trellis was invented by Steven Meyers in 1999, who describes it as a "redeemed Bridg-It"; despite similarities between the two games, the double-move mechanism defeats the Bridg-It symmetry strategy. See Steven's Trellis web site for details: http://home.fuse.net/swmeyers/trellis.htm.

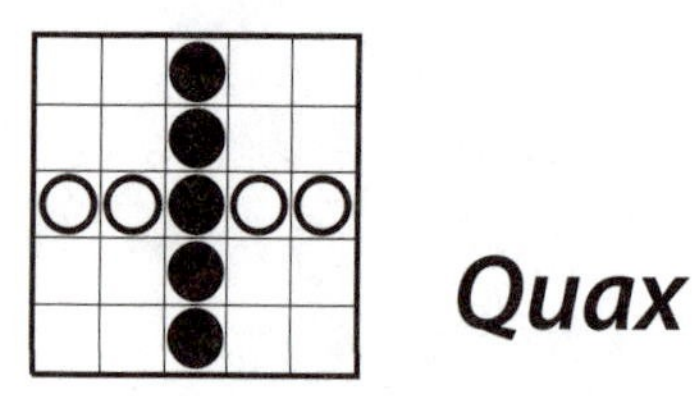

Quax

Quax is a connection game that uses a bridging mechanism to transcend its square grid and avoid deadlocks.

Rules

Quax is played on the intersections of an 11 x 11 square grid. Two players, Black and White, own alternating sides of the board that bear their color. The board is initially empty.

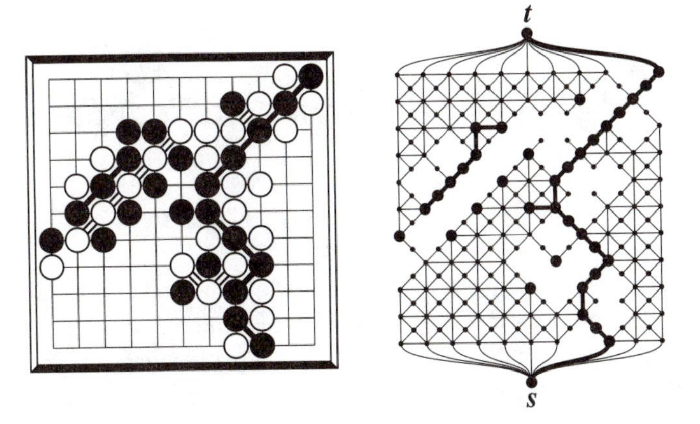

Figure 7.36. An 11 x 11 game won by Black and its game graph.

Players alternate taking turns, in which the current player may place either

- a piece of his color on an empty intersection, or
- a bridge of his color between two diagonally adjacent pieces of his color (provided it does not cross an enemy bridge).

Black wins by connecting the top and bottom edges with a black chain, while White wins by connecting the left and right sides with a white chain. Pieces are connected if they are orthogonally (squarely) adjacent or joined by a bridge. Figure 7.36 (left) shows a game won by Black.

Exactly one player must win. A single-move swap option is recommended.

Notes

Figure 7.37 demonstrates how bridges resolve deadlocks on the square grid. The importance of bridges means that Quax has a strong diagonal bias, as evident in Figure 7.36. Once a player starts pushing his connection along a diagonal, careful planning and a bit of room is required to defend effectively. For this reason Quax does not scale down well, and games on smaller boards do not allow much scope for strategic play.

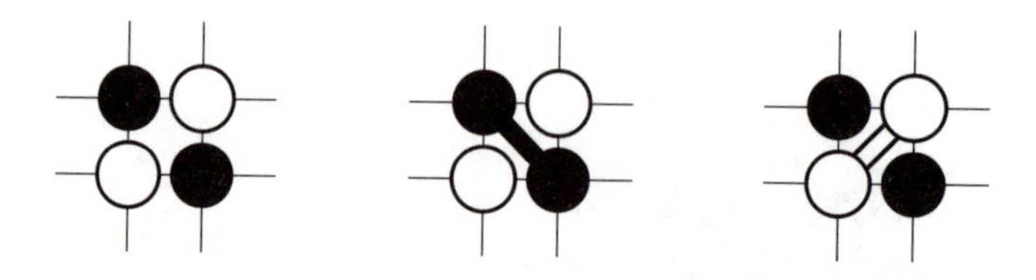

Figure 7.37. A temporary deadlock, a black bridge and a white bridge.

At first glance Quax appears to be both a Shannon game on the vertices (piece placement) and on the edges (bridge placement). However, Quax is effectively played on the Archimedean 4.8.8 grid if diagonal crossings are treated as another type of piece move. This grid is trivalent and games cannot be tied.

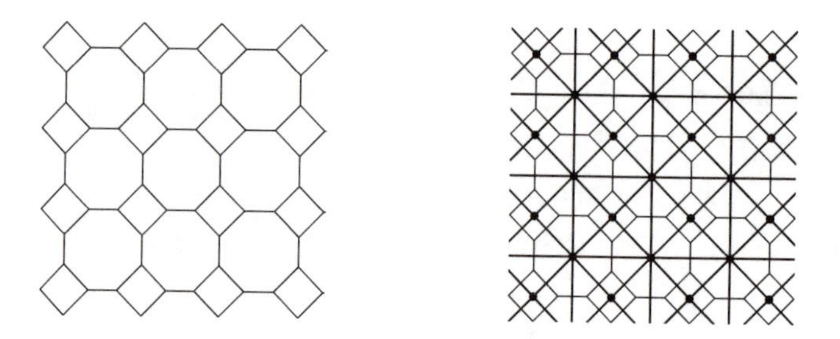

Figure 7.38. A 4.8.8 tiling and the Quax adjacency graph.

Quax has a lot in common with Twixt. For instance, it is not possible to stop the two pieces shown in Figure 7.39 from connecting (in isolation). These pieces are a knight's move apart, the basic connective step in Twixt.

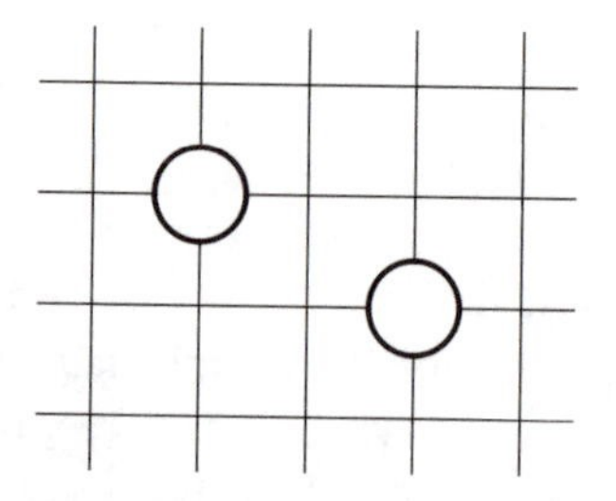

Figure 7.39. A strongly connected pair.

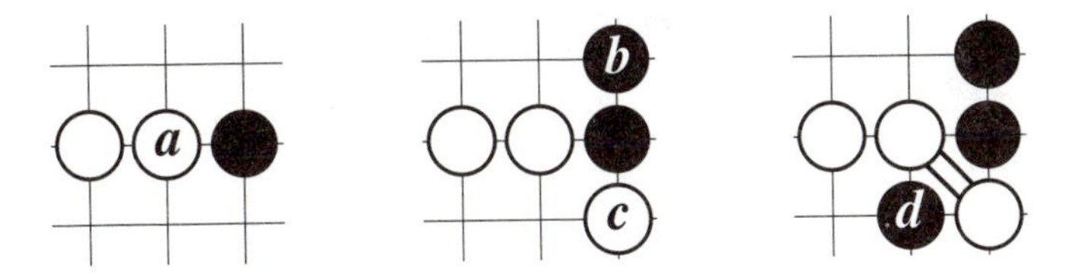

Figure 7.40. White can force his connection.

Strategy and Tactics

Figure 7.40 demonstrates a strong play by White. Piece ***a*** ensures that White can connect around the single black stone. If Black tries to block upwards with move ***b***, then White can play piece ***c*** to create an immediate threat. Black's reply ***d*** is ineffectual as White can play a bridge to secure the connection.

Notes from Bill Taylor

Placing pieces orthogonally so that they connect is always better than placing pieces diagonally.

In relation to the diagonal bias of the game, the two main diagonals have the same significance as the various important (knightrider) diagonals in Twixt. For example, whoever gets to play next in Figure 7.41 is guaranteed to connect to his edge if he plays at the main diagonal point ***x***. This would not be so if this system was one row or column closer to either edge. Whoever had the overlap onto the main diagonal would then win even against the move.

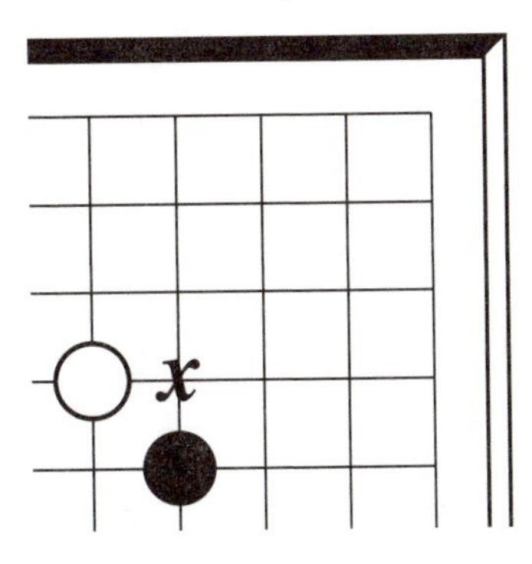

Figure 7.41. The player to move at ***x*** connects.

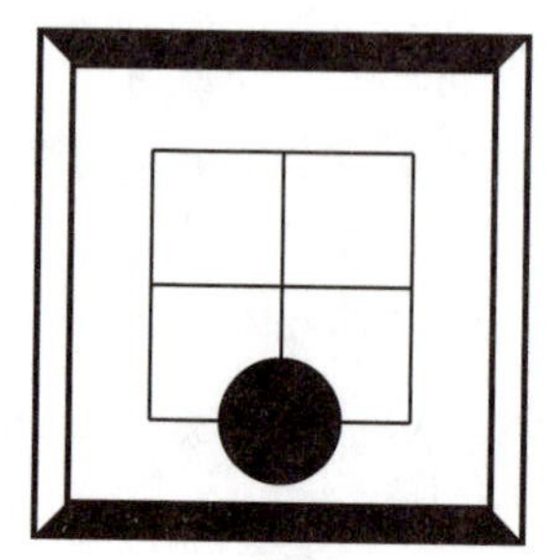

Figure 7.42. Puzzle by Bill Taylor: Black started, who wins?

History

Quax was invented by Bill Taylor in 2000. The name Quax is a contraction of Quadrangular Hex. According to Bill, this is also the sound made by the winner as he makes the killer move. Much of the above material was derived from João Neto's Quax page: http://www.di.fc.ul.pt/~jpn/gv/quax.htm.

Variants

The following entries show the Quax move mechanism applied to related games on the square grid. This mechanism constitutes a metarule that allows some hexagonally-based connection games to be successfully ported to the square grid.

ZeN

ZeN is played on a Quax board wrapped around a torus; connections wrap between the left and right sides of the board, and between the top and bottom sides of the board. Rules are as for Quax except for the winning conditions.

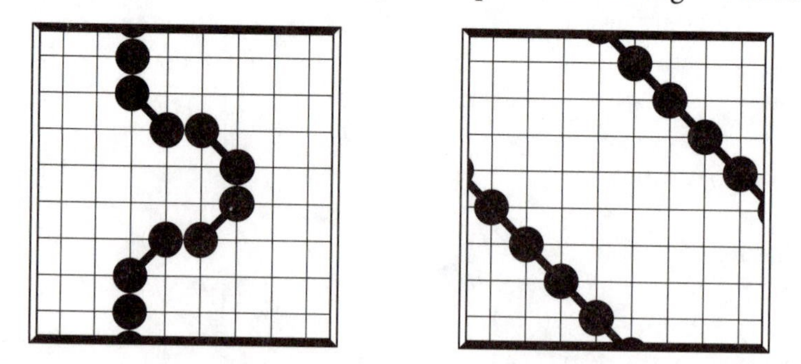

Figure 7.43. Black aims for a vertical loop (left) or a NW-SE spiral (right).

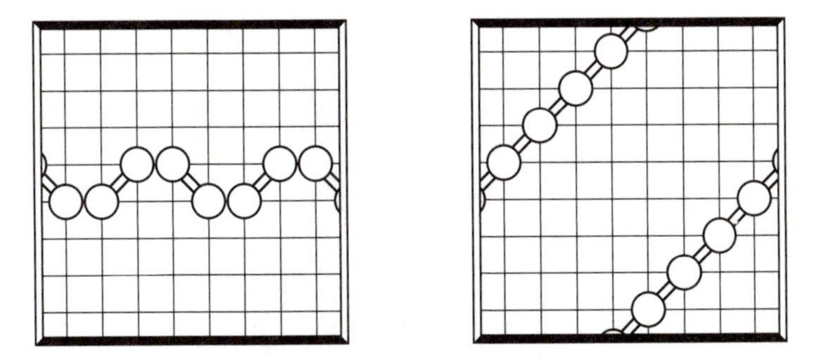

Figure 7.44. White aims for a horizontal loop (left) or a SW-NE spiral (right).

Black (N) wins by completing either a vertical loop or a NW-SE spiral (\) around the torus, as shown in Figure 7.43. White (Z) wins by completing either a horizontal loop or a SW-NE spiral (/) around the torus, as shown in Figure 7.44. Note that edge pieces and bridges projecting off the board are partially drawn at the corresponding point on the far side of the board where their connections wrap. This aids board interpretation.

As in standard Quax, exactly one player must win. This would not be possible without the tie-breaking bridge moves.

ZeN was designed by Bill Taylor in 2003. The name refers to the fact that the stroke directions in the letters *Z* and *N* correspond to the players' winning goal directions.

Three-Player ZeN

The winning conditions must be slightly modified to incorporate a third player, Gray:

- Black wins by completing a vertical loop around the torus,
- White wins by completing a horizontal loop around the torus, and
- Gray wins by completing either a NW/SE spiral (\) or a SW/NE (/) spiral around the torus.

Figure 7.45 shows a game won by Gray, who has completed a SW/NE spiral. Note that it's possible to connect pieces at diametrically opposed corners, for instance the bottom left black piece is connected to the top right black piece via a wrapped bridge.

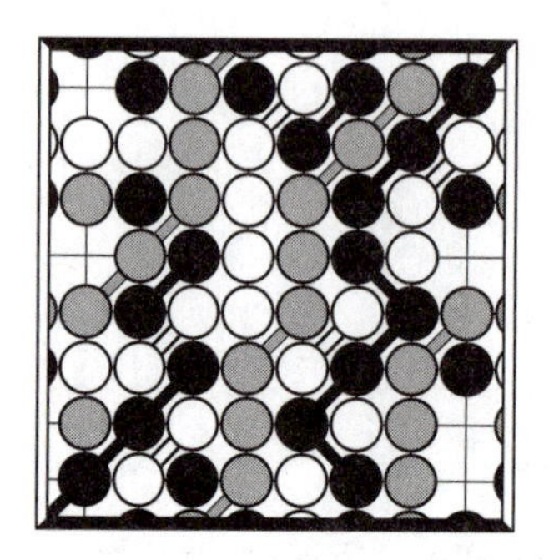

Figure 7.45. A game won by Gray.

Gray, with two winning conditions, has the easier task and should play last. However, any imbalance within the game tends to be masked by three-player dynamics (petty diplomacy) in which the two losing players cooperate to pull back the leading player.

Three-Player ZeN was designed by Bill Taylor in 2003, and was originally called Porus Torus. ZeN is one of the few connection games that translates easily to a three-player version that works almost as well as its two-player counterpart.

Unlur

Unlur is a connection game with unequal goals that incorporates an extended swap sequence as an integral part of the game.

Rules

Unlur is a two-player game played on a hex hex board, typically with six cells per side. The board is initially empty.

> **Contract Phase:** Players take turns placing a black piece on any empty hexagon (except along the edge). A player may elect to pass, in which case he becomes Black, the opponent becomes White, and the contract phase ends.

Combat Phase: Players take turns placing a piece of their color on any empty hexagon. Players are no longer permitted to pass.

White wins by completing a *line* connecting any two opposite edges, and Black wins by completing a *Y* connecting any three alternating sides (Figure 7.46). If a player achieves the opponent's goal with his pieces without achieving his own goal, then the opponent wins. A player wins if both winning conditions are achieved at the same time (Mover Wins).

Notes

Black's position should steadily build up in the contract phase. Edge cells are forbidden in this phase because some moves along the edge would actually worsen Black's position, resulting in excessively long contracts. This also stops a player from achieving a winning condition before the contract phase has ended.

A most unusual aspect of Unlur is the inequality between players' goals, as it's much easier to achieve a line than a Y. The contract phase was introduced as a balancing mechanism to address this inequality, and players must be careful to end this extended swap option as soon as the balance tips in Black's favor. Judging this point can be difficult, and this decision usually decides the outcome of the game.

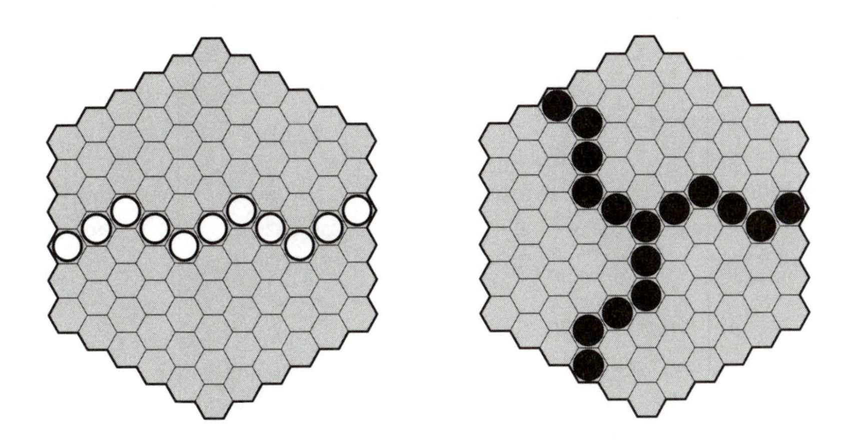

Figure 7.46. A white line and a black Y.

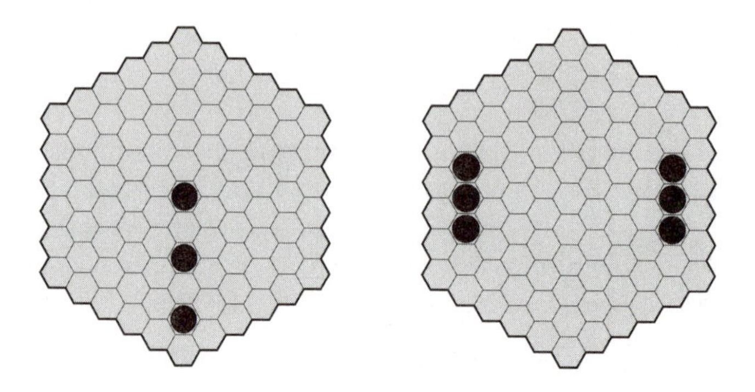

Figure 7.47. A center opening and a block opening.

There appear to be two common types of opening contracts:

- the *center* opening based upon a black piece on the central hexagon (Figure 7.47, left), and
- the *block* opening consisting of blocks of black pieces on opposite sides of the board while the center remains empty (Figure 7.47, right).

The nature of the contract has a significant impact on the game's development. Center contracts are typically short (3 to 5 pieces) and White is forced to find a way around the central obstacle. Block contracts are typically longer (8 to 10 pieces) and White will generally take a central hexagon as his first move and build outwards from there.

Unlur is a borderline inclusion in the Absolute Path category. It shares some traits with members of the Path Race category in that winning conditions may coexist for both players (when taken out of context) and winning paths made of opponent's pieces are possible. However, the danger of handing the win to the opponent by achieving his goal first reduces this problem. The basic nature of Unlur is very much that of an Absolute Path game.

Strategy and Tactics

Players can exploit the fact that the opponent will not want to achieve the player's goal. For instance, Figure 7.48 shows a board position that looks good for Black, who has a virtual connection between three alternating sides. If White intrudes into this virtual connection with move *1* then

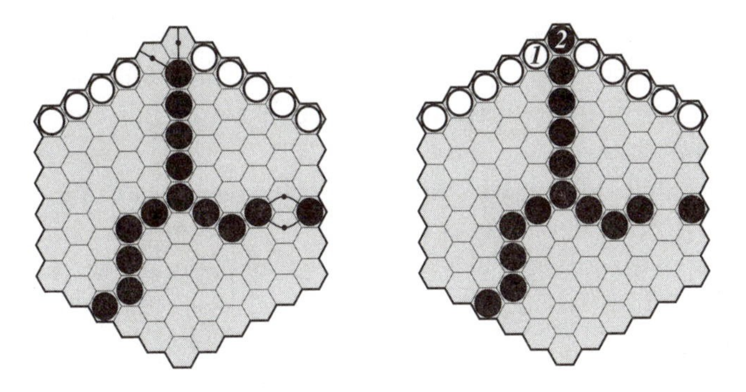

Figure 7.48. White can force a win.

Black must reply with move *2* to stop White making a line next turn; however this move itself achieves a line, giving White an immediate win.

This example suggests a general strategy for Black: *avoid the corners if possible*. Corner plays leave Black open to forced line attacks.

Players should be aware of the three basic win forms described by Taral Guldahl Seierstad [Arrausi et al. 2002] as shown in Figure 7.49:

1. White must eventually win if he surrounds two opposite edges,
2. Black must eventually win if he surrounds three nonadjacent edges, and
3. White must win if both players each surround an opposite edge with touching groups.

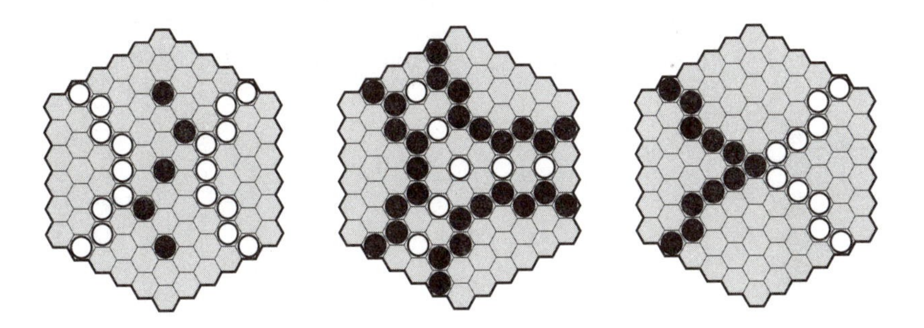

Figure 7.49. The three basic win forms: (1) White wins, (2) Black wins, and (3) White wins.

In each case the player will win even if he tries not to, as the opponent is eventually forced to achieve the player's goal without achieving his own. The third case is quite subtle; Black can achieve a Y but cannot do so without first completing a line.

Players should strive to achieve their goal directly, but must also be aware of the three basic win forms as an alternate way to stop the opponent if play develops accordingly. The first case provides a simple way for White to stop Black from winning, however the second case tends to occur most frequently in play.

History

Unlur was invented by Jorge Gómez Arrausi in December 2001. It was the winner of the Unequal Forces Game Design Competition hosted by *Abstract Games* magazine [Gómez et al. 2002]. The original board size of eight cells per side was reduced to six to encourage new players to try the game. Six cells per side has become the standard size, but it is recommended that experienced players seeking a deeper game use the larger size.

This section was drawn largely from notes submitted by Taral Guldahl Seierstad.

Variants

Misère Unlur

Suggested by Taral Guldahl Seierstad, who points out that if the contract phase is ignored, then the misère version of Unlur is in fact equivalent to the standard game with the roles reversed. In the misère version White wins by creating a Y and loses by creating a line with no Y, while Black wins by creating a line and loses by creating a Y with no line.

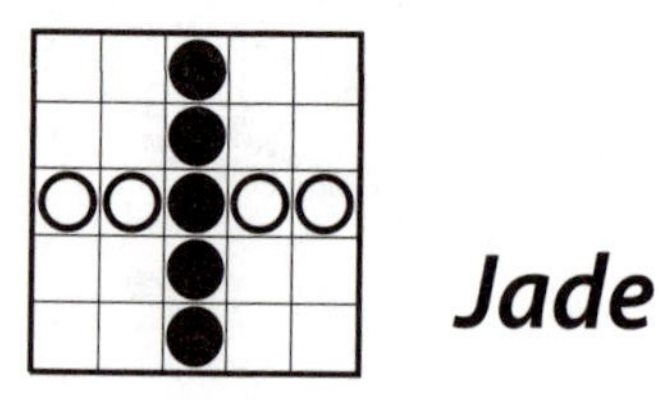

Jade

Jade is a connection game with unequal goals that introduces a simple but brilliant new move mechanism: players choose which color to play each turn.

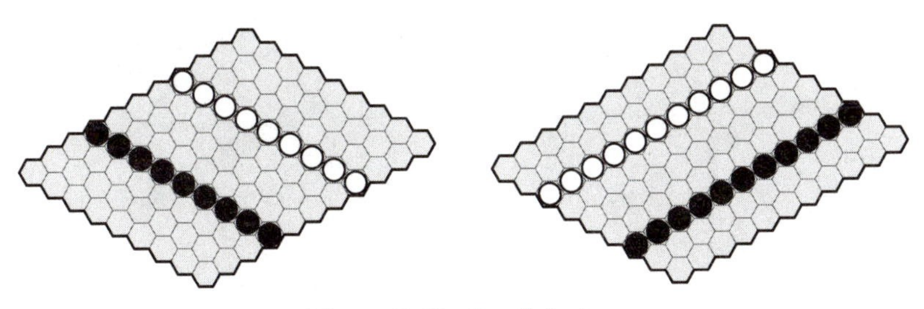

Figure 7.50. Parallel wins.

Rules

Jade is played on a hexagonally tiled rhombus similar to a Hex board, except that the number of cells along each side should be odd and the two dimensions should differ in size. The standard board size is 9 x 11. The board is initially empty. Two players, Cross and Parallel, take turns placing either a white piece or a black piece on an empty board cell. Cross plays first.

Parallel wins by connecting two opposed sides with both a white chain and a black chain (Figure 7.50).

Cross wins by connecting all four sides with either a white chain or a black chain (Figure 7.51). As in Hex, corner cells are deemed to belong to both adjacent sides. Exactly one player must win.

Notes

Cross has the easier task of the two players, and the luxury of choosing either color for his winning chain. It is in Cross's interest to delay committing to either color for as long as possible. Swapping the emphasis from one color to the other in mid-attack can throw Parallel into disarray and is one of the true joys of the game.

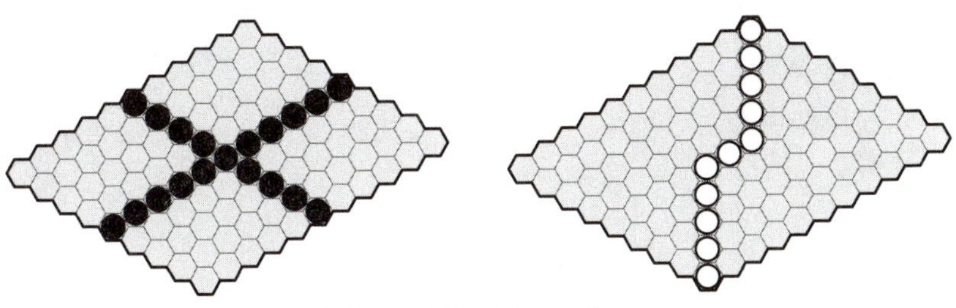

Figure 7.51. Cross wins.

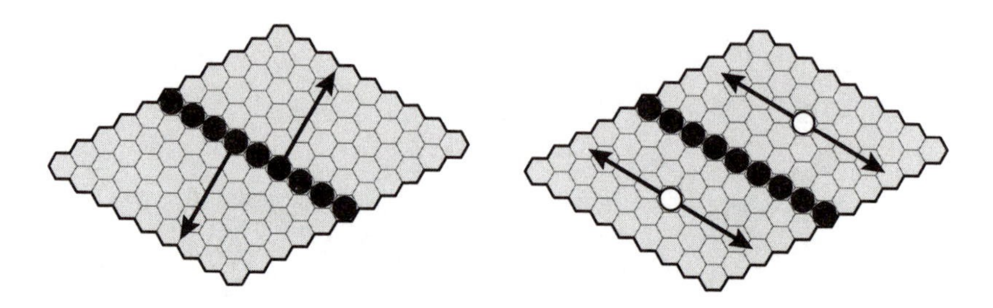

Figure 7.52. Cross has the easier task.

Just imagine that a black chain connects two opposed sides as per Figure 7.52. This is half of the winning condition for both players, but it is much easier for Cross to complete his connection than for Parallel to do so.

It is therefore surprising that Cross gets first move. However, making Cross move first gives Parallel the opportunity to play the opposite color to the opening move, starting the game with an even balance of colors. Cross, moving second, would almost certainly play a piece of the same color as the opening move, creating a dominant color and making a Cross win even more likely.

Note that it is easier for Parallel to win along the shorter ($n = 9$) direction rather than the longer ($n = 11$) direction. The larger the discrepancy between the number of cells in each direction, the better for Parallel. A good compromise is 9 x 11.

The game graph for Jade can be constructed in a similar manner to that of Hex. Unlike Hex, however, two graphs of the same board are required, one in each direction, as shown in Figure 7.53. Both graphs are from the perspective of the same piece color (not player).

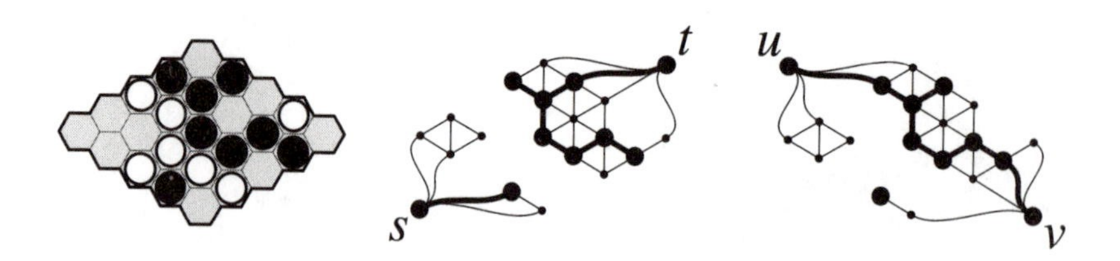

Figure 7.53. A 5 x 5 game won by Parallel and its game graphs.

The left figure shows a 5 x 5 game won by Parallel. The middle figure shows the game graph in the (*s*, *t*) direction cut from black's perspective, and the right figure shows the game graph in the (*u*, *v*) direction joined from black's perspective.

All possible winning conditions can be described as follows:

- Parallel win = (exactly one of (*s*, *t*) or (*u*, *v*) cut) AND (exactly one of (*s*, *t*) or (*u*, *v*) joined), or
- Cross win = (both (*s*, *t*) and (*u*, *v*) cut) OR (both (*s*, *t*) and (*u*, *v*) joined).

Jade is more subject to cold wars than most other connection games. They do not occur often, but ruin the game when they do. A cold war occurs when there exist points on the board that will guarantee an eventual win for the opponent if either color is played there. Players are forced to play elsewhere, filling up the board, until no more safe moves remain and one of the danger spots must be taken, handing the win to the opponent.

Figure 7.54 shows two cold wars of the same form on different board sizes. If either player places a piece of either color on any cell marked *x*, then the opponent has a guaranteed win. Cold wars are discussed in Section 4.4.5.

Jade is difficult to analyze as it bends the usual connection game rules, and has yielded many surprises. For instance, having an extra piece on the board can be a disadvantage because pieces do not belong to either player. Hence the usual strategy-stealing argument (see Appendix D) does not apply to Jade.

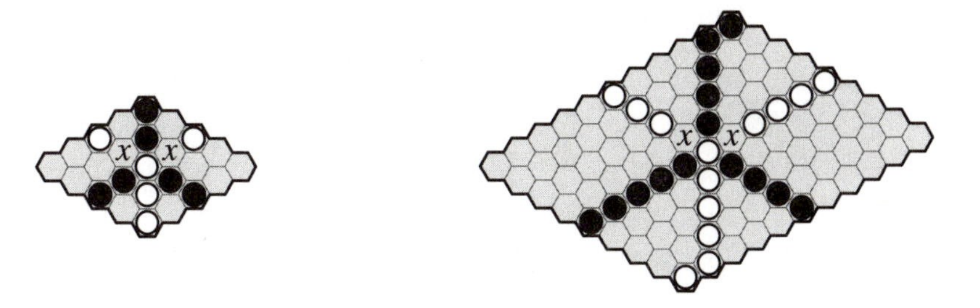

Figure 7.54. Two cold wars of different sizes.

Strategy and Tactics

As mentioned above, it is usually to Parallel's advantage to have approximately equal number of black and white pieces on the board. Cross would prefer one color to dominate.

Paul van Wamelen has shown that symmetry strategies guarantee a win for a particular player on all but odd-and-unequal-sided boards (see Appendix E).

Cross tends to adopt the more attacking role of opening up possibilities and laying traps, while Parallel's role is more defensive, closing down Cross's options in case of a possible color switch. Cross is arguably the more interesting role to take. Jade has some similarity to Quoridor, as avoiding commitment is an important strategy in both games.

History

Jade was invented by Mark Thompson in 2001. The original rules had each player placing both a black and a white piece per turn (presumably only one on the first move). Claude Chaunier suggested the current rule of placing only one piece of the player's choice each turn.

Shortly after Jade's invention, an intense period of testing and analysis revealed the various biases in Cross's favor and symmetry strategies. Several artificial rules were tried to address these, however simply restricting the game to odd-and-unequal-sided boards proved an elegant solution.

Chameleon, a variant of Jade, is discussed in its own section.

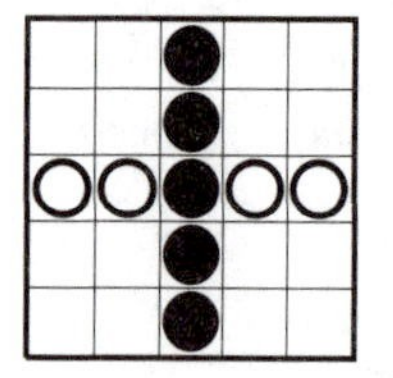

Gaia

Gaia has a similar layout and theme to Split but introduces an effective tie-breaking mechanism.

Rules

Gaia is a game in which two players, Black and White, attempt to connect opposed points on a rounded board (Figure 7.55). The board is irregularly tessellated and contains six five-sided cells; all other cells are six-sided. The board is initially empty.

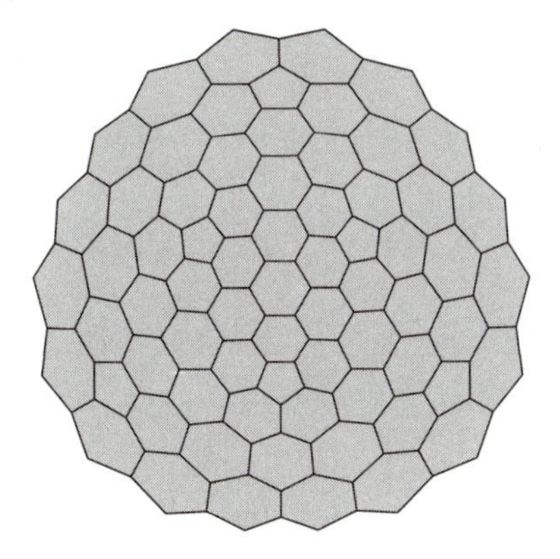

Figure 7.55. The Gaia board.

Players take turns placing a piece of their color on an empty cell. If a piece is played on an edge cell, then another piece of the same color is also placed on the maximally removed edge cell on the far side of the board.

A player wins by either

1. completing a chain of his pieces across the board between two opposed edge cells, or
2. forming a Y that renders any Type 1 connection by the opponent impossible.

Every game must have exactly one winner. A single-move swap option is recommended. Edge cells may be marked by letters or some appropriate symbol to explicitly show opposed pairs; the original board design used astrological symbols.

Notes

Figure 7.56 (left) shows a game won by White, who has completed a chain of white pieces between two opposed white edge pieces. Figure 7.56 (right) also shows a game won by White. This is a Type 2 connection in which the white Y stops all possible black connections across the board. Note that every piece occupying an edge cell has a matching piece of the same color on the far side of the board.

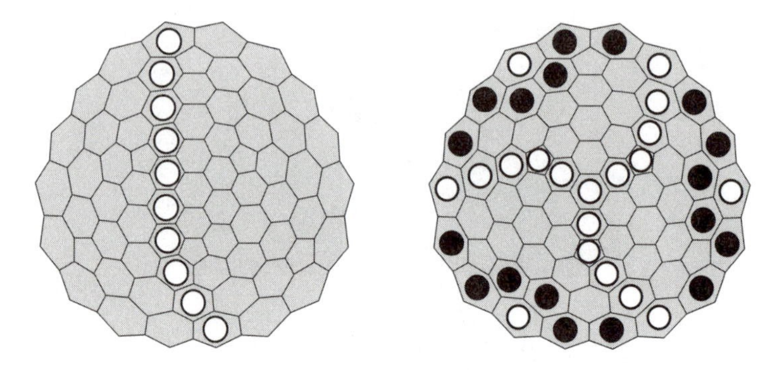

Figure 7.56. Two games won by White.

The winning conditions can also be described as follows. Any edge cell that is part of a chain is called an *anchor*. The object of the game is to create a chain with at least two anchors, in which the gap between every pair of consecutive anchors (going around the board) is less than half the board's perimeter.

The game graph for Gaia is more complicated than that for Split, and involves a separate graph for each pair of antipodal cells (which form the terminals for that graph). If any such pair is joined then the player wins; if all such pairs are cut then the opponent wins.

Notes from Mark Thompson

Although the center is still somewhat more valuable than the perimeter, since a path through the center can take advantage of more two-way stretches, opening play needs to gain influence over large areas of the board. Any part may be ultimately important, especially on larger boards.

Note that either a straight or curved path linking two opposite cells requires about the same number of cells. But ladders around the circumference are also important, and the player unlucky enough to be on the inside of such a ladder can only win it by running into a friendly stone, prudently placed ahead of time.

History

Gaia was invented by Mark Thompson in January 2002.

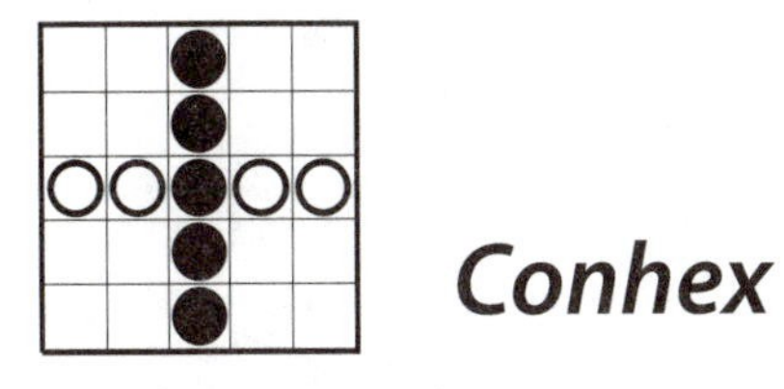

Conhex

Conhex is an entertaining and well-balanced connection game in which players battle to color a map.

Rules

The Conhex board consists of 41 cells, each with a hole at each interior corner. Edge cells have three holes, the central cell has five holes, and all other cells have six holes. The top and bottom sides of the board are colored black and the left and right sides are colored white. The board is initially empty.

Two players, Black and White, take turns placing a peg of their color in an empty hole. The first player to occupy at least half of a cell's holes then owns that cell for the rest of the game. Cell ownership can be shown by coloring the cell or placing a marker piece on it.

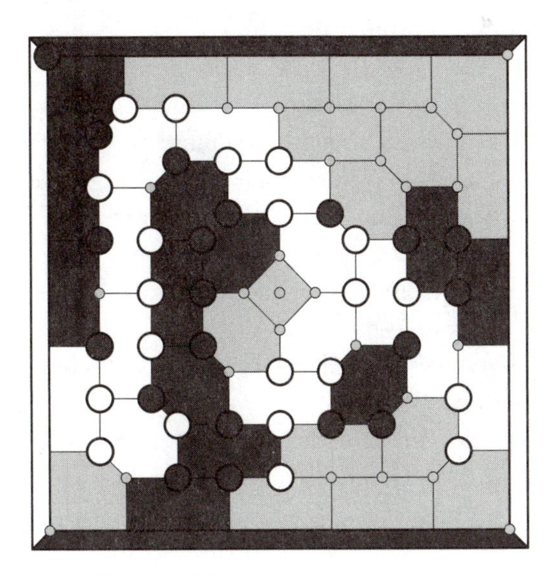

Figure 7.57. A game won by White.

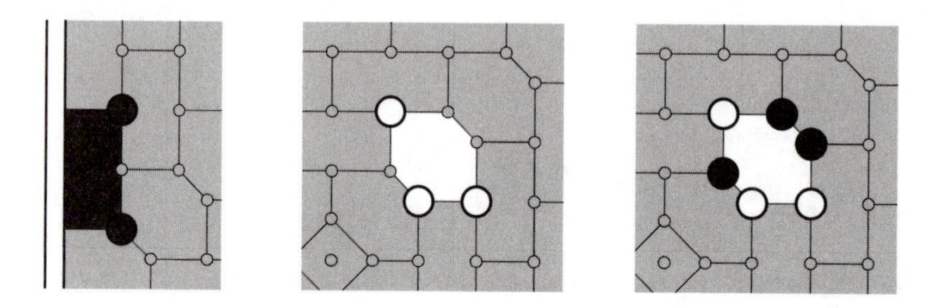

Figure 7.58. Cell ownership.

The first player to complete a chain of his cells between his sides of the board wins. Figure 7.57 shows a game won by White. Exactly one player must win; the Conhex board is a trivalent grid.

Notes

Conhex is one of the best recent connection games. It has proved popular with a variety of players, including some who generally avoid connection games—a rare feat. At the heart of Conhex is the idea of map-coloring, which was the original inspiration behind Hex.

The fact that the five-sided central cell is relatively weak and the edge and corner cells are relatively strong (only requiring two pegs for ownership) means that cell strengths are more evenly spread across the board than for most other connection games.

Figure 7.58 shows the mechanics of cell ownership. Black occupies two of the three holes on the edge cell (left) and therefore claims it. Similarly, White occupies three of the interior cell's six holes (center) and therefore claims it. White owns this cell for the rest of the game, even if Black places pegs in the remaining three holes (right).

This system of peg voting for cell ownership may seem to introduce a nonconnective element into the game. However, if we examine its game graph we see that Conhex is a Pure Connection game at its most fundamental level, and that the player's actual moves (the peg placements) are removed from this level by a layer of abstraction.

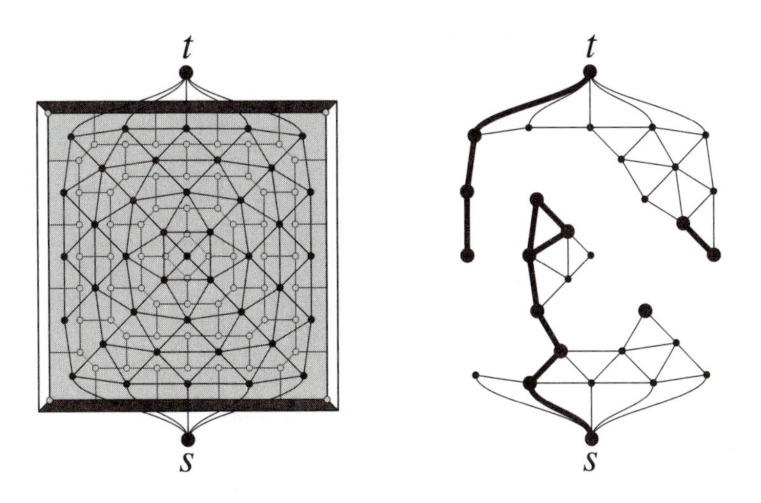

Figure 7.59. The Conhex game graph.

Figure 7.59 (left) shows the game graph of an empty board from Black's perspective. As usual, the adjacency graph is given by the dual of the board's tiling. Figure 7.59 (right) shows the game graph of the game shown in Figure 7.57; it can be seen that White has won this game as the graph has been cut so that the black terminals s and t lie in disconnected halves.

The Caeth metarule, in which players choose an edge rather than a cell corner each turn, is the dual of Conhex. Conhex itself forms a metarule, as the peg voting mechanism for establishing cell ownership can be successfully applied to other connection games.

Strategy and Tactics

As in most connection games, efficiency is critical in Conhex. Players should keep in mind that each peg influences three incident cells (except the center and corner holes) and should place their pegs to influence as many cells as possible. Playing a peg around a cell owned by either player is a wasted move in that direction.

For instance, Figure 7.60 shows the territory influenced by the black pieces. Lightly shaded cells are two pegs away from Black ownership, and darkly shaded cells are one peg away from Black ownership.

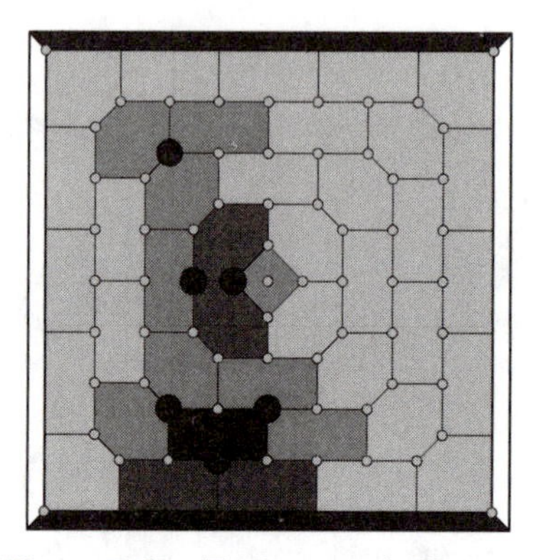

Figure 7.60. Territory and efficiency.

Black's two central pieces have secure virtual connections to both the top and bottom sides of the board; White cannot stop them from connecting to win this game. Notice however that the lower connection involves some unnecessary overlap, and that two of the bottom-most black pieces could have been put to use elsewhere.

Players should spread their territory by placing single pegs in areas not yet challenged by the opponent. Cells can be reinforced as the opponent attacks them.

It is good to work to an overall strategic plan but to keep alternative paths open. Players should have contingency plans and be prepared to abandon their preferred line if a local battle along its path does not fare well.

Most games between players of equal skill will involve meandering paths that take up most of the board, emphasizing what a well-balanced game Conhex is; most other connection games are over before a third of the board is filled.

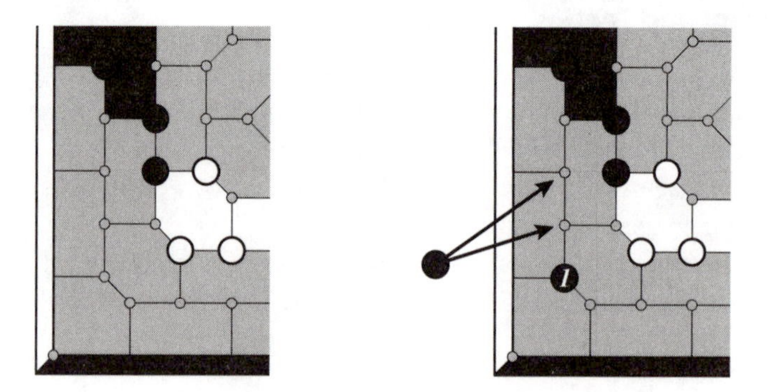

Figure 7.61. A fork by Black.

Figure 7.61 shows a good tactical move *1* by Black whose connection to the bottom is then guaranteed by a fork. If White moves in either of the holes indicated then Black just plays in the other one to secure the connection.

Just as important as efficiency is tempo. If two players have an equal number of pegs around a cell, then the player with the move in hand will win that cell unless his opponent makes a fork or forcing move. For instance, Figure 7.62 (left) shows a situation that looks almost hopeless for White, who has connected to the left but faces an imposing black gauntlet to the right. However, White has the move in hand and the tempo.

White plays a series of forcing moves *1*, *3*, and *5* to which Black has no choice but to reply as shown with moves *2*, *4*, and *6*. White's move *7* wins the top right corner and sets up a fork back to the main white group, and move *9* completes the winning path.

The seemingly innocuous white piece in the top right corner becomes a game-winner. This demonstrates the even distribution of cell strengths and that acute corners are as vital as any other hole. In fact, Michail Antonow points out that it can be beneficial to occupy the corner spaces, and that players should always look for opportunities to push along the edges of the board. Edge cells provide excellent opportunities for forcing moves.

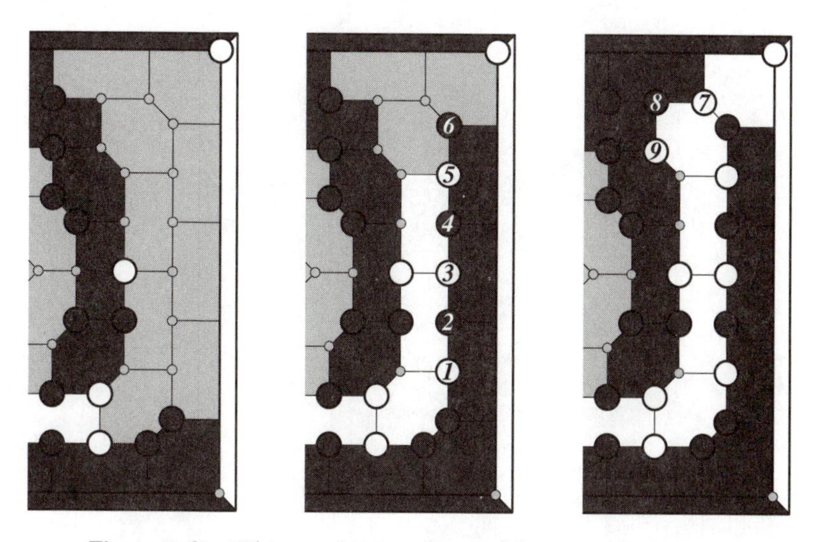

Figure 7.62. White to play a sequence of forcing moves to win.

This example demonstrates that *gaining territory is important, but saving tempo is decisive*. Players should not waste moves defending cells that are not yet attacked by the opponent.

History

The Conhex board and rules were designed by Michail Antonow in 2002. Michail has experimented extensively with other board shapes and sizes and found the official board shown in Figure 7.57 to provide the most balanced game. Larger boards significantly increase the complexity of the game and the time required to play it, whereas smaller boards provide unsatisfactorily short games that unduly favor the first player.

The Conhex board is similar to the Square Hex board, which in turn is similar to a square Poly-Y board shown in Schensted and Titus's *Mudcrack Y and Poly-Y* [1975, 181]. However Conhex was derived from a different route, via an earlier game of Antonow's called Pula.

Figure 7.63 shows the Pula board, which consists of a hex hex board with holes at interior corners. The game is played in the same way as Conhex except that the winner is the player with the most hexagons at the end of the game. Pula was published in 1996 by Norbert Dinter Holzspielwaren.

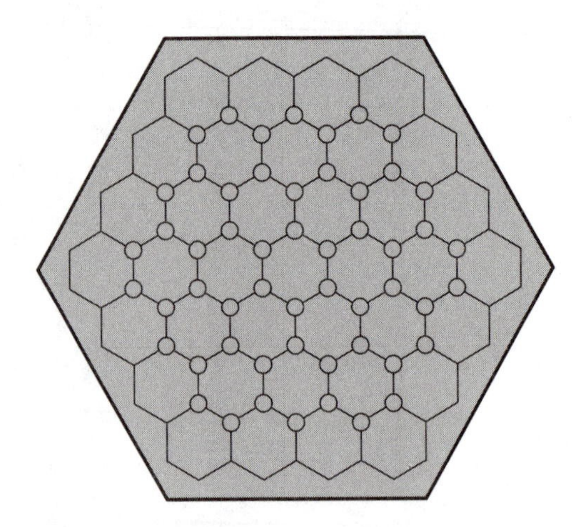

Figure 7.63. The Pula board.

A more sophisticated version of the game, Pula 2, is identical to Pula except that players score points as follows:

- 1 point for connecting two neighboring sides of the board with a chain of their color,
- 5 points for connecting two opposite sides of the board with a chain of their color, and
- 3 points for connecting two nearly opposite sides of the board with a chain of their color.

The winner is the player with the most points at the end of the game. Pula and Pula 2 are not Pure Connection games as their goals are score-based, however they do feature strictly Connective Play. Conhex was derived directly from Pula 2.

Variants

Noc

Noc is a metarule in the same family as Caeth. Noc uses an edge-voting mechanism similar to Conhex, but may be played using the board and rules of almost any parent game with a planar adjacency graph. Noc is discussed in the Caeth section.

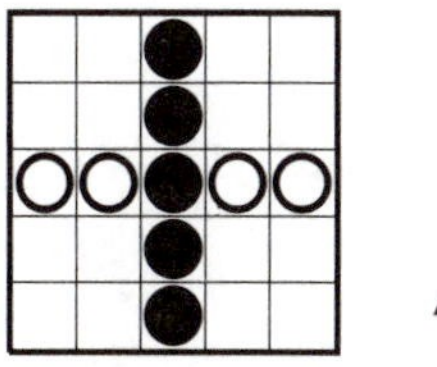

Akron

Akron is a three-dimensional connection game played with marbles on a square grid. Players may stack marbles to step over enemy blocks and resolve deadlocks.

Rules

Akron is played on a square grid of holes, typically 9 x 9. Two players, Black and White, own alternating sides of the board that bear their color. The board is initially empty.

Each player owns a pool of 40 marbles of his color. Players alternate taking turns, in which the current player may either

1. place one of his marbles on an empty hole on the board, or
2. move one of his marbles to a valid empty point that touches a piece connected to it.

A point is valid if it is empty and either on the board surface, or supported by a flat stable square of four marbles. Once a player's pool of marbles runs out he can only play Type 2 moves for the remainder of the game. A single-move swap option is recommended.

Moving a marble may cause higher-level marbles to drop. However, a marble that directly supports more than one piece on the level above cannot be moved. Marbles that drop during the move cannot be used as support pieces for the moving marble.

Two marbles are connected if they touch, in other words if they are orthogonally adjacent and on the same level, or if one rests directly upon the other from the level above. If a connection crosses over an opponent's connection at any point then the upper-most connection prevails; the lower connection is cut until the upper one is broken or removed (*over/under rule*).

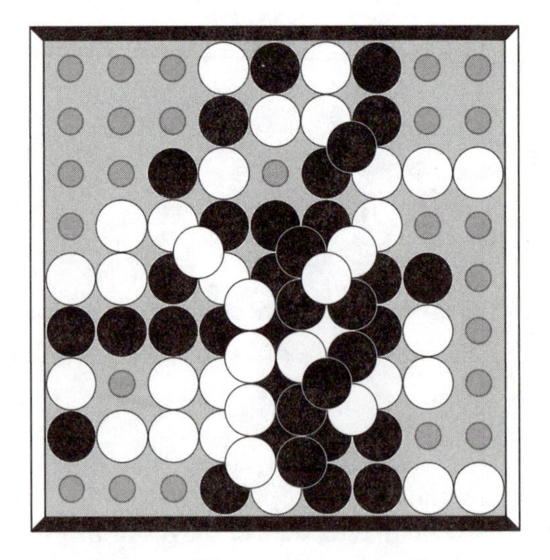

Figure 7.64. A game won by Black.

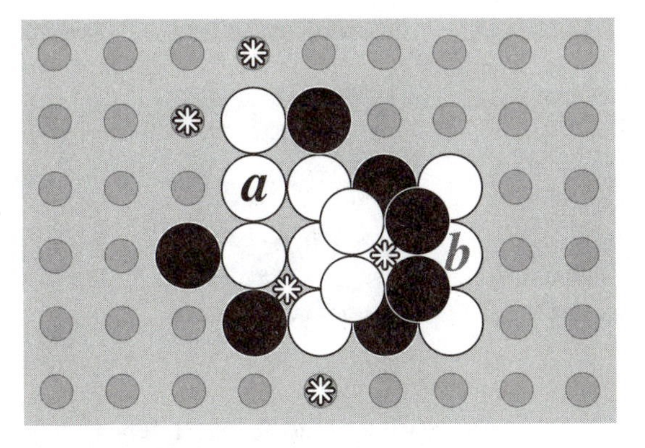

Figure 7.65. Valid moves for white marble ***a***.

A player wins by connecting his sides of the board with a chain of his marbles (see Figure 7.64). The win only counts if the connection still exists after the opponent's replying move (win-on-opponent's-turn metarule).

Notes

Figure 7.65 demonstrates valid moves for white piece ***a***. All valid empty points touching a piece connected to ***a*** are marked with an asterisk; ***a*** can move to any of these points. Note that the white marble ***b*** is disconnected from ***a*** due to the over/under rule. Also note that ***b*** cannot move as it is pinned by two marbles resting directly upon it.

Figure 7.66 shows how the removal of white support marble ***a*** causes a *cascade* of two black marbles to drop. Players are therefore able to shift opponent's pieces within a limited scope, allowing some subtle tactical plays.

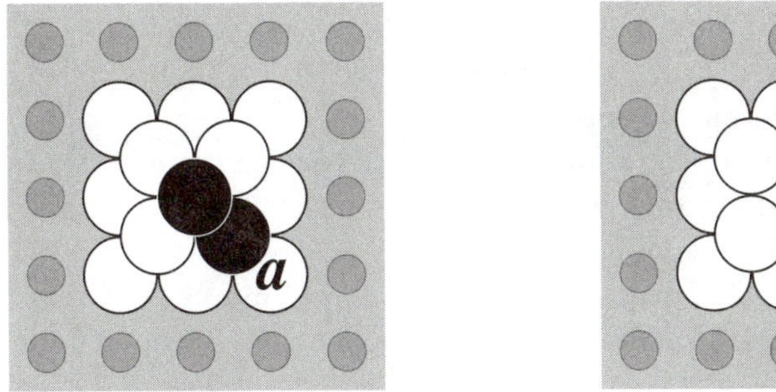

Figure 7.66. Pulling out a support piece causes marbles to drop.

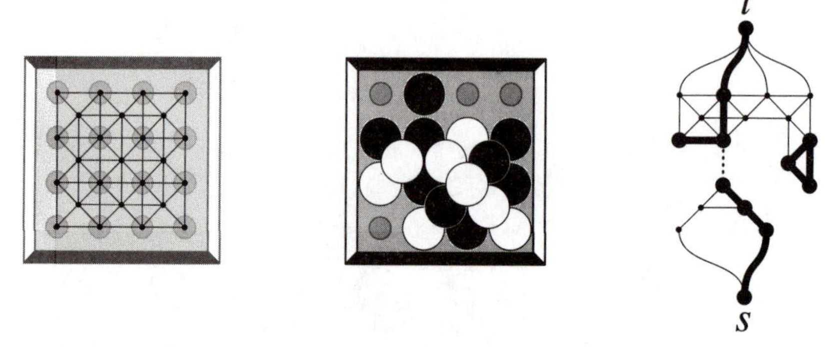

Figure 7.67. The adjacency graph and game graph of a completed game.

The recommended number of marbles for each player for a board of size *n* is given by *n* x *n* / 2 (rounded down).

The possibility of paths crossing above or below each other would seem to preclude the Cut/Join property, however the over/under rule makes it viable. Connection is based only on the upper-most pieces at each point, effectively projecting the game to two dimensions viewed from above. For instance, Figure 7.67 (left) shows the planar adjacency graph of a small 4 x 4 board.

The game graph (right) can then be represented in the standard Cut/Join format typical of Absolute Path games. It can be seen that White has won this game as the graph is cut with a terminal in each disconnected half. The dotted line indicates a buried black connection cut by the over/under rule.

Akron violates the Macro Reduction property common to most connection games (see Appendix F). Removing a row or column of a completed connection may also involve removing higher-level pieces that depend upon them, which may cut higher-level connections and even reveal a win for the opponent.

Contrary to most other connection games, having an extra piece on the board can actually harm a player's position in some cases. This nullifies the strategy-stealing argument for Akron (see Appendix D).

Although marbles may move in Akron, this movement is strictly connection-based; a marble may only move to connected valid empty points.

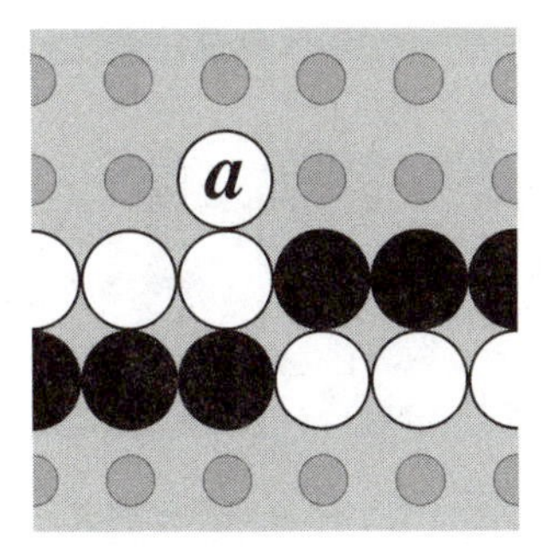

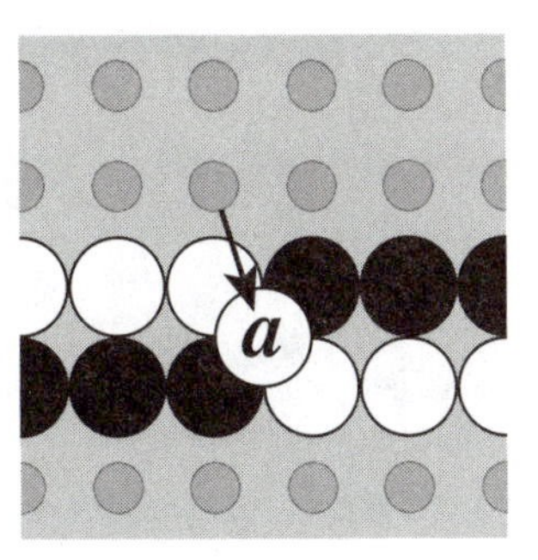

Figure 7.68. Using up a freedom to break a deadlock.

Strategy and Tactics

The *freedom* is probably the most important tactical concept in Akron. A freedom is a marble connected to a chain that will not break the chain's connection if moved.

For instance, Figure 7.68 shows a freedom ***a*** promoted to complete a white connection across a temporary deadlock. Note the cost of using up that chain's last remaining freedom; White must now eat into this connection in order to move any further pieces connected to it. The number of freedoms available to a player indicates his attacking potential.

Diagonally placed pieces of the same color form a strong virtual connection if the player has a move in hand. Even if the opponent occupies the opposing diagonal points to create a temporary deadlock, then the player can promote a freedom onto the four deadlocked marbles to win the connection.

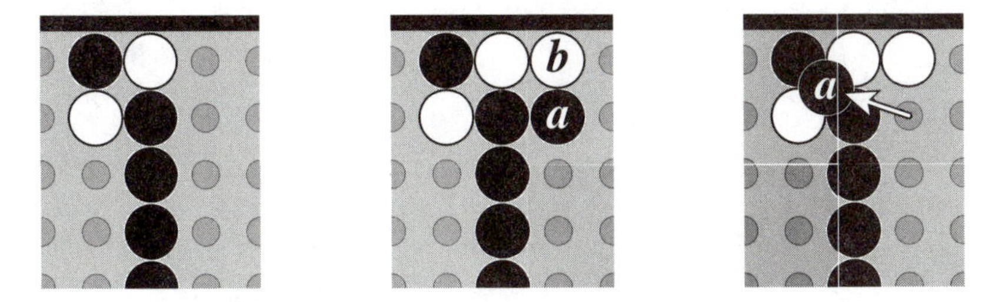

Figure 7.69. Creating a freedom with a forcing move.

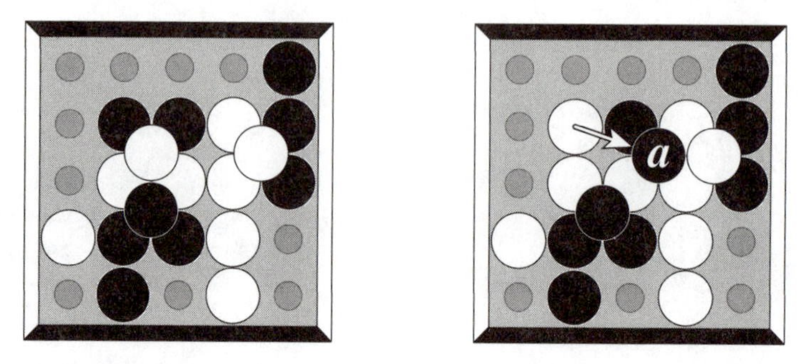

Figure 7.70. Black to play and win (left) with killer move *a* (right).

Forcing moves are a good way to create additional freedoms. Consider Figure 7.69, in which Black has almost connected to the top side of the board, but has run out of freedoms (left). Black marble *a* threatens a direct connection and White is forced to reply with move *b* (middle). Black can then promote his newly created freedom to complete the connection (right).

This demonstrates the importance of efficiency; players should strive to make each move serve more than one purpose. *Drop and threaten* is an especially potent combination.

Figure 7.70 (left) shows a 5 x 5 puzzle with Black to play and win. Move *a* (right) is a killer move which simultaneously

- drops an enemy piece,
- threatens a strong vertical connection (via virtual diagonal connection), and
- pins two enemy pieces.

The complete solution proving that this move leads to a win for Black is complex.

Higher-level marbles are generally more powerful than lower-level ones. However, this example highlights another useful tip: *don't trust enemy foundations*. Players should be wary of relying upon any piece that can be shifted by the opponent.

It's generally good to cluster pieces where possible, as larger chains provide more opportunities for movement. Small chains or isolated pieces are sitting targets with limited movement, and may require several moves to connect into something useful. This balances the usual desire in connection games to spread pieces out and claim territory.

Ladders play a limited role in Akron as a wall of enemy pieces can usually be built over.

History

Akron was invented by Cameron Browne in 2002, and incorporates rule improvements suggested by Steven Meyers. Akron is a Greek word meaning "highest extreme." For a more complete history of Akron see Browne [2003].

The win-on-opponent's-turn rule is a recent addition that improves the game as it

- extends games so that boards fill up more,
- reduces the race-to-edge element,
- promotes higher-level development, and
- simplifies the rules as players no longer need to test for revealed wins when moving.

Variants

Dipole

Dipole is played on an Akron board. Players share a common pool of *dipoles*, each consisting of a white marble joined to a black marble. Players take turns placing a dipole on the board to sit either

1. *flat* such that both marbles occupy adjacent points on the same level, or
2. *tilted* such that one marble rests on one level and the other marble rests at an angle on the level above. Tilted pieces can only rest upon diagonally adjacent pairs of pieces (Figure 7.71).

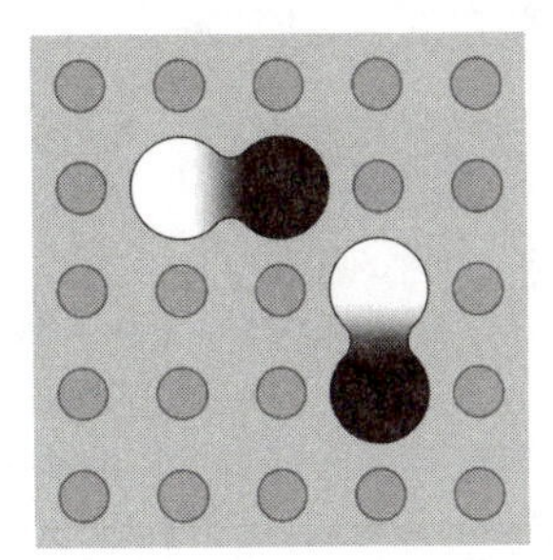
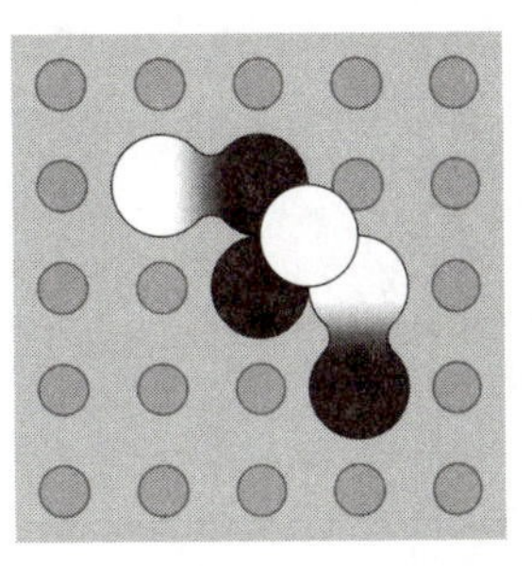

Figure 7.71. Flat dipoles (left) and a tilted one resting upon them (right).

The first player to connect his edges with a chain of his color wins. The over/under rule is in effect.

Dipole has the interesting aspect that a player must introduce an opponent's marble with every move. Each move is therefore a balance between maximizing the impact of the friendly marble and minimizing the impact of the enemy marble.

Dipole was designed by Cameron Browne in November 2003.

Lakron

Lakron stands for Lipped Akron. It is played as per standard Akron on a board containing a lip around its perimeter, as suggested by Dale Watson. The lip is typically one ball diameter in size and is equivalent to playing the game with a row of marbles along each home edge, however larger lips can be used.

Marbles are deemed to connect to a home edge only if they rest on top of the lip. This tends to prolong games and promote higher-level play, and means that deeper games can be played on smaller boards.

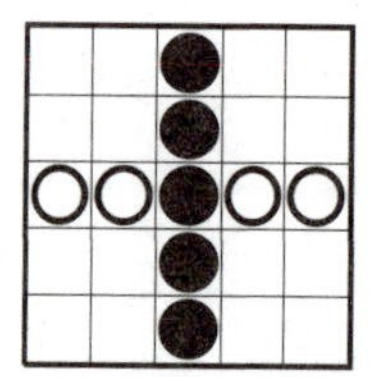

Antipod

Antipod is a connection game with unequal goals played on the two halves of a sphere.

Rules

Antipod is played on two equally sized hex hex boards, typically with six cells per side. The board is set up with a black piece in the center of each half, forming Black's *antipodal goals*, and white pieces placed at the six corners of each board, forming White's *handicap pieces* (Figure 7.72).

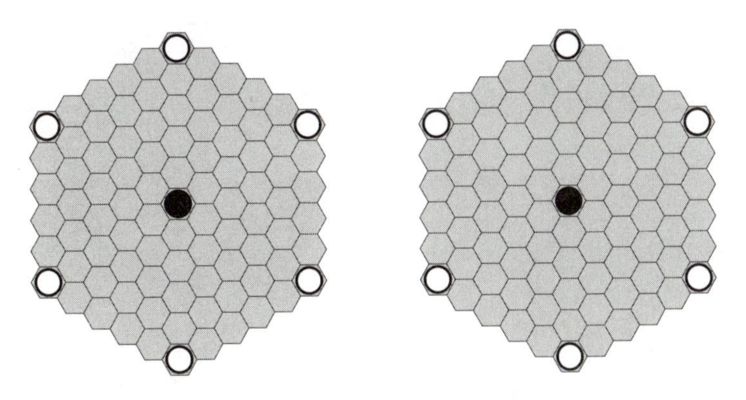

Figure 7.72. The Antipod board and its initial set-up.

Two players, Black and White, take turns placing a piece of their color on an empty cell. If any piece is placed on an edge cell of one board, then a duplicate piece of the same color is automatically placed at the same cell on the other board (which must be empty). Such duplicate pairs actually represent the same piece, allowing continuous connections between the two boards.

Black wins by forming a chain of black pieces between the antipodal goals of each board half. For instance, Figure 7.73 shows a win by Black who has completed a chain between the antipodal goal pieces ***a*** and ***e***, via edge crossings at ***b***, ***c***, and ***d***.

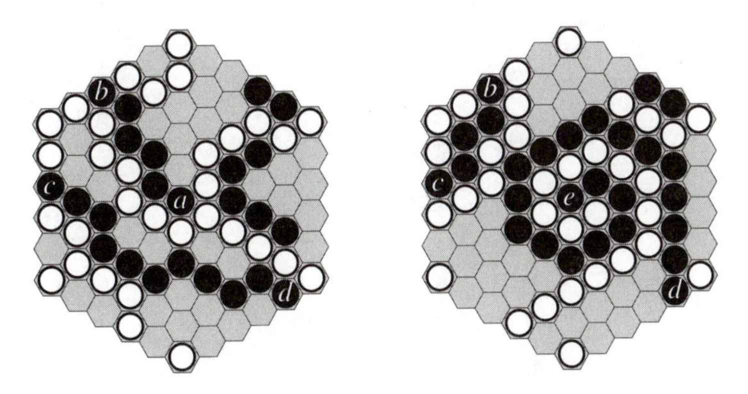

Figure 7.73. A game won by Black.

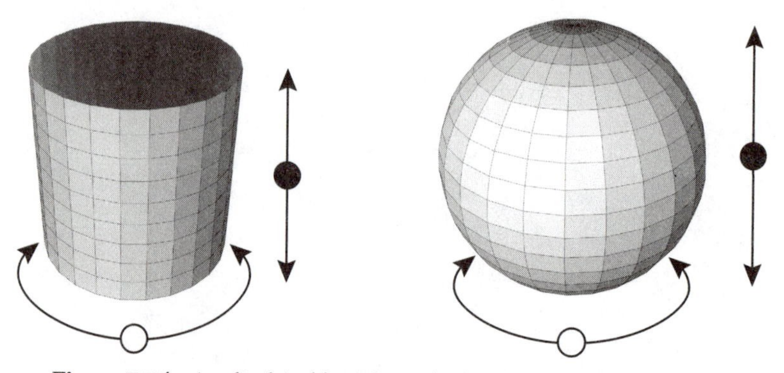

Figure 7.74. A cylindrical board pinched to form a spherical board.

White wins by preventing a Black win. This amounts to completing a cycle of white pieces around the board, disconnecting Black's antipodal goals. Exactly one player must win.

Notes

Antipod is closely related to Cylindrical Hex. Figure 7.74 illustrates how the Cylindrical Hex board (left) can be turned into a punctured sphere (right) by pinching the cylinder at each end. In both cases Black aims to establish a chain between top and bottom, while White aims to establish a cycle around the circumference.

The two Antipod board halves each correspond to a hemisphere of this pinched sphere, and Black's two antipodal goal points correspond to its two poles. The Antipod board can be projected onto two hemispheres to show this explicitly (Figure 7.75).

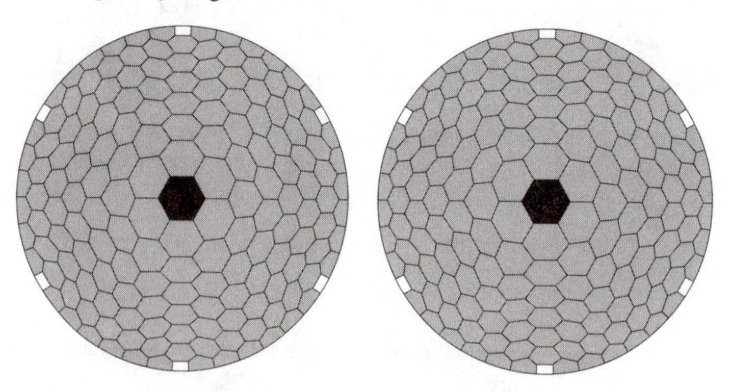

Figure 7.75. Antipod is played on two hemispheres.

It now becomes clear why edge pieces must be duplicated. They actually represent a single piece played on the equator but viewed from both the top hemisphere (left) and the bottom hemisphere (right).

A completely hexagonal tiling of the sphere is not possible, and six of the equatorial cells are of necessity four-sided rather than six-sided. The six white handicap pieces are placed at these four-sided equatorial cells for a number of reasons:

- to address an inherent imbalance between the players' unequal goals (Black's task is easier than White's),
- to eliminate the weaker four-sided cells from play,
- to orient the game in space (if the game were ever played on an actual sphere then there would be no doubt as to the whereabouts of the Black antipodal goals),
- for aesthetic purposes, and
- to defeat potential symmetry strategies.

The last point is the most important; Antipod would otherwise be subject to the same symmetry strategy that solves Cylindrical Hex (see Appendix E).

Figure 7.76 shows how the Antipod board gives a planar adjacency graph. Note that pieces on the right hemisphere must be reflected about its vertical axis to achieve the correct adjacencies. This is equivalent to viewing the right hemisphere from behind (that is, from the inside of the sphere looking out) rather than viewing it from in front.

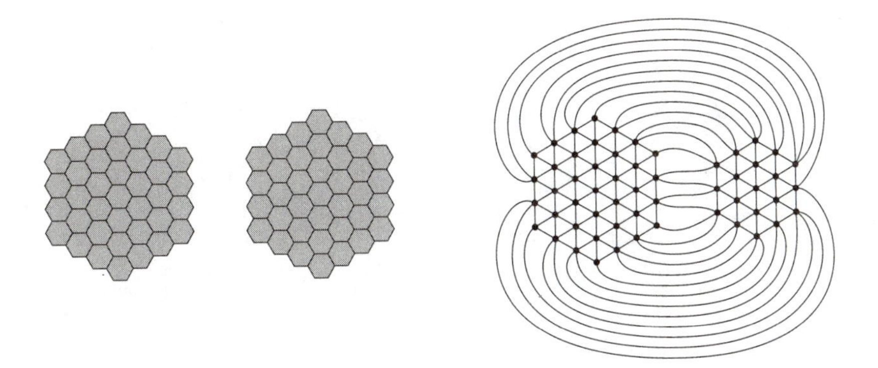

Figure 7.76. The adjacency graph of a small Antipod board.

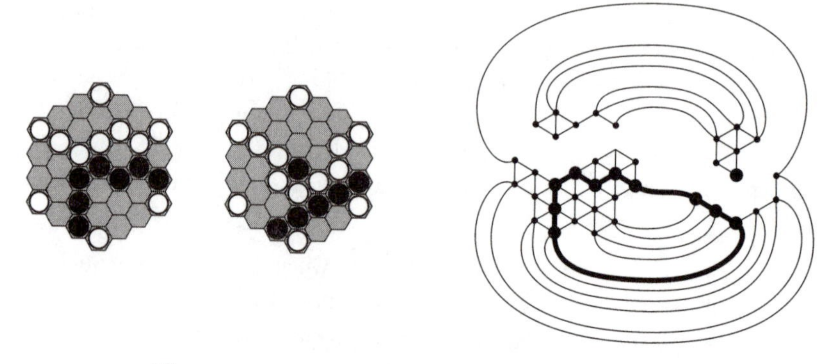

Figure 7.77. A completed game and its game graph.

Figure 7.77 shows a completed game won by White and its game graph. There can be no doubt that White has won since the game graph has been cut into two disconnected components with an antipodal goal in each; Black cannot possibly connect these now.

Strategy and Tactics

Antipod allows some interesting strategic and tactical play due to the spherical nature of the board. For instance, Figure 7.78 shows how Black can exploit two longitudinal attacks in tandem to achieve a connection. Black move ***a*** is a forking move that wins the game; White can block one longitudinal attack with either move ***b***, but cannot stop both at once.

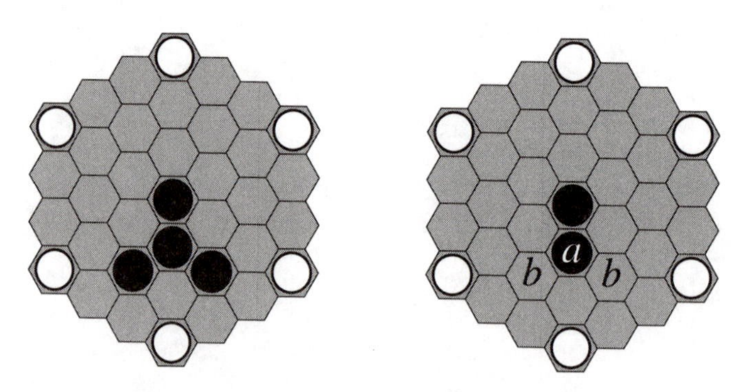

Figure 7.78. Black wins with forking move ***a***.

Smaller board sizes favor Black, while larger board sizes favor White. The standard board size achieves a reasonable balance.

History

Antipod was invented by Cameron Browne in 2003. It was devised while investigating projective Hex games with Bill Taylor, and was originally called Corpus Callosum.

Variants

Escape Hex

Escape Hex is similar to Antipod but played on a single hemisphere. Black aims to achieve a path of black pieces connecting the central cell with any equatorial cell, while White aims to achieve a cycle of white pieces that prevents this. Jack van Rijswijck proposed Escape Hex as a hypothetical variant of Antipod.

The Escape Hex board is equivalent to the Cylindrical Hex board with one end pinched to a point, and unfortunately suffers from the same symmetry strategy as Cylindrical Hex. White handicap stones around the equator do not stop this winning strategy unless they completely occupy the equator to form a cycle, which defeats the purpose of the game.

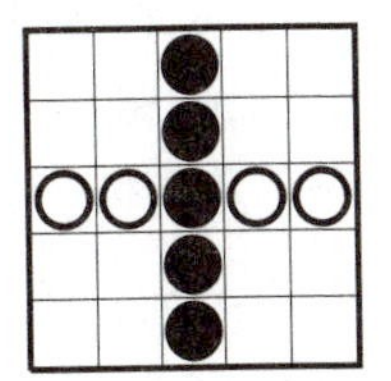

TriHex

TriHex is Y with a twist—or more accurately, a tear. The board is topologically equivalent to a Y board split along one seam and with a rotated copy of itself inserted in the break.

Rules

TriHex is played on a hex hex board. Each cell, except the center, has a diametrically opposed partner on the far side of the board. The board is initially empty.

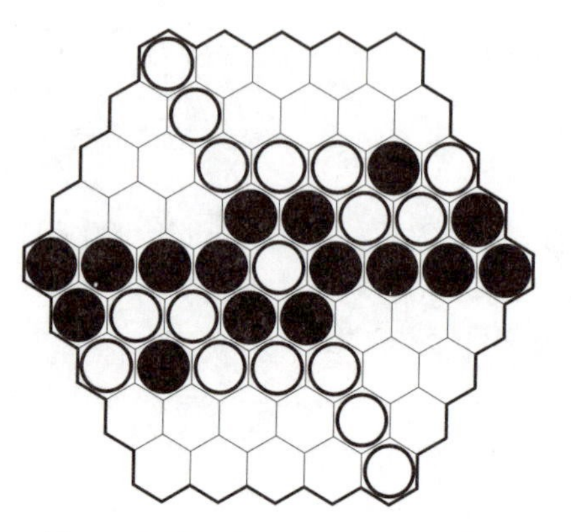

Figure 7.79. A game won by White.

Two players, Black and White, take turns placing either

- a single piece of their color on the empty central hexagon, or
- two pieces of their color on a matching pair of empty cells.

A player wins by connecting at least three consecutive sides with a chain of his pieces. Exactly one player must win.

Notes

Figure 7.80 shows what is meant by *diametrically opposed partners*. Each cell in the top half of the board has a matching counterpart in the bottom half (and vice versa). For instance, if White plays either piece ***a*** then its matching partner ***a*** must also be played on the other side of the board, and similarly for black piece ***b***. The board center ***c*** does not have a partner.

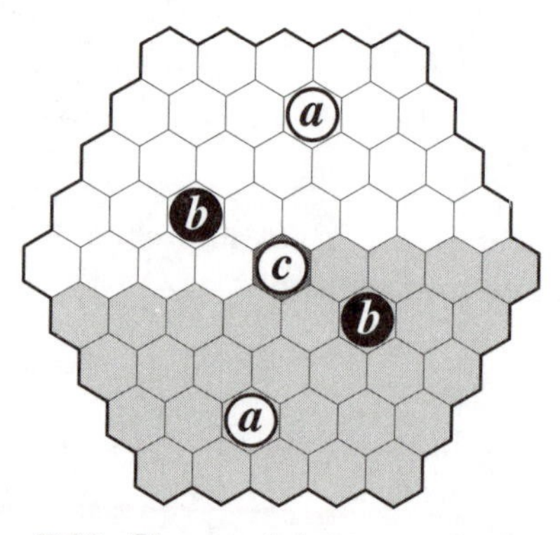

Figure 7.80. Pieces and their rotated counterparts.

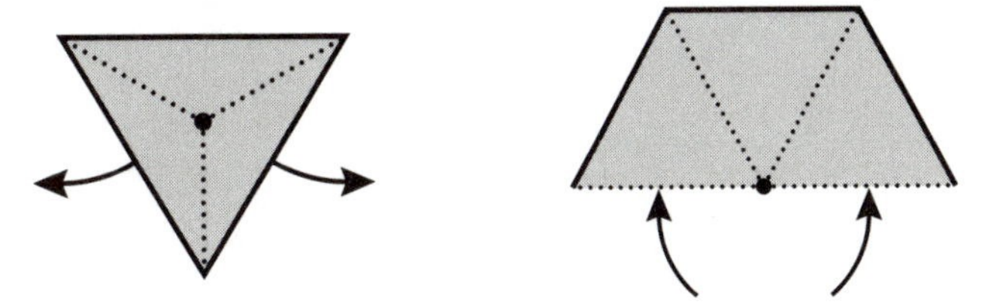

Figure 7.81. A Y board unfolded to give a TriHex board.

This method of piece pairing reflects the fact that TriHex is effectively played on a Y board split along a seam, and a rotated copy inserted in the gap. Figure 7.81 shows a Y board (left) unfolded to give the top half of the TriHex board (right), which is then duplicated and rotated to give the bottom half. The pivotal point around which the unfolding occurs forms the singleton central cell.

TriHex can be played without paired moves using the single unfolded board half shown in Figure 7.82. However, players must visualize the implied adjacencies across the seam (dotted lines). This can detract from the clarity of the game, and it's generally better to play TriHex with paired pieces on the full hexagonal board.

This unique board design makes TriHex equivalent to playing a game of Y on the three exposed faces of a tetrahedron (Figure 7.83) with the aim of connecting the three bottom edges ***a***, ***b***, and ***c***.

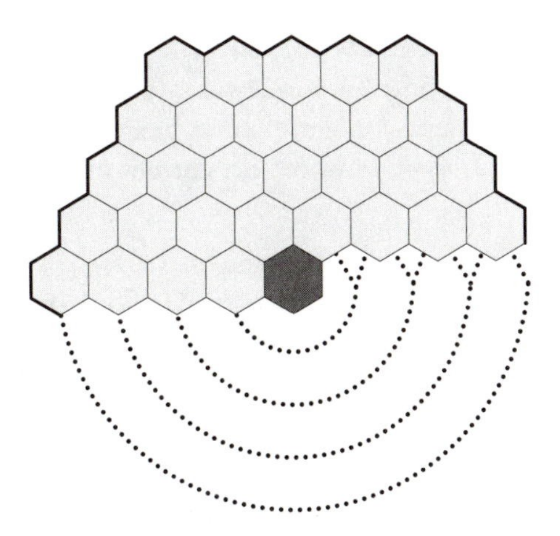

Figure 7.82. Implied adjacencies across the unfolded seam.

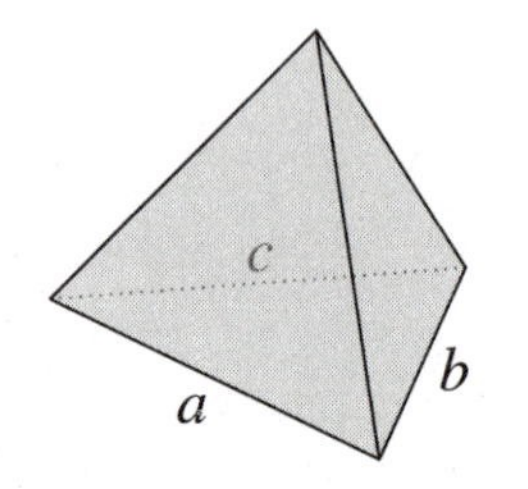

Figure 7.83. TriHex is Y played on the exposed faces of a tetrahedron.

History

TriHex was invented by Scott Huddleston in 2003.

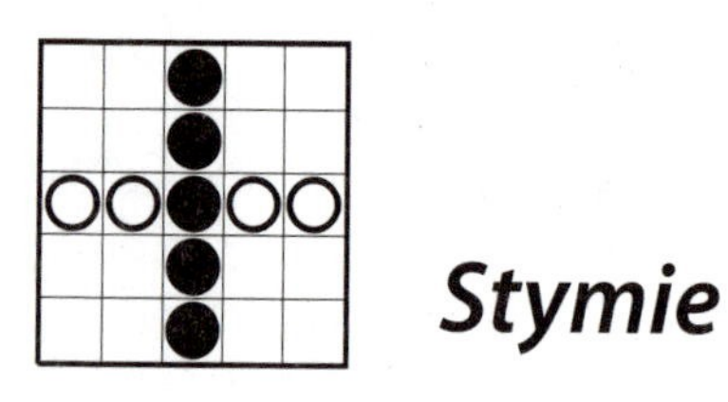

Stymie

Stymie is played on a field of squares and octagons. It is unusual in that single moves may be more powerful than multiple moves.

Rules

Stymie is played on a field of octagons, typically 8 x 8, with interstitial squares. Two players, Black and White, own alternating sides of the board that bear their color. The board is initially empty.

White plays first and claims a single cell (first-move equalizer). Thereafter, players alternate taking turns, in which the current player may either

1. claim a single cell,
2. claim an adjacent square and octagon, or
3. claim three cells (either an octagon and two adjacent squares, or a square and two adjacent octagons) but only if the opponent's previous move was a single move.

A player wins by connecting his edges of the board with a chain of cells in his color. Exactly one player must win. Figure 7.84 shows a game won by White.

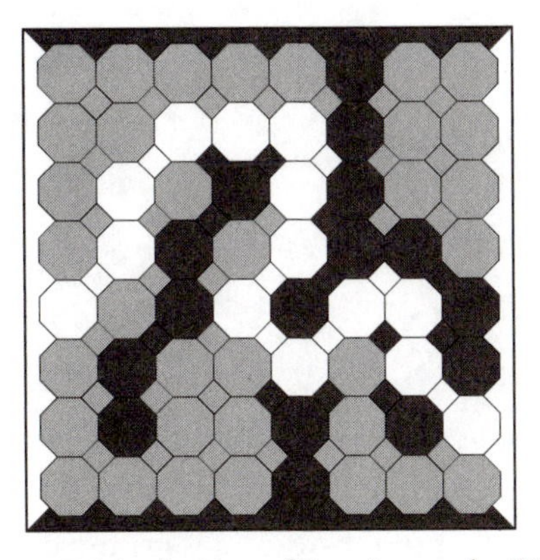

Figure 7.84. A game of Stymie won by White.

Notes

Figure 7.85 demonstrates the various types of move available to the players. Note that all moves involving more than one cell form a connected chain in themselves and include at least one square cell.

The Stymie board is the 4.8.8 Archimedean tiling (see Appendix H). Deadlocks will not occur as this tiling is trivalent. Like its close relatives Octagons and Quax, Stymie encourages the inclusion of weaker square cells–in this case by making all multiple moves include at least one square.

The paradox of Stymie is that single moves (even single squares) can be quite powerful in some circumstances, and must be balanced out by allowing the opponent a triple move next turn.

Figure 7.85. Singleton moves, double moves, and triple (special) moves.

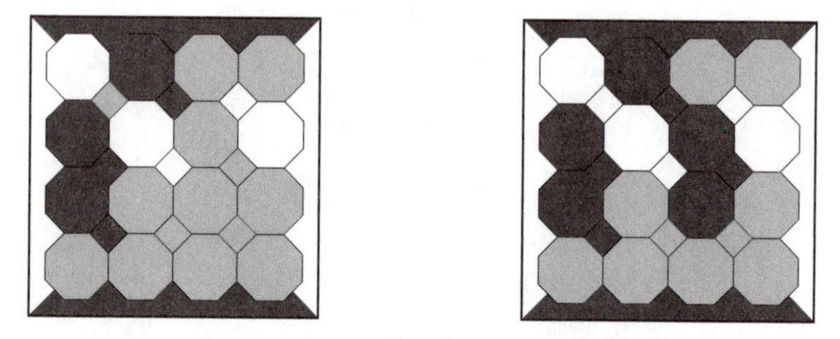

Figure 7.86. A single cell stymie (left) and a bad resolution for White (right).

This point is illustrated in Figure 7.86 (left) in which it is White's turn to play. Note that both White's and Black's strongest connections are stymied by the empty square cell at the top left of the board (hence the game's name).

If White makes the disastrous decision to claim the single stymying cell (Figure 7.86, right) then Black's triple-move reply sets up a win. White should instead attack Black's connection towards the bottom of the board.

Strategy and Tactics

Players are advised to take an octagon rather than a square on the opening move, and to take pairs of octagons rather than pairs of squares on triple moves, unless circumstances dictate otherwise.

Each time a key cell is stymied, the players engage in a game of chicken in which the first player to crack and take the cell gives the opponent a valuable triple move. Players are not forced to make a triple move following an opponent's singleton, but it is usually wise to do so.

History

Stymie was created by Randall Bart in December 2003, who was inspired by his bathroom floor.

Stymie is essentially the same game as Square Board Connect (described in Section 8.1) played on a more complex tiling and with the triple move limited to a conditional bonus.

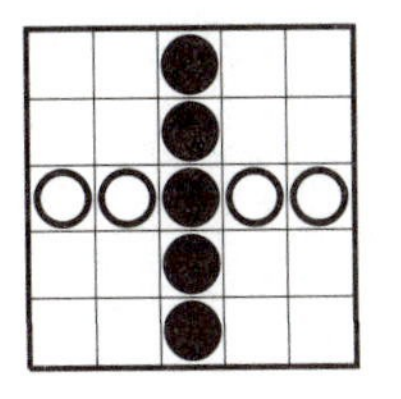

Druid

Druid is a three-dimensional block building connection game played on the square grid. Deadlocks are resolved by letting players build blocks over enemy pieces.

Rules

Druid is played on a square grid, typically 10 x 10. Two players, Black and White, own alternating sides of the board that bear their color. The board is initially empty.

Each player owns a number of the following types of stones of his color:

- *sarsens* (single unit cubes), and
- *lintels* (three unit cubes long).

Players take turns placing one of their stones on the board. Stones must always be aligned with the board squares. A single-move swap option is recommended.

Sarsens can be placed directly on the board surface, or on top of any same-colored stone.

Lintels cannot be placed on the board surface, but must be elevated by stones at both ends so that they lie flat and sit directly upon exactly two units of same-colored stone. Figure 7.87 shows some legal lintel placements by way of example.

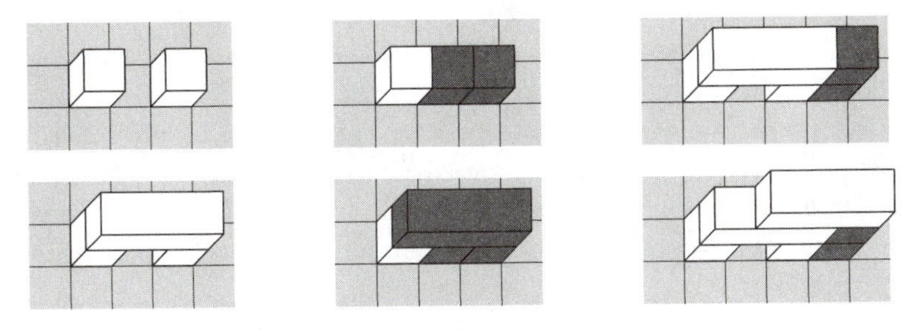

Figure 7.87. Legal lintel placements.

Figure 7.88. A game won by Black.

The game is won by the player who completes a connected set of their stones between their sides of the board (see Figure 7.88).

Two stones are connected if they are the same color and occupy orthogonally adjacent board squares, irrespective of their relative heights. Only the top-most stone at each board square counts in any connection; upper connections cut lower connections (*over/under rule*).

Notes

The rules of lintel placement imply that if a lintel does not bridge a gap with its central unit, then it must sit on exactly one unit of enemy stone. This rule not only stops players from stacking lintels (an overly strong play) but means that players can build over enemy stones. This is especially important on the square grid, which is susceptible to deadlock. The triple-unit lintels provide an escape route over enemy blocks just as stacked marbles do in Akron. It has not been proven that all local Druid deadlocks can eventually be defeated, but this appears to be the case so far.

Like Akron, the over/under rule allows the game to be projected to a two-dimensional game graph, greatly simplifying matters. The game graph must contain terminal pairs for both players in the Join/Join style due to the nature of the square grid and temporary deadlocks, as discussed in Section 3.3.

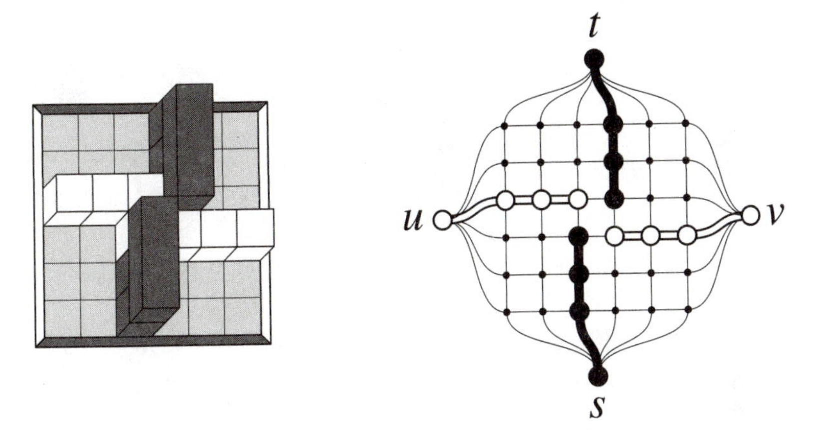

Figure 7.89. A temporarily deadlocked 6 x 6 game and its game graph.

Figure 7.89 shows a temporarily deadlocked 6 x 6 game (left) and its game graph (right). If the white terminals were not present then this graph would appear to be cut and hence a win for White, but nothing is further from the truth; Black can force a win no matter who plays next.

Druid is not limited to the square grid, and can be played on any regular or semiregular tiling using prismatic extrusions of the board tiles (irregular tilings make lintel design difficult). Lintels of different shapes and sizes can also be devised. Figure 7.90 shows some possible sarsen and lintel arrangements on nonsquare tilings.

Strategy and Tactics

Druid is a rich tactical game. The fact that lintels threaten to build over enemy stones means that forcing moves are common, and traps readily set for an unwary opponent.

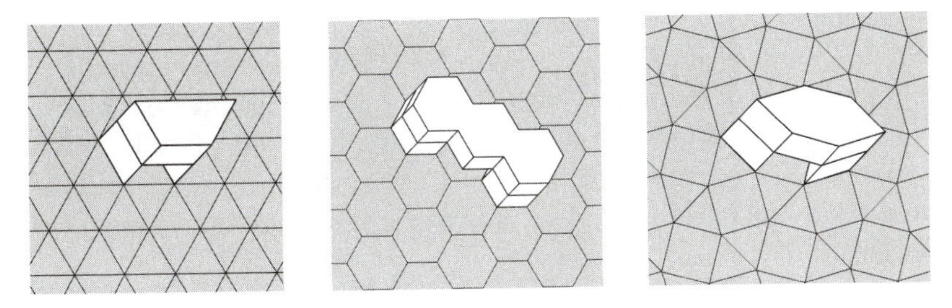

Figure 7.90. Druid played on other tilings.

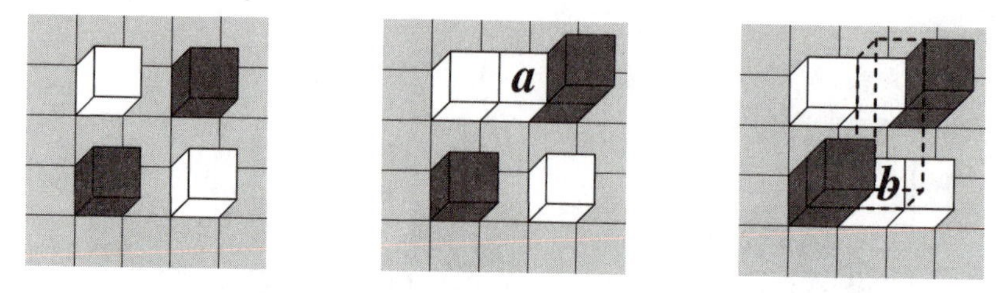

Figure 7.91. Forcing move ***a*** and forking move ***b*** yield a connection.

It is usually good to begin the game by staking out a staggered path of well-spaced sarsens across the board, since sarsen placement is generally the only possible move in the early stages anyway.

Placing friendly sarsens separated by an empty space is a good compromise between threatening an immediate lintel placement and gaining territory. However, pieces separated one empty space from enemy stones may be vulnerable to forcing moves, as shown in Figure 7.91.

Figure 7.91 (left) shows a typical sparse formation after the early stages of the game, with White to move. White move ***a*** forces Black to build his sarsen up a level or risk having a lintel built over it. White has gained a square of territory while Black has been forced to waste a move without gaining any territory. This is a common play.

White move ***b*** is a similar forcing move that gains territory. It is also a fork that guarantees connection, and hence a very efficient play. If Black again decides to save his besieged sarsen by building upwards as shown, then White can place a lintel between ***a*** and ***b*** to connect his two groups (dotted line). Note that Black could have performed a similar attack if it was his turn; as with most connection games, the momentum afforded by having the move in hand (tempo) can be decisive.

Building upwards is an obvious way to block an opponent's lintel placements, but lower-level blockers can be just as effective. For example, Figure 7.92 (left) shows a situation in which White wishes to connect to the left side of the board, with Black to move. Black piece ***a*** blocks the higher-level White connection. Black would like to build a lintel along the edge to remove the dangerous White sarsen; however doing so would raise the lower-level block ***a*** and give White an immediate lintel placement to connect.

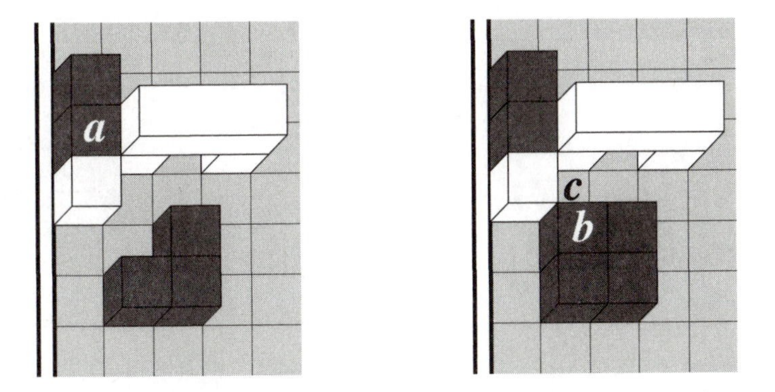

Figure 7.92. Black blocks White to the left.

Instead, move ***b*** (Figure 7.92, right) is a much better play for Black. It forms a strong 2 x 2 group that threatens to clobber any enemy piece placed alongside it. White would like to have played at ***c*** next turn, but this is no longer a viable move and Black can play there instead to complete the block.

Figure 7.93 (left) shows a common sequence of forcing moves that guarantee connection to an edge. Move ***a*** forces White to block upwards, move ***b*** forces an immediate block to the edge, and move ***c*** guarantees connection for Black. If White tries to block the immediate connection then Black can place a lintel to restore it. Move ***a*** was deceptively subtle; if White were not forced to first build upwards then piece ***c*** would be under threat of a White lintel.

This example is similar in nature to Akron (see Figure 7.69) but instead of using a forcing move to create a freedom, in this case the forcing move

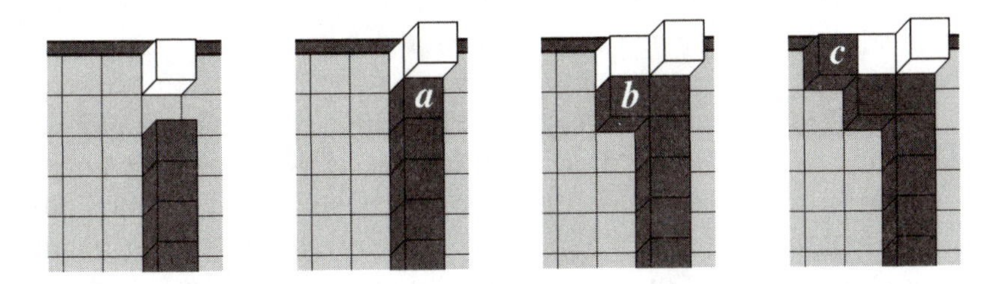

Figure 7.93. Black forces a connection to the top.

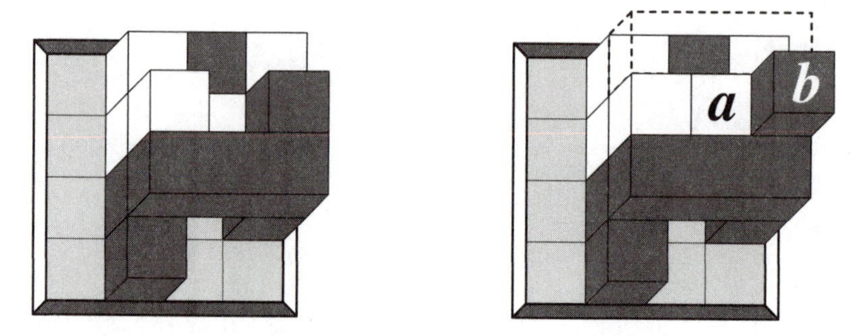

Figure 7.94. White to play and win (left) with forcing move ***a*** (right).

prepares a platform for future lintel placement. Akron and Druid share many similarities, both being three-dimensional connection games on the square grid with similar goals.

Figure 7.94 (left) shows a mini puzzle with White to play and win. Black appears to have the stronger position, but the turn in hand again proves decisive and Figure 7.94 (right) shows a killer move ***a*** that leads to a win for White. Black is forced to build upwards with ***b***, preventing any defense against White's next lintel move (dotted).

History

Druid was invented by Cameron Browne in 2004.

Variants

Ryan's Game

Ryan's Game is played on a 15 x 15 square grid with a number of pegs placed in a symmetrical pattern [Ryan 1974]. Players own alternating sides of the board that bear their color, and a number of tiles of their color each covering three square units and with two holes drilled into them. Some tiles are straight and some are L-shaped.

Players take turns placing one of their tiles on the board. The tile's holes must fit over board pegs, which then become owned by the player; the opponent cannot use those pegs for the rest of the game. Tiles may stack up to three levels in height, and may be placed such that the square unit without a hole passes over an enemy connection. This does not cut the enemy connection, as Ryan's Game does not include the over/under rule.

A player wins by connecting his edges of the board with a path of his tiles (which may be buried at points). Druid was created independently without prior knowledge of Ryan's Game.

Druid's Walk

Druid's Walk is played as per standard Druid, except that the first time a friendly piece is played along a player's home edge, a token called the "Druid" is placed upon that stone. Each turn thereafter, the current player may optionally move his Druid after making the usual stone placement.

Druids may move to any adjacent stone of the same color, regardless of height. Stones may not be placed on top of any Druid and no Druid may step into gaps left underneath stones. Each player has exactly one Druid. The game is won by the first player to move his Druid to the opposite edge from which it was entered.

Druid's Walk has the effect of extending the standard game. A player might complete a connection between his sides but still require several moves to get his Druid home, giving the opponent time in which to cut the connection and keep the game alive.

Span

Span is played as per Druid except that the playing grid is rotated through 45 degrees relative to the sides of the board (see Figure 7.95) and a player wins by connecting either pair of opposite board sides with a chain of his color. The Span rules are by Cameron Browne and Pin International Co., Ltd.

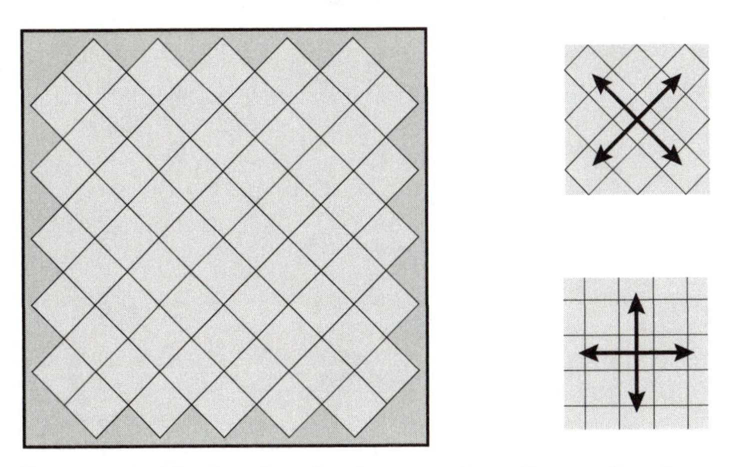

Figure 7.95. The Span board and a comparison of strong directions.

The rotation of pieces relative to the sides of the board actually has a significant impact on play. In Druid each piece has a single strongest direction of connection to each side (bottom right), whereas in Span there are two equally strong directions of connection to each side (top right).

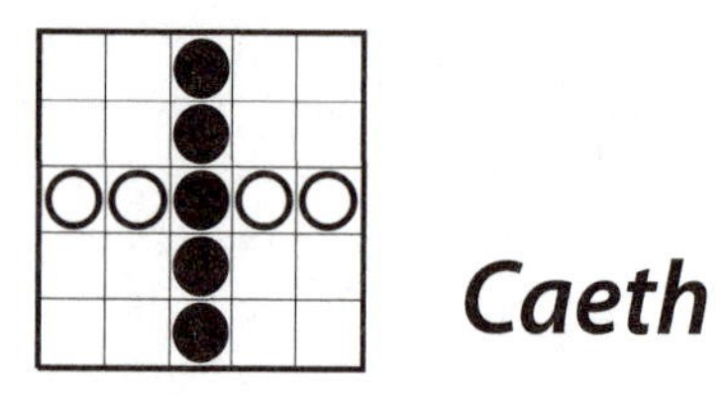

Caeth

Caeth is a metarule that adds a layer of tactical richness to many connection games.

Rules

Caeth is a metarule rather than any particular game. It is played on the edges of an adjacency graph on which two players, Black and White, take turns claiming an unoccupied edge. The first to claim at least half of the edges incident with a vertex then claims that vertex.

For instance, Figure 7.96 shows an interior vertex with six neighbors won by White (left), a side vertex with four neighbors won by Black (middle), and a corner vertex with two neighbors won by White (right). Vertices remain permanently colored even if the opponent claims the remaining edges.

This Caeth mechanism of edge voting is called the *undergame*. The game to which it is applied is called the *overgame*. The winning condition

Figure 7.96. The Caeth metarule in action.

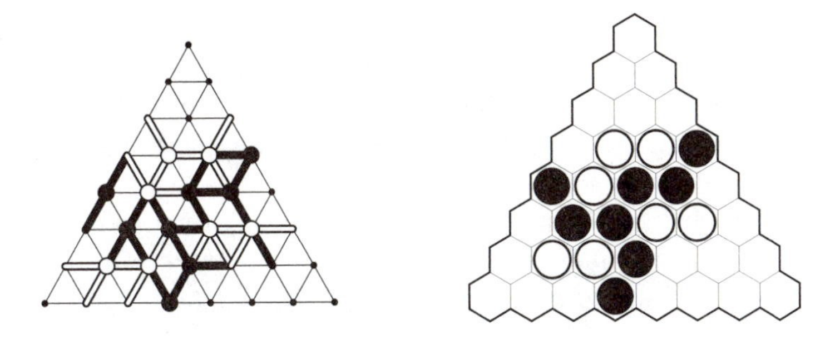

Figure 7.97. A game of Caeth Y won by Black.

for any Caeth game is the same as that for the overgame played on the vertices. A single-move swap option may be used if the overgame does not have a specific first-move equalizer.

For instance, Figure 7.97 (left) shows a game of Caeth Y won by Black, who has established a path of adjacent black vertices between all three sides of the board. The standard Y equivalent of this game is shown on the right.

Notes

Although several connection games include ideas that can be applied successfully to other games, Caeth is the only game truly intended as a universal metarule. Without a parent game Caeth has no specific board shape or winning condition.

The adjacency graph on which the Caeth undergame occurs must not be confused with the overgame's graph. Figure 7.98 demonstrates that these are quite distinct.

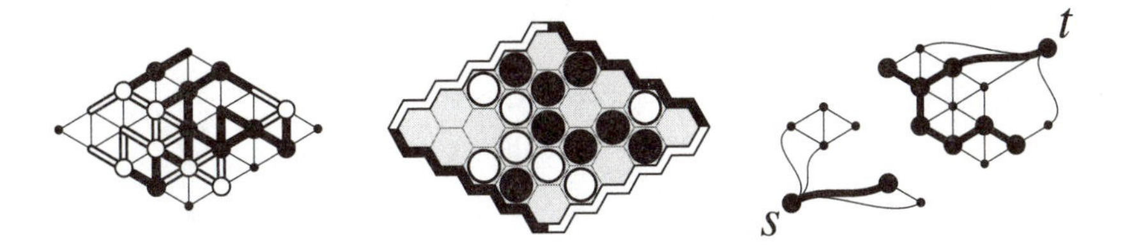

Figure 7.98. A game of Caeth Hex won by White and its game graph.

Figure 7.98 (left) shows a game of Caeth Hex won by White, who has established a path of white vertices between the white edges of the board. Figure 7.98 (middle) shows the standard Hex representation of this game. Figure 7.98 (right) shows the game graph proving that White has cut the graph to win.

Although players choose an edge each turn, the Caeth undergame is *not* a Shannon game played on the edges, as the winning connection is defined strictly by the vertices. The edges of the Caeth undergame are used strictly for voting and are a level of abstraction removed from the overgame. Caeth is played on the edges but won on the vertices.

Caeth can potentially be applied to any connection game with a simple adjacency graph. It is less applicable to games with nonplanar adjacency graphs, or in which pieces move from one board point to another.

However, the Caeth undergame is robust enough to incorporate the piece-surround capture rule of Go and Gonnect. Figure 7.99 shows a black vertex in peril (left) in Caeth Gonnect. White move *e* claims the final surrounding vertex (center), capturing the surrounded black vertex (right). Capture in this context involves resetting the surrounded vertex and its incident edges back to the neutral state.

Caeth is a layer of tactical complexity laid over an existing game, and for this reason tends to suit games with simple rule sets especially well. Recursive subdivision of edges into smaller majority-vote units was considered (a sort of inverse Y Reduction) but this double level of abstraction proved too complex for even the simplest overgames.

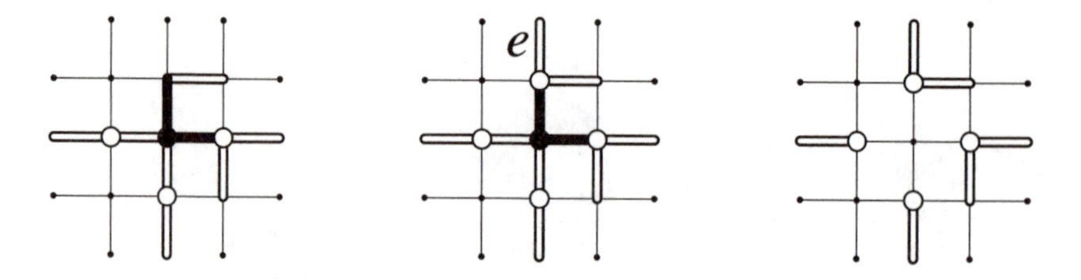

Figure 7.99. Piece-surround capture using the Caeth metarule.

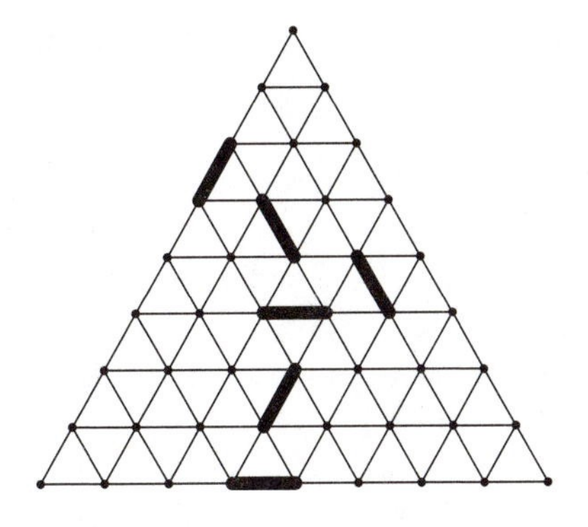

Figure 7.100. An unbeatable connection for Black in Caeth Y.

Strategy and Tactics

Caeth games are more tactical in nature than their parent games, as every vertex must be fought for rather than simply played in. These mini battles for each vertex seep into each other and combine beautifully to give a very rich overall game. Players should always try to make efficient moves in which both ends of each edge serve a purpose.

Figure 7.100 shows an unbeatable position for Black in a game of Caeth Y. No matter where White intrudes along this connection, Black has a safe alternative play that restores it. A player's virtual connection is unbeatable if he is ahead on every vertex along the path and the opponent has no moves that encroach upon two of these vertices at once.

Figure 7.101 (left) shows one such double intrusion by White, called the *splitter* by Phil Bordelon. This is both a strong defensive move that

Figure 7.101. Basic defensive plays by White: the splitter and the slingshot.

stops an immediate Black connection and a strong attacking fork, which threatens both vertices and gives White a good chance of winning at least one.

Figure 7.101 (right) shows a strong marching line development by Black, which cannot be broken by the opponent as all vertices along its length (except the head) have two colored edges, none of which are subject to forking attacks. White's only option here is to block the advancing connection head with a *slingshot* move that gives White a head start on whichever side the Black head tries to veer around.

A *swing vertex* is a vertex with an equal number of edges of both colors that is not yet controlled by either player. A *swing edge* is an edge between two swing vertices. Players should avoid leaving swing edges (potential forks) between two swing vertices that they wish to own.

Variants

There are several different varieties of Caeth. The term "Caeth" above describes the standard variety, which is properly called "Caeth Anu."

Caeth Det

This variant is played as per Caeth Anu except that players own a vertex whenever they own more incident edges than their opponent. A vertex belongs to neither player if they both own the same number of incident edges, except that the first player to own half or more of the incident edges permanently owns the vertex (as in Anu).

Caeth Cha

This variant is played as per Caeth Anu except that players own a vertex by controlling more incident edges than their opponent, and ownership does not revert to neutral if the players subsequently tie in the number of incident edges they control.

Caeth Cha requires a larger board than other Caeth variants as each move is effectively a double move; every edge move captures two vertices (unless the player is forced to waste an end). It is easy for a players to frog-hop their connection across smaller boards before the opponent can react.

Noc

Noc is a metarule like Caeth but is its dual; it is played on the dual of the adjacency graph at the intersection points where three edges meet. Players

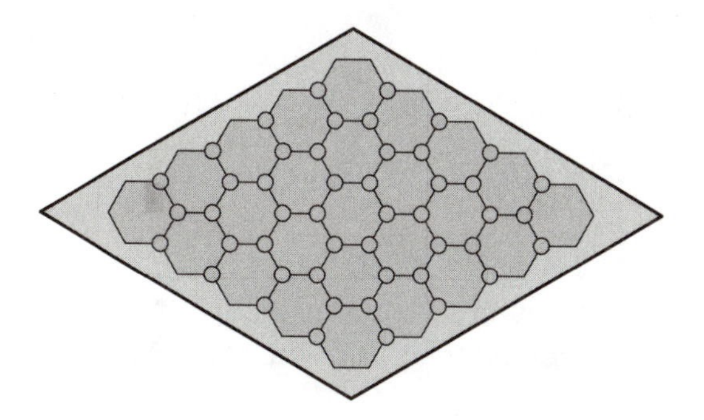

Figure 7.102. A Noc Hex board.

take turns coloring an intersection point, and as soon as a player owns at least half of a cell's intersection points he then owns that cell. Again, the game's winning condition is borrowed from the overgame to which it is applied.

Noc makes explicit the close relationship between Conhex and Caeth, which are effectively duals of each other. Compare the 5 x 5 Noc Hex board shown in Figure 7.102 with the Pula board shown in the Conhex section, and indeed the names Noc Hex and Conhex.

History

The Caeth system was devised by Phil Bordelon in 2004. Caeth Det was designed first but found to have too fast a tempo. Caeth Cha was then devised, then the clearer and calmer Caeth Anu, which has remained the standard.

Note on terminology: Caeth games are described by the metarule, the overgame and the undergame modifier. For example, the various Caeth forms of Hex are as follows:

- Caeth Hex (or Caeth Hex Anu),
- Caeth Hex Det,
- Caeth Hex Cha,
- Noc Hex (or Noc Hex Anu),
- Noc Hex Det, and
- Noc Hex Cha.

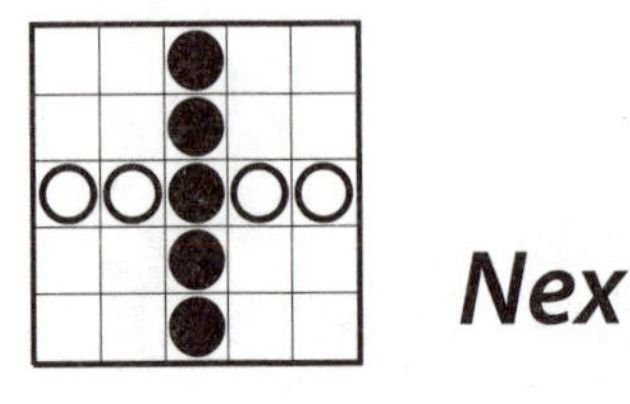

Nex

Nex is a Hex variant that introduces a neutral piece mechanism that may have widespread application to other abstract board games.

Rules

Nex is played on a standard Hex board. As in Hex, two players, Black and White, aim to connect the alternating board sides that bear their color with a chain of their color.

The novelty of Nex lies in the inclusion of neutral pieces (shaded gray and marked with question marks in the figures) and a special move mechanism. On their turn players may either

1. place a piece of their color and a neutral piece on empty cells, or
2. convert two neutral pieces to their color, and convert a different piece of their color to neutral.

Figure 7.103 (left) shows the three types of Nex pieces. Figure 7.103 (middle) shows a Type 1 move by White, and Figure 7.103 (right) shows a Type 2 move by Black.

Figure 7.104 shows a game won by Black, who has established a chain of black pieces connecting the black sides of the board. Neutral pieces do not count in any connection. As with Hex, it is probably the case that exactly one player must win in Nex, but this has not been proven.

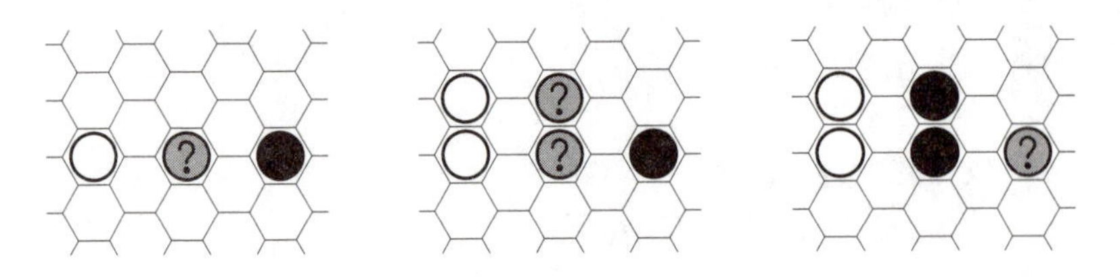

Figure 7.103. The three piece types, a Type 1 move by White, and a Type 2 move by Black.

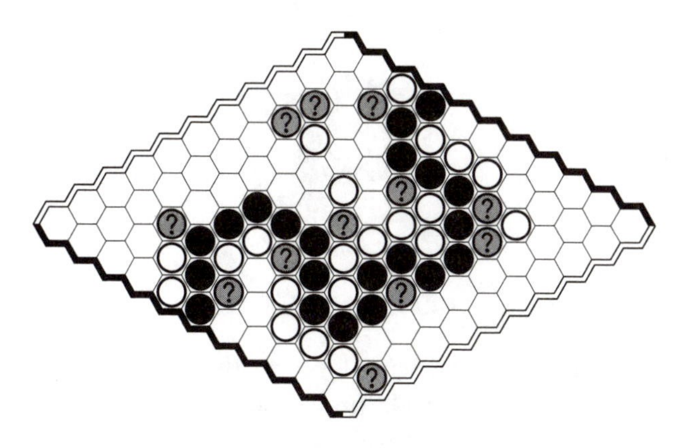

Figure 7.104. A game won by Black.

Notes

The two move types are nicely balanced and mean that there will always be an equal number of black and white pieces at the end of each round. The unusual Type 2 moves allow a form of piece recycling; unwanted pieces can be returned to the neutral state in exchange for converting another piece in a more useful location.

Figures 7.105 and 7.106 illustrate how neutral pieces can affect connectivity during play.

Figure 7.105 (left) shows a pair of white pieces in the familiar bridge formation. This connection is not guaranteed in Nex, since if Black intrudes into both key points with a Type 1 move (Figure 7.105, right)

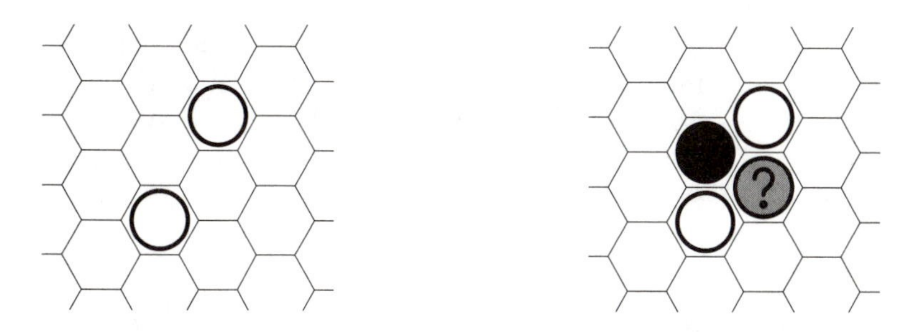

Figure 7.105. Bridges can be exploited by the opponent.

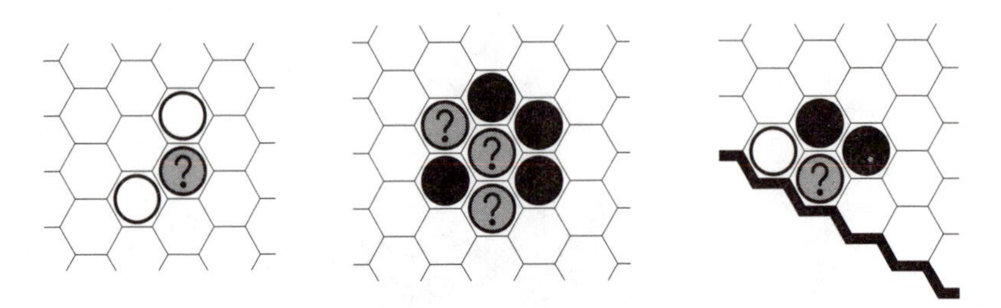

Figure 7.106. Safe connections in Nex.

then White must convert the neutral piece with a Type 2 move next turn to save the connection. This can be problematic if White has no free pieces to convert, and can even be impossible if there is no other neutral piece to convert, although this is not likely to occur during play.

By comparison, the white bridge shown in Figure 7.106 (left) looks vulnerable but is in fact a strong connection. The black connections (middle and left) are also strong.

Strategy and Tactics

A key tactic unique to Nex has emerged: *play forcing moves to stop the opponent converting*. As long as the opponent is forced to place an extra piece of his color he is unable to make Type 2 moves, and hence is unable to convert any pieces already on the board.

Some passages of play tend to have an element of chicken, when players delay making a conversion that may also benefit the opponent. In such cases players tend to play Type 1 moves until absolutely necessary.

History

Nex was invented by João Pedro Neto in 2004. The name Nex is a contraction of "Neutral Hex." Originally, Type 2 moves did not require that a friendly piece also be converted to neutral, making them much stronger than Type 1 moves. The current rules make the moves more balanced, and ensure that there will be an equal number of white and black pieces at the end of each turn.

Variants

Like many Hex variants, the neutral piece mechanism of Nex actually constitutes a metarule that can be successfully applied to other games. For

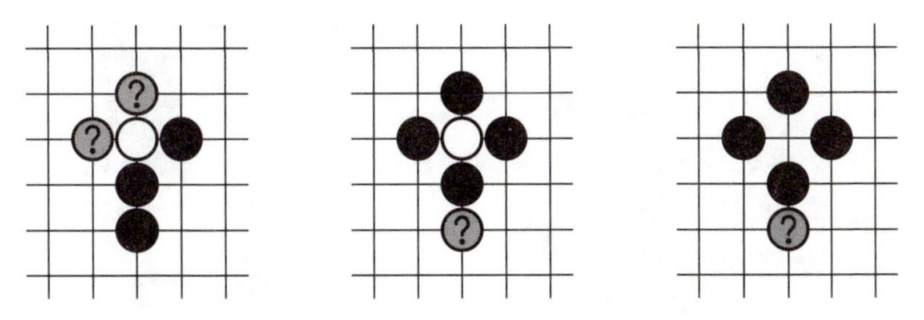

Figure 1.07. Piece-surround capture using Nex moves.

instance, Figure 7.107 demonstrates how the Nex neutral-piece metarule can be applied to such games as Go, Gonnect, and Tanbo that involve piece-surround capture. Neutral stones do not count as capturing pieces (left) but must be converted to a color (middle) before any surrounded pieces are captured (right). Note that the bottom-most black piece is reverted to neutral as part of the move.

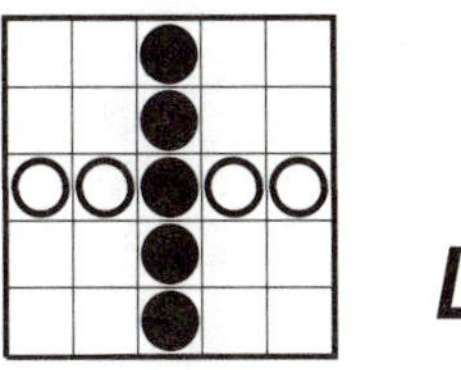

Lynx

Lynx is a simple connection game based on Dots & Boxes that introduces a novel way to avoid deadlocks on the square grid.

Rules

Lynx is played on a square grid of pegs, typically 8 x 8. Two players, Black and White, own alternating sides of the board that bear their color (Figure 7.108, left). The board is initially empty.

Players have access to a common pool of links exactly long enough to connect two orthogonally adjacent pegs. Player take turns placing a link between any two pegs, provided that no peg ever has four links attached to it. If the link just played encloses an area, then the player claims that area by marking it with his color.

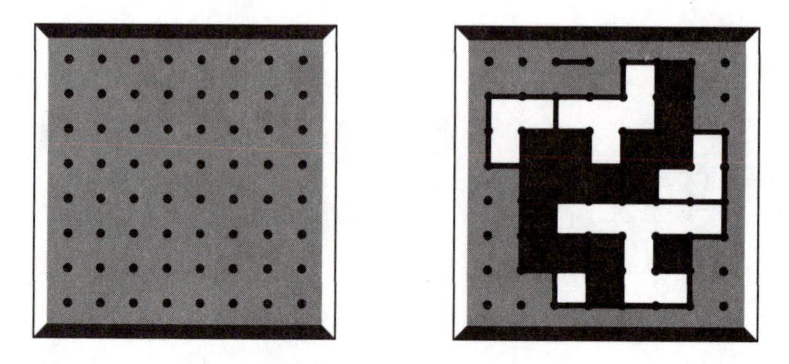

Figure 7.108. The Lynx board (left) and a game won by Black (right).

A player wins by establishing a path of his color between his edges of the board. For instance, Figure 7.108 (right) shows a game won by Black. Exactly one player must win.

Notes

The restriction that no peg may ever have four links ensures that the adjacency graph for any possible combination of enclosures remains trivalent. The usual deadlock condition on the square grid (Figure 7.109) will never occur, and every game will have a winner. See Section 3.3 for a discussion of deadlocks on the square grid.

Strategy and Tactics

There may be several potential moves that enclose an area (Figure 7.110, left). However, for any such enclosure there will always be one optimal move

Figure 7.109. The deadlock condition that will never occur.

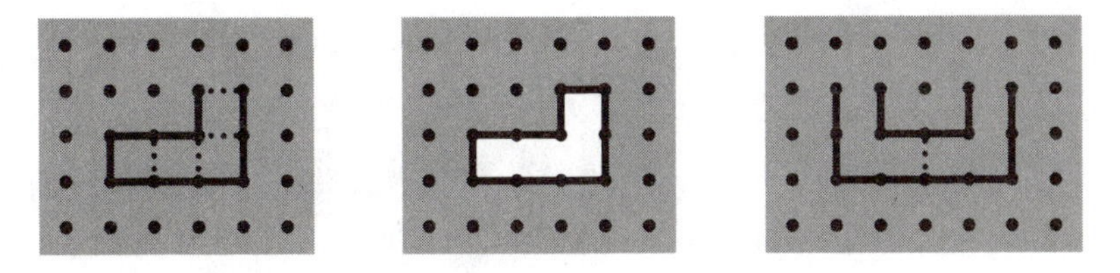

Figure 7.110. Possible enclosures, the optimal move, and a potential subdivision.

that will enclose the most area (middle). The optimal move should always be made, unless the player is trying to trick the opponent with an especially subtle trap.

Such traps may be laid by *subdividing* a run of edges with two openings into two enclosable areas (Figure 7.110, right). If the player makes the move indicated by the dotted line then the opponent claims three units next turn, but the player is then guaranteed three units the following move.

A player who especially wants to claim a particular unit may choose a subdivision point that tempts the opponent with a larger but less critical area (see Section 5.9, Sacrifice). Such subdivisions add tactical richness to the game. The Dots & Boxes rule of awarding a free turn for each capturing move is not adopted, as this would make such sacrifices redundant.

History

Lynx was invented by Dan Troyka and Cameron Browne in May 2004. The simple but brilliant link restriction mechanism that guarantees a trivalent tiling of the square grid was devised by Dan Troyka for a computer game called Random Links. This mechanism was then integrated with the rules of the classic game Dots & Boxes to produce a playable board game. Lynx has not been tested extensively yet.

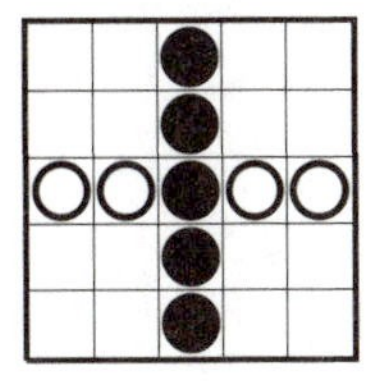

Warp and Weft

Warp and Weft is a state-based connection game played on a woven board design.

Figure 7.111. A game won by Warp (black) on the standard 4 x 4 board.

Rules

Warp and Weft is played on the tiling shown in Figure 7.111. The vertical *threads* belong to Warp (black), and the horizontal threads belong to Weft (white). The *patches* between threads may be claimed by either player.

Players exist in either of two states: Threads or Patches. While in the Threads state, a player may claim empty threads in their direction. While in the Patches state, a player may claim empty patches adjacent to at least one thread of their color.

The game starts with both players in the Threads state. Warp moves first, then players take turns moving. Each turn, the current player must either claim a cell according to his current state, or change states.

Warp wins by connecting the top and bottom sides of the board with a chain of black cells. Weft wins by connecting the left and right sides of the board with a chain of white cells. For instance, Figure 7.111 shows a game won by Warp. A player also wins if the opponent has no legal move.

Exactly one player must win. A single-move swap option is recommended.

Notes

The Warp and Weft board can be obtained by reshaping the cells of the 4.8.8 grid. The two grids are similar in a connective sense if the octagons on the 4.8.8 grid are assigned alternating directions, as shown in Figure 7.112.

Figure 7.112. An alternating 4.8.8 grid and the Warp and Weft grid.

This pattern of connection results in a sparse adjacency graph (Figure 7.113, left). The game graph can be represented in the usual Cut/Join format (Figure 7.113, right); in this case, however, a cut graph indicates an eventual win for Weft rather than an immediate win.

Strategy and Tactics

Since changing states takes up a turn, state changes must be judged carefully and made as seldom as possible. Changing states too early may leave a player with limited options, forcing a temporary change back that wastes two moves. However, changing states too late may be catastrophic if it puts the player one move behind in a race for critical patches.

Players should generally try to avoid playing two threads adjacent to the same patch; such plays represent wasted board influence.

Figure 7.113. The adjacency graph, a completed game, and its game graph.

History

Warp and Weft was invented by Phil Bordelon in August 2004.

Variants

The standard game is described as *Thread-Primary*. In *Patch-Primary* Warp and Weft, players begin in the Patches state and may claim empty patches without constraint while in that state, but may only claim empty threads adjacent to at least one patch of their color while in the Threads state. Claiming threads in this variant is merely a formality; the game is won and lost on the patches.

Light Cycle

Light Cycle is not so much a variant as a general metarule inspired by the idea of state transitions. It can be applied to most games whose boards can be divided into *phase spaces* (described in Appendix H).

A marker indicates which phase (or state) each player is currently in, and the player may only play on cells of that type for that turn. The marker may cycle automatically, or may require a move on behalf of the player to change states, as in Warp and Weft.

Light Cycle was devised by Phil Bordelon in August 2004.

7.1.2 Pure Connection > Absolute Path > Edge-Based

In contrast to playing on the vertices of the game graph, playing on the edges involves choosing an unclaimed edge (or connection) each turn. These are usually implemented as bridges that the players place between board points.

Edge-Based games are either equivalent or directly related to the Shannon game on the edges, in which players alternately claim or cut an edge of a graph each turn (see Section 3.2.1).

The icon for this group shows a chain of black edges cutting two white edges.

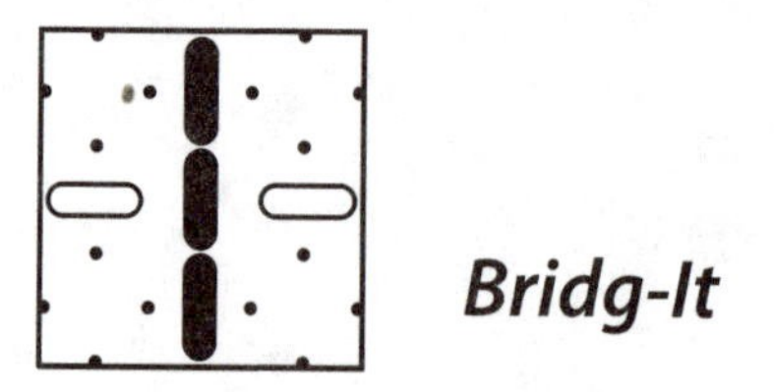

Bridg-It

Bridg-It, the quintessential Shannon game on the edges, is one of the most fundamental connection games. Unfortunately a winning strategy has been known almost since the game's invention.

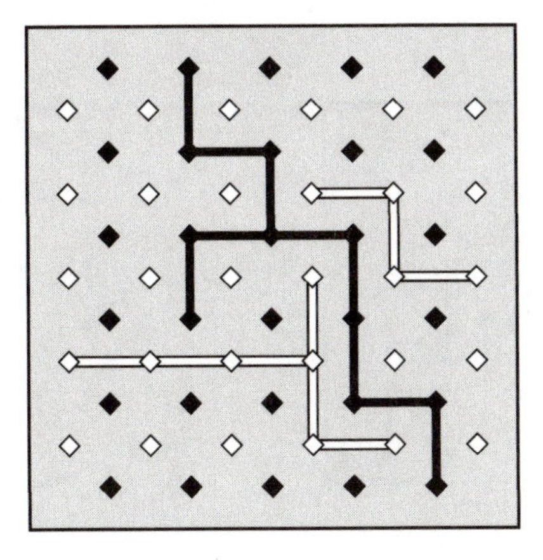

Figure 7.114. A game of Bridg-It won by Black.

Rules

The board consists of two staggered 5 x 6 square grids of differently colored pegs. Two players, Black and White, take turns placing a bridge of their color between squarely adjacent pegs of their color. The board is initially empty. Black moves first.

Black wins by completing a chain of black bridges connecting the top and bottom sides, and White wins by completing a chain of white bridges connecting the left and right sides. One player must win.

Notes

Figure 7.115 (left) shows the adjacency graph for Bridg-It overlaid on a section of the board. Note that vertices exist corresponding to black pegs

Figure 7.115. Edge operations on the adjacency graph.

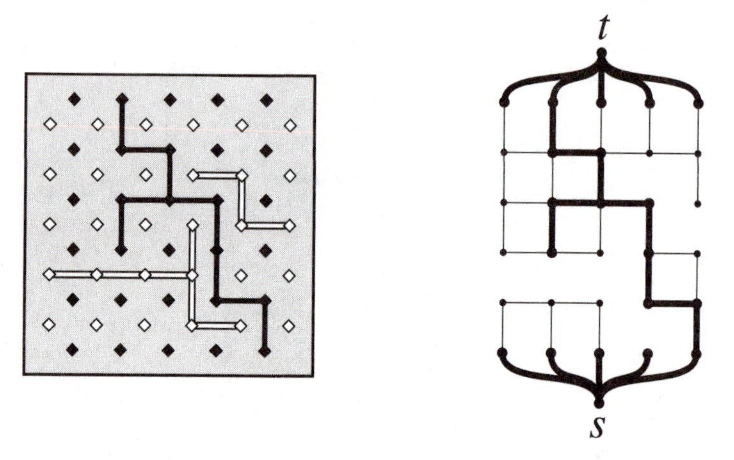

Figure 7.116. A game won by Black and its game graph.

but not white pegs; the game fundamentally consists of two overlapping grids, but only one is required for the game graph. Each Black move claims a potential edge, while each White move cuts a potential edge (Figure 7.115, right).

Figure 7.116 shows a completed game won by Black and its game graph. Recall that Black wins by connecting terminals *s* and *t*, while White wins by cutting the graph so that *s* and *t* are permanently disconnected.

Strategy and Tactics

A winning strategy first discovered by Oliver Gross in the early 1960s is described in Appendix E. Lehman [1964] provided a rigorous proof of a solution to the Shannon game on the edges, and hence Bridg-It, using matroid theory (see Appendix B).

History

Claude Shannon created a robot Bridg-It player in 1951 using a resistor network corresponding to the game graph, however this was not widely known until much later [Gardner 1961]. Bridg-It was independently reinvented by David Gale in the late 1950s. It is also known as Gale, Birdcage, Connections, and Connexxions.

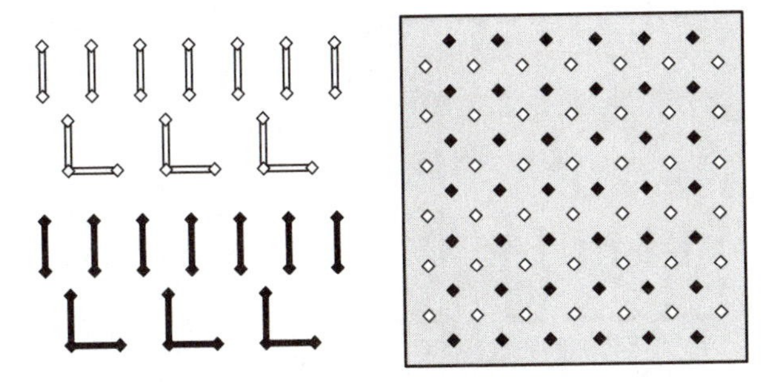

Figure 7.117. The original September set.

Variants

Various forms of Bridg-It have been patented over the years, including Goetz [1962], Christy [1968], Minty [1978], Shimizu [1980], and McNamara [1993]. Christy and Shimizu's games are notable in that edges may cross under some circumstances, Minty's game is notable for featuring directional bridges, and McNamara's game is notable in that players may also win by surrounding an opponent's peg or bridge with a cycle of bridges.

Web

Web is played identically to Bridg-It using the same board, except that the player with the largest connected chain of bridges at the end of the game wins. Web is Joris game #40 [2002] and is not a Pure Connection game.

September

September is played on a 6 x 7/7 x 6 Bridg-It board as shown in Figure 7.117. Each player has exactly those pieces shown on the left, some of which are double bridges. Like Bridg-It, players take turns placing a piece of their color between pegs of their color. Unlike Bridg-It, players must move an existing piece to a new location when they run out of pieces. The winning conditions are identical to those for Bridg-It. The double-edge pieces mean that Gross's winning strategy for Bridg-It does not apply to September.

September was invented by Danny Kishon and first released in 1985 [Cornelius and Parr 1991]. Paradigm Games published a larger version in 1986, played on a 10 x 11/11 x 10 board with the expanded piece set shown in Figure 7.118.

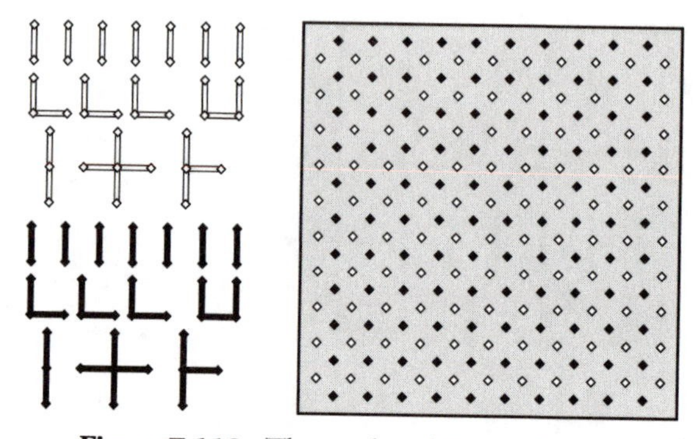

Figure 7.118. The marketed September set.

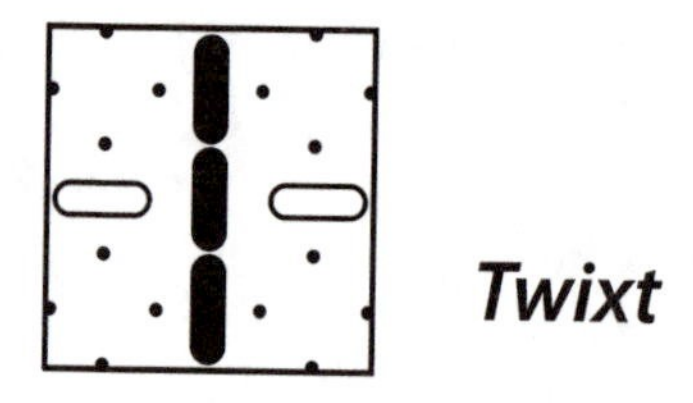

Twixt

Twixt is one of the more well-known connection games. Overlapping knight's-move adjacencies make this an interesting Edge-Based game.

Rules

Twixt is played on a 24 x 24 square grid of holes, minus the corner holes. Two players, Black and White, own the holes along alternating sides of the board that bear their color (Figure 7.119). The board is initially empty.

Each player has an ample supply of *pegs* and *links* of their color. Pegs fit into the board holes, and links are exactly long enough to connect two pegs a knight's move apart (opposite corners of a 2 x 3 rectangle). Each player typically has 50 pegs and 50 links, which is sufficient for most games.

Players alternate taking turns. Each turn the current player makes the following moves:

1. place a peg of his color in an empty hole (except in the opponent's goal areas), and
2. remove any number of existing links of his color, and place any number of links of his color between pairs of his pieces spaced a knight's move apart (both optional). No link can ever cross an existing link.

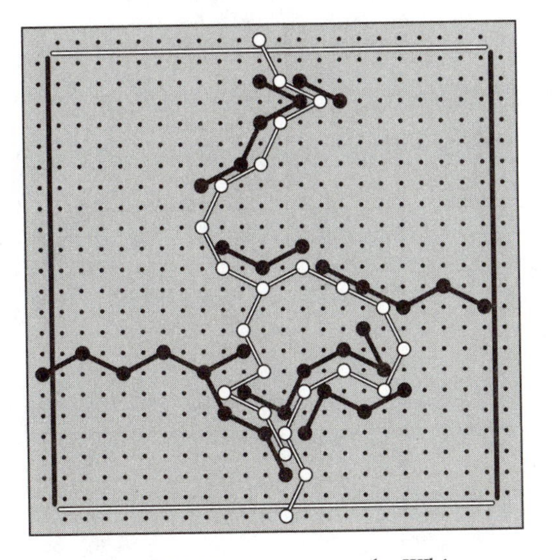

Figure 7.119. A game won by White.

A player wins the game by establishing a chain of links connecting his goal areas. The game can be tied. At most one player can win. A single-move swap option is recommended.

Figure 7.119 shows a game won by White. Note that links must actually cross over the colored boundary to connect to a goal area; the right-most black link is not connected to the right goal area.

Notes

Figure 7.120 shows the five essential bridge formations in Twixt. In each case the paired pegs have two possible link paths (indicated) that cannot both be blocked in one move by the opponent. From left to right, these are called: Beam, Mesh, Coign, Tilt, and Short.

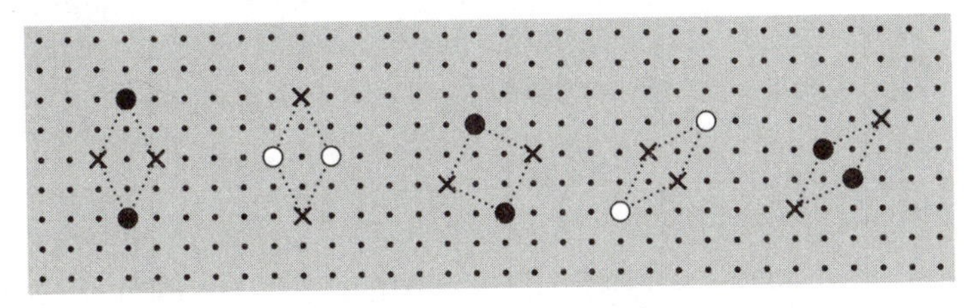

Figure 7.120. The five essential bridge formations.

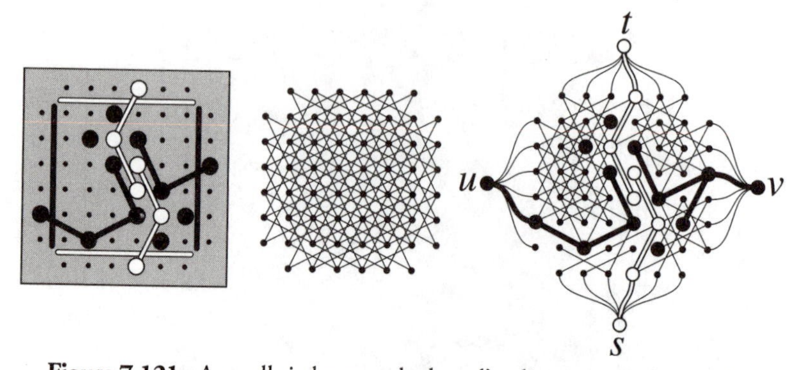

Figure 7.121. A small tied game, the board's adjacency graph, and the complete game graph.

Figure 7.121 (left) shows a tied game on a small board; neither player can win from this position without the opponent's cooperation, which is unlikely. Figure 7.121 (middle) shows the rather complex nonplanar network of adjacencies for this small board, hinting at the tactical richness afforded by Twixt.

Figure 7.121 (right) shows the complete game graph for this game. Twixt is a Shannon game on the edges, but the possibility of draws makes it a Join/Join game rather than a Cut/Join game, hence the game graph must encode both players' colors simultaneously.

Each peg move colors a vertex, and each link move colors an edge and cuts all neutral edges that cross it. Removing a link restores that edge (and all edges that previously crossed it) to the neutral state. It can be seen that the game shown in Figure 7.121 is tied, as it is not possible to trace a path between either player's goals that does not pass through a vertex of the opponent's color, even if friendly links are removed and replaced elsewhere.

Twixt is one of the most popular and widely marketed of all connection games. This demonstrates that the no-tie property common to most connection games is a theoretical nicety but not essential to a game's success.

Strategy and Tactics

The following advice is distilled from material by David Bush [2000, 2001].

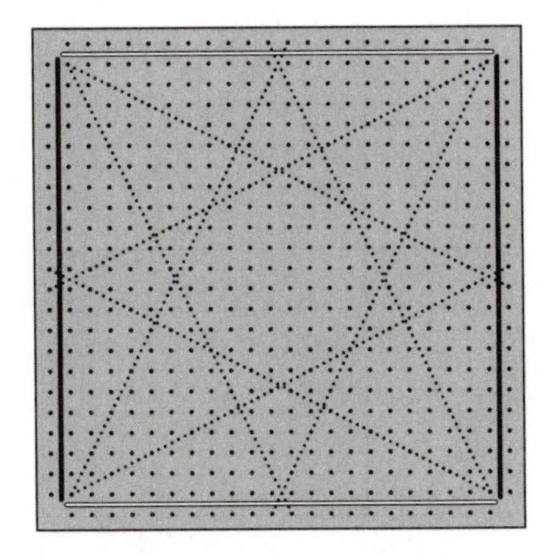

Figure 7.122. The crucial diagonals of the Twixt board.

Experienced players will generally spread their pegs out around the board in the opening phase of the game, in an effort to influence as much area as possible and maximize the number of potential link threats.

Players should always keep the whole board in mind, and abandon local battles in favor of superior strategic moves.

Players should usually place all possible links as each piece is played. It is rare to remove links, but this can occasionally be necessary to free up some elbow room.

The *crucial diagonals* of the Twixt board (marked in Figure 7.122) provide convenient landmarks for players to orient themselves on the large Twixt board, and mark the line at which a player is likely to win a race to their home edge. For instance, Figure 7.123 (left) shows a race to connect, with Black to move. White has reached the crucial diagonal and wins the race (right).

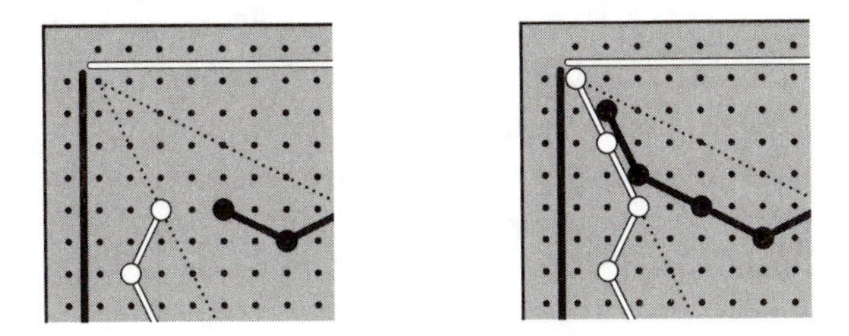

Figure 7.123. White has reached the crucial diagonal and wins the race.

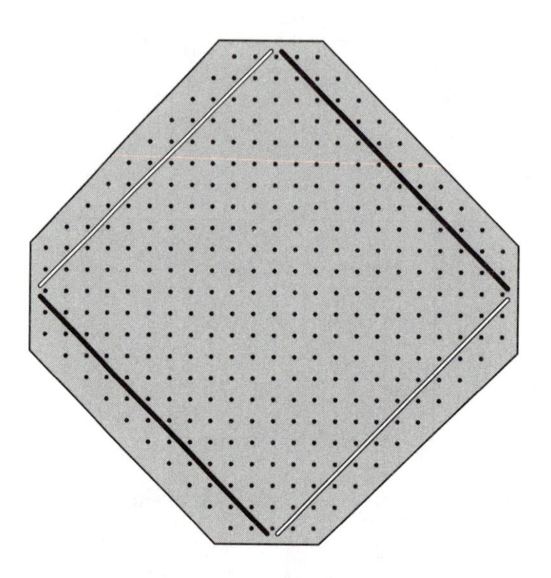

Figure 7.124. The Diagonal Twixt board.

Even though a game can be tied, good defense is still often the best offense.

History

Twixt was invented by Alex Randolph in 1961. It has been marketed by various companies over the years: 3M (1962), Avalon Hill (1976), Schmidt (1979), Selecta (1979), Klee (1990), and Kosmos (1998).

Variants

Diagonal Twixt

Diagonal Twixt is played on the board shown in Figure 7.124 according to the rules of Twixt. It is described by Mark Thompson on his Twixt web page: http://home.flash.net/~markthom/html/twixt.html.

The rationale behind rotating the grid alignment through 45 degrees relative to the board axes is to reduce the overwhelming importance of the crucial diagonals. Thompson emphasizes that this variant has not been tested.

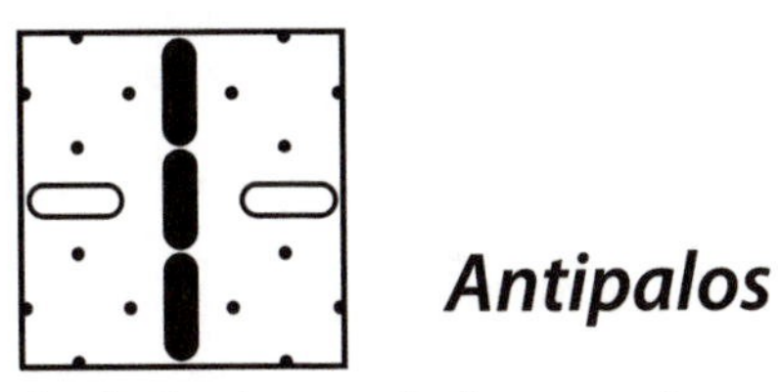

Antipalos

Antipalos is a tactical connection game played on the diagonal edges of a square grid. Players must be vigilant for phase problems.

Rules

Antipalos is played on a 17 x 17 square grid of pegs. Two players, Black and White, own alternating sides of the board which bear their color. The board is initially empty.

Players take turns placing a bridge (presumably of their color) between two diagonally adjacent pegs. The bridge cannot lie on top of or cut across existing bridges. The first player to connect his edges of the board with a chain of his bridges wins the game. Paths may cross where they meet at a peg. The official set comes with a fixed number of bridges. Players who run out of bridges must pick up an existing bridge and play it elsewhere.

Notes

The pegs on the board correspond directly to the vertices in the adjacency graph (Figure 7.126). Note that the graph is nonplanar and that each neutral edge is crossed by a perpendicular edge. Playing on an edge colors it and deletes its perpendicular cross-edge. This is similar to other Shannon games on the edges, but Antipalos has one important difference: endpoint vertices are not colored when an edge is colored. This allows players' connections to cross at a vertex.

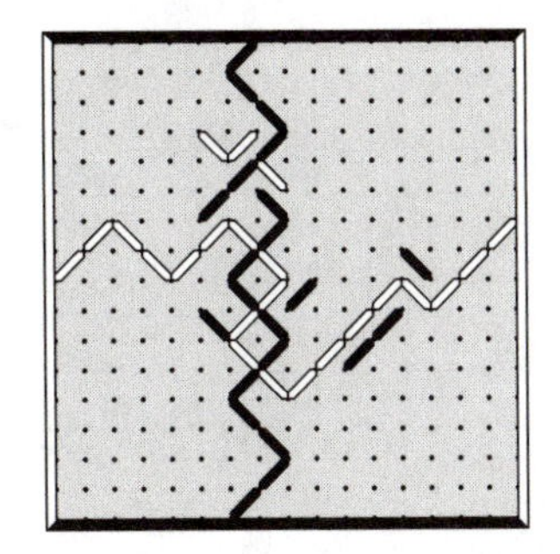

Figure 7.125. A game won by White.

Figure 7.126. Crossing paths share vertices.

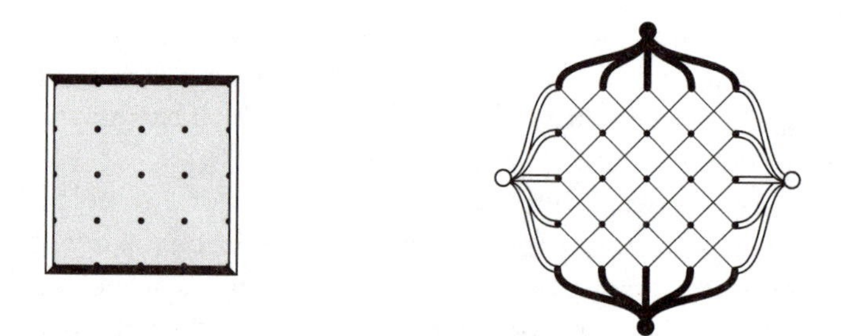

Figure 7.127. A small 5 x 5 game and its graph.

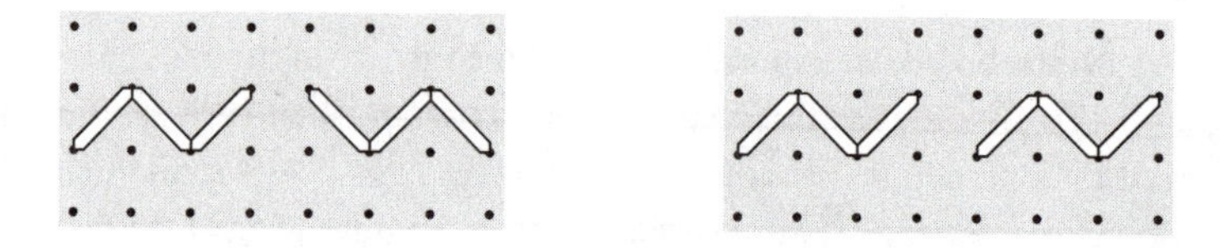

Figure 7.128. Two subpaths out of phase (left) and in phase (right).

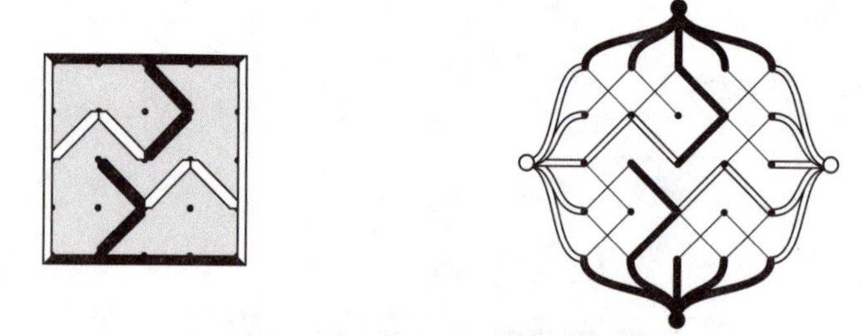

Figure 7.129. A deadlocked game.

The complete game graph for a small 5 x 5 board is shown in Figure 7.127. It is not possible to cut the graph in either direction since at least one edge of each perpendicular pair will always exist; hence separate terminal pairs exist for each player in a Join/Join format.

Paths may exist in either of two phases depending on which edge is chosen from each perpendicular pair. This is a crucial point, as subpaths that are out of phase can never be joined together. Figure 7.128 (left) shows two white subpaths out of phase with each other. These paths can never be joined. Figure 7.128 (right) shows two paths in phase that can be joined. Players should endeavor to make all paths the same phase if possible.

Many games of Antipalos may end in deadlock unless players are conscious of phase, as shown in Figure 7.129. The fact that players can move existing bridges when their stockpiles run out helps alleviate potential deadlock problems, and means that a temporarily tied game may eventually have a winner; however this is not certain.

Antipalos is a rather tactical game; it is difficult to develop strategies and plan ahead. Instead, most of the game is spent in low-level combinatorial play addressing the opponent's developing connection.

Antipalos is rare in that connections may pass through each other. Blocking open-end paths is difficult and can require a lot of room, which may be the reason for the large board size. Once the central connection is decided, the game usually becomes a race to the edges.

History

Antipalos was designed by Martin Collier and published by Antipalos Games in 1982.

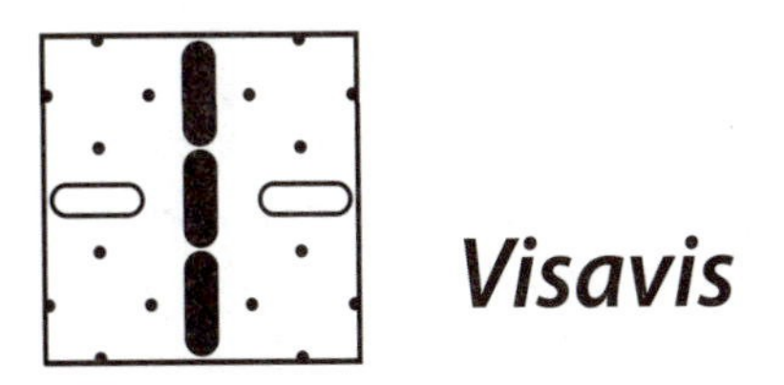

Visavis

Visavis is a tactical connection game played on the edges of a cleverly designed board. The fact that each move cuts two potential opponent moves makes Visavis less susceptible to the analysis used to solve Bridg-It.

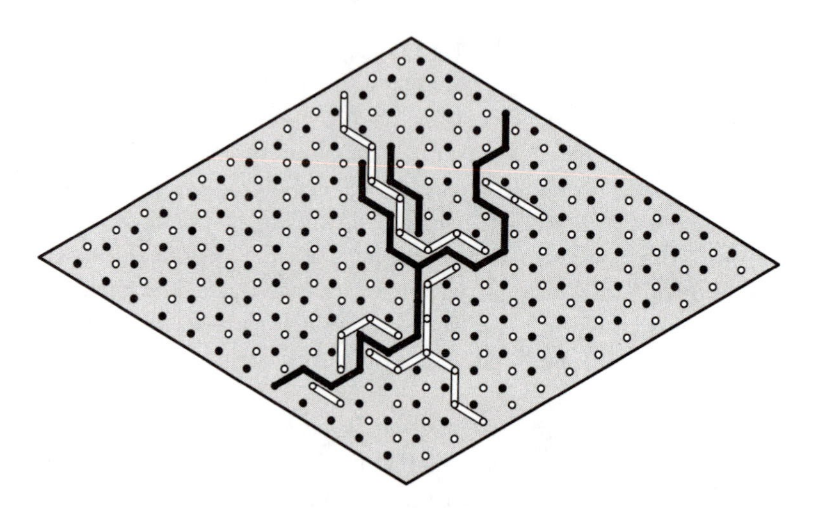

Figure 7.130. A game won by Black.

Rules

Visavis is played on a 12 x 12 grid of white pegs interleaved with a 12 x 12 grid of black pegs. The board is initially empty of pieces.

Two players, Black and White, each own 60 bridges of their color. Black starts, then players take turns placing a bridge of their color between two adjacent pegs of their color, such that no bridge crosses any other. Players who run out of bridges must pick up a bridge of their color from the board and play it elsewhere.

Black wins by completing a chain of black bridges connecting the black sides (bottom left and top right). White wins by completing a chain of white bridges connecting the white sides (top left and bottom right).

Notes

At most one player can win (winning paths cannot coexist for both players). The finite number of pieces and dynamic movement of bridges means that a game is not guaranteed to end, although this will generally happen in practice.

As with Bridg-It, only one player's grid of pegs is required to define the game graph for this board. Figure 7.131 (left) shows a section of the board with the game graph relative to Black overlaid on top.

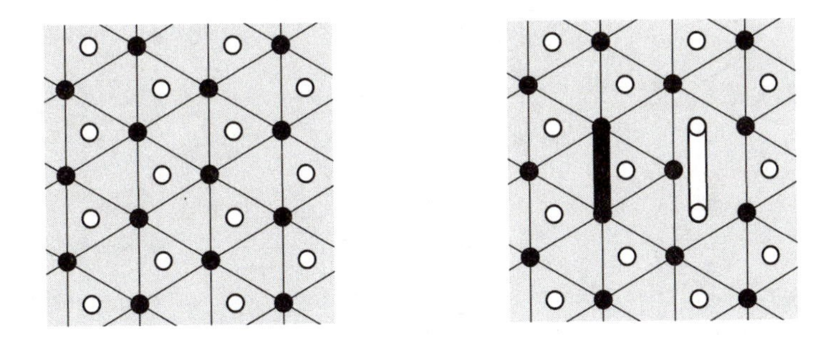

Figure 7.131. Game graph overlaid on the board, and edge operations for each player.

Figure 7.131 (right) shows the edge operations corresponding to a move by each player. Placing a black bridge will claim one potential edge in the game graph. Placing a white bridge will cut (remove) two potential edges in the game graph.

Visavis is a variation of the Shannon game on the edges with the important difference that *two* of the opponent's potential connections are cut with each move. Note that potential edges may be reset to their initial states if pieces are moved later in the game.

Figure 7.132 shows the complete game graph for a small 5 x 5 board. The terminals *s* and *t* are permanently disconnected, hence White has won this game.

It may appear that Visavis is not much different to the solved game Bridg-It, and hence susceptible to the same analysis. Fortunately, the fact that each move cuts two potential opponent edges rather than one means that Gross's winning strategy for Bridg-It [Gardner 1966] and Lehman's

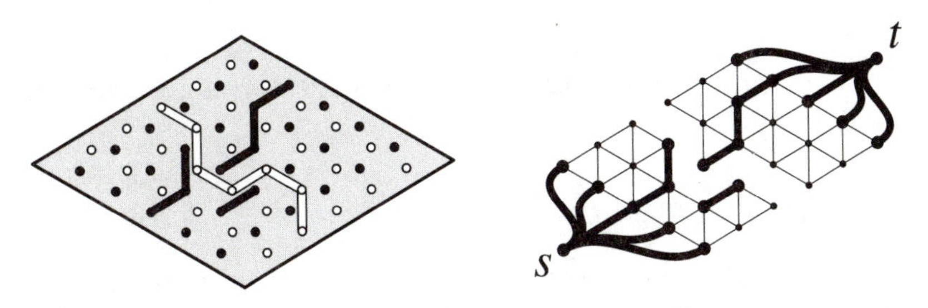

Figure 7.132. A 5 x 5 game won by White and its game graph.

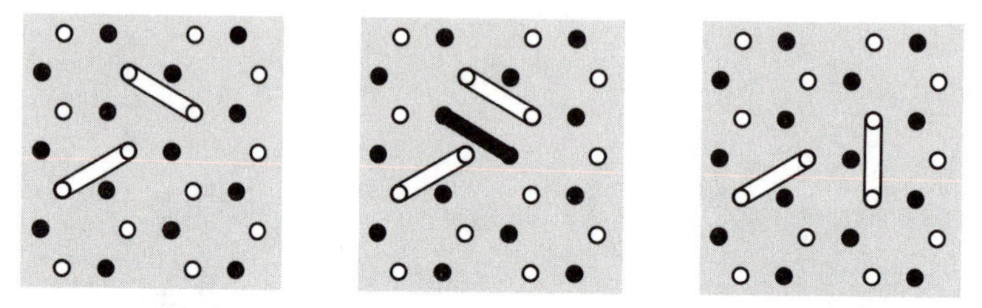

Figure 7.133. Two white pieces in blockable and unblockable relationships.

[1964] solution for the Shannon game on the edges do not directly apply to Visavis.

Strategy and Tactics

Each peg may threaten in six potential directions. Players should keep in mind that three of these threats can be immediately blocked by the opponent and three cannot, as illustrated in Figure 7.133.

Figure 7.133 shows how two white pieces in close proximity (left) can be blocked by a single black piece (middle). However, the next potential connection in a clockwise direction (right) cannot be immediately blocked by Black and is virtually connected.

Michail Antonow suggests that players should defend from a distance, as with most connection games. This is especially important in Visavis where a lot of space is required to mount a successful block, hence the unusually large board size. Play tends to be combinatorial and tactical rather than strategic in nature.

Antonow also points out that placing bridges in parallel is generally weaker than placing bridges that change direction. Nonparallel bridges threaten more potential connections.

History

Visavis was invented by Michail Antonow in 1995, and was originally called Nexus.

7.2 Pure Connection > Path Race

Path Race games are those in which players race to complete their connection before their opponent completes theirs. Unlike in Absolute Path games, a player's path does not preclude all opponent's paths. Goals are not complementary, and instead tend to be either equivalent or symmetrical.

Some of the elegance and mathematical purity of Absolute Path games is lost in the race; a player who gets one move ahead is usually hard to stop. Defensive play is generally not as important as it is in Absolute Path games (where good defense equals good attack) and race games tend to reward attacking play.

Path Race games fall into two main categories:

- Separate Paths, and
- Shared Paths.

7.2.1 Pure Connection > Path Race > Separate Paths

Players in Separate Path games have pieces or tiles distinguished from their opponents' and race to complete paths of their color only.

These games are typically characterized by the fact that it is possible to construct board positions showing winning connections for both players at once. To stop this from happening during play, additional rules must be introduced, such as awarding the win to the mover if both players' goals are achieved on the same move.

The icon for this group shows a pair of different colored paths racing across the board in parallel.

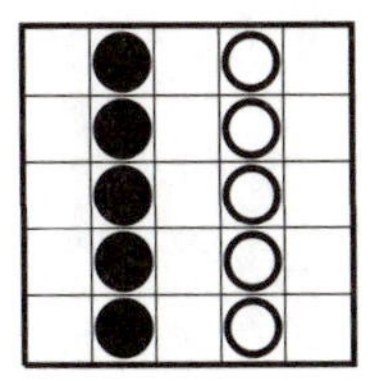

Lightning

Lightning is a historical curiosity; a connection game marketed 50 years before the invention of Hex. It differs from most other connection games in that players have individual boards, either of which may be played on each turn.

Rules

In Lightning, two players each own a separate 8 x 17 square grid, and share a common pool containing 30 of each of the tiles shown in Figure

Figure 7.134. The five types of Lightning tiles.

7.134. The boards are initially empty and the tiles placed face down on the table.

Players take turns choosing a tile and playing it on either their board or their opponent's board. The first tile on each board must connect with the board's starting edge (a short edge that should be facing its owner). Thereafter tiles must be placed adjacent to the last tile placed on that board, so as to extend the active path.

If a player is unable or unwilling to place the current tile on either board, then that tile is placed face up in a free tile pool. Either player may choose a tile from this pool in lieu of turning over an unknown tile.

If the path hits a board edge and may not continue to either side it is said to have *terminated*, and the player must then branch the path from the most advanced junction at which the subsequently drawn tile fits. For instance, Figure 7.135 shows a path that has terminated at the lower edge (left) and a possible continuation from the most advanced junction (right).

The first player to complete a line from his starting edge and terminating at the far edge wins the game. Figure 7.136 shows a completed game of Lightning given by Polczynski [2001], in which the winning connection starts at the left and terminates at the right. If it becomes impossible for one player to complete his connection, then the game continues until the opponent wins. If it becomes impossible for both players to complete their connection then the game is drawn.

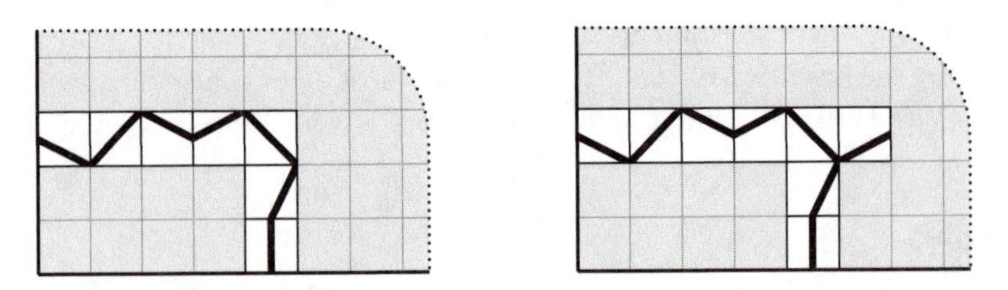

Figure 7.135. Blocked paths may branch at the most advanced junction.

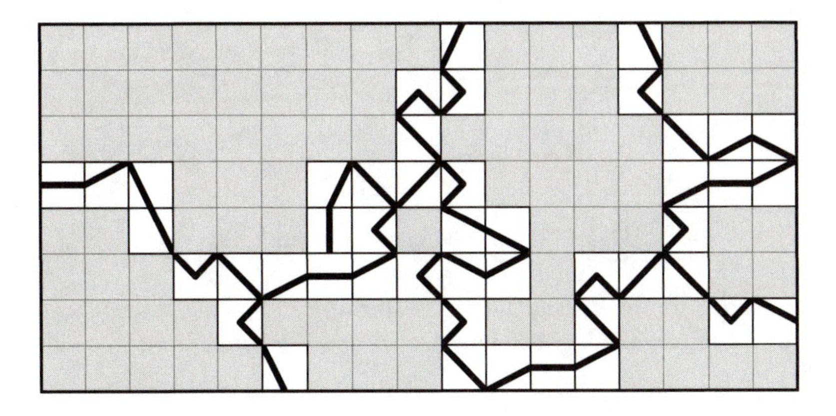

Figure 7.136. A completed game of Lightning.

Notes

A path reaching the far edge of the board must terminate to win. For example, Figure 7.137 (left) shows a path that touches the right (target) edge but does not terminate. Figure 7.137 (middle) shows a continuation of play that terminates at the top edge, and Figure 7.137 (right) shows a continuation from the most advanced junction that eventually terminates at the target edge to win.

Strategy and Tactics

Lightning is an unusual connection game; not only do players have their own boards, but they may choose which board to play upon each turn.

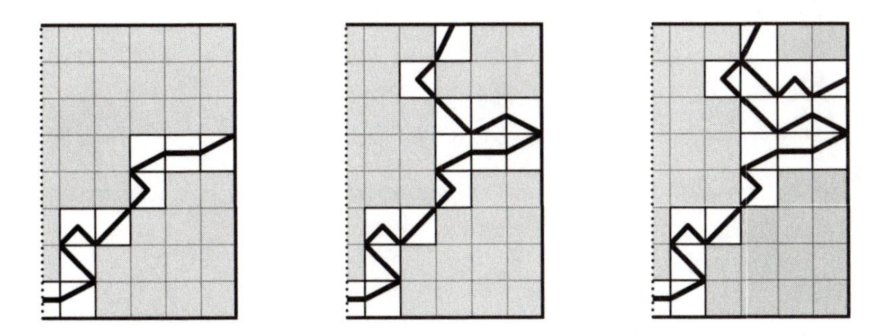

Figure 7.137. A line must terminate at the target edge to win.

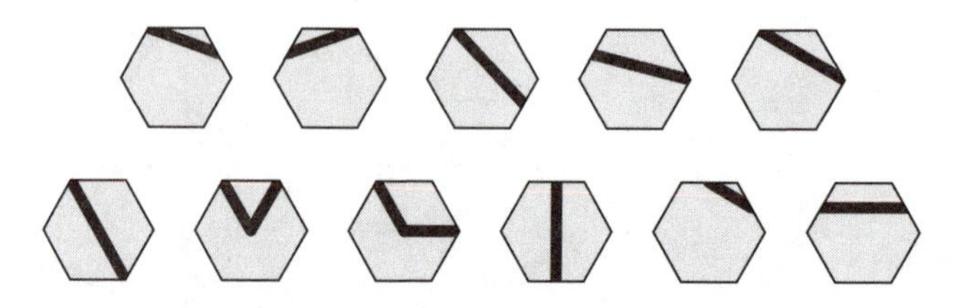

Figure 7.138. The Hex-Lightning tiles.

This allows some tactical plays such as turning an opponent's connection in on itself, terminating it at an edge, or choosing a suboptimal branch.

History

Lightning was patented by H. Doty on March 29, 1892, making it the first connection game by several decades [1892]. Jim Polczynski describes the game in further detail [2001].

The original Lightning patent fails to list the fourth tile in Figure 7.134, and also includes a blank tile that may be used to block the opponent.

Variants

Hex-Lightning

Jim Polczynski suggests the game of Hex-Lightning played according to the rules of Lightning, but using the tiles shown in Figure 7.138 on a hexagonally tiled board [2001]. This game has not been tested.

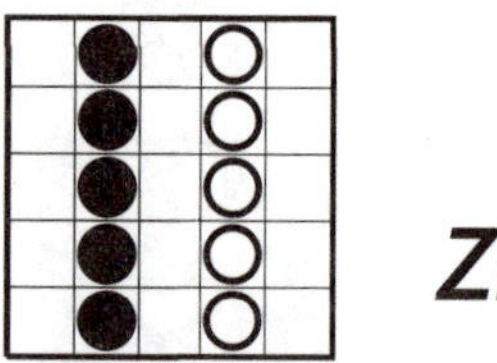

Zig-Zag

Zig-Zag is a pioneering connection game that predates Hex and features double moves controlled by dice rolls.

Rules

Zig-Zag is played on a 12 x 12 grid of squares, each marked with a number from 1 to 6. While the exact numbering pattern on the official board

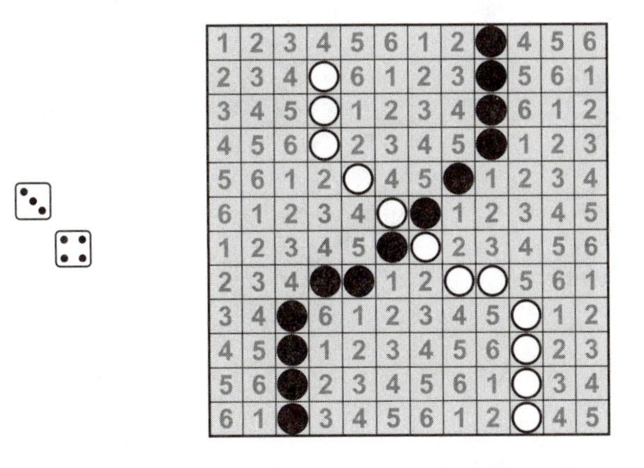

Figure 7.139. A game of Zig-Zag won by Black.

is not known, an ordered numbering of board squares cycling from 1 through 6 is used in the following examples.

The game may be played by two players, Black and White, or four players in two teams of two. The board is initially empty.

In the two-handed game, both players throw a pair of dice to determine who starts. The player with the highest roll places a single piece of his color on any board square. Thereafter players take turns rolling the two dice and placing two pieces of their color on empty squares corresponding to the numbers rolled. Doubles do not confer any special privileges. Presumably, if there are no available squares matching either die, then that move is forfeited.

A player wins by completing a chain of his pieces between the top and bottom sides of the board. The chain may include orthogonal and diagonal connections. For instance, Figure 7.139 shows a game won by Black. The dice show three and four pips; hence the last two black pieces must have been placed on squares labeled 3 and 4.

Notes

A game of Zig-Zag cannot be drawn. Diagonal connections mean that deadlock cannot occur, even though the game is played on a square grid.

Figure 7.140 shows a small 6 x 6 game won by White and its game graphs. A separate graph is required for each player as they are both aiming to connect the same two sides.

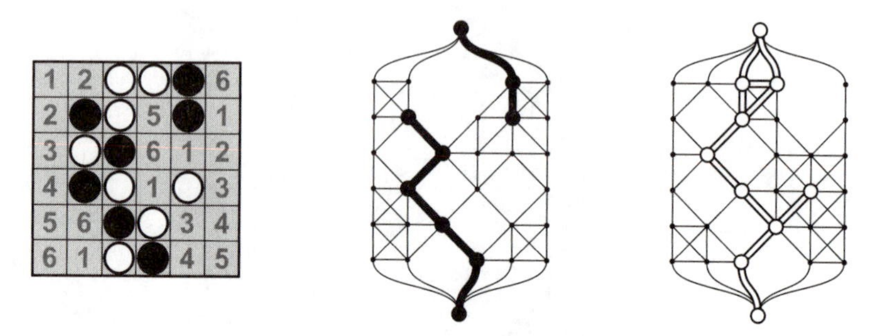

Figure 7.140. A 6 x 6 game won by White and the corresponding game graphs.

The unconstrained placement of a single piece on the first move is an elegant first-move equalizer.

As Zig-Zag is a race game, the two-player version may actually benefit from using a rectangular rather than a square board. A longer and narrower board such as 6 x 18 may encourage more interference between players' paths and provide a more engaging game. Note that board sides should generally be multiples of six units in length.

Strategy and Tactics

The inclusion of dice rolls does not disqualify Zig-Zag from being a Pure Connection game; it merely constrains players to choose from a subset of available board squares each turn. Dice rolls introduce a probabilistic element that has since been largely overlooked in connection games (with the exception of Round the Bend and games involving random tile or card selections).

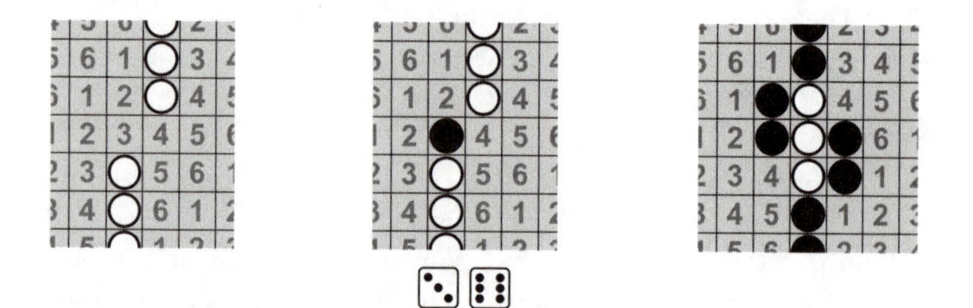

Figure 7.141. Virtual connections are probabilistic rather than certain.

For instance, Figure 7.141 (left) shows what appears to be a strong virtual connection for White, having alternative paths through squares labeled **3** and **4**. However, Black has a 1 in 18 chance of rolling {3, 4} or {4, 3} to break this connection next turn. If Black instead rolled {3, 6} and took the **3** square as shown in Figure 7.141 (middle), White would then have an 11 in 36 chance of rolling at least one 4 to save the connection.

Figure 7.141 (right) shows a more favorable virtual connection for Black. White has only a 1 in 36 chance of rolling a double 4 to sever this connection.

Zig-Zag seamlessly combines connection strategy with the probability and risk estimation of dice games such as Backgammon. The introduction of the doubling cube could make Zig-Zag a rewarding gambling game.

History

Zig-Zag was published by Parker Brothers in 1932, making it the first known piece-based connection game. Zig-Zag is mentioned in passing in Martin Gardner's article "Four Unusual Board Games" [1963].

Zig-Zag is remarkable for the number of typical connection game traits it introduced decades before these traits became common knowledge. These include

- the concept of chains of pieces connecting sides of the board,
- a first-move equalizer years before Nash's strategy-stealing argument demonstrated why this was necessary,
- diagonal connections to avoid deadlocks well before hexagonal tilings became commonplace for such games, and
- a double-move mechanism, over half a century before Trellis, Octagons, Square Board Connect, and Stymie put this idea to good use.

Combined with the clever use of dice, these elements make Zig-Zag a truly pioneering game. It is a shame that the game and the name of its inventor have been forgotten over the years.

Variants

Criss-Cross Zig-Zag

Criss-Cross Zig-Zag is played according to the same rules as standard Zig-Zag, except that players own the alternating sides of the board that bear

Figure 7.142. A game of Criss-Cross Zig-Zag won by White.

their color, and win by connecting their sides of the board with a chain of their pieces. This game is essentially Hex on the square grid, but was invented over a decade before Hex was conceived.

Figure 7.142 shows a game won by White, whose last two pieces must have been placed on squares labeled 2 and 5.

Even though players have perpendicular goals, Criss-Cross Zig-Zag is still a race game as the existence of diagonal connections means that those goals are not complementary.

With the benefit of hindsight, Criss-Cross Zig-Zag looks to be a more interesting game than standard Zig-Zag, as perpendicular goals mean that players are more likely to engage in local battles where their paths cross.

Four-Handed Zig-Zag

Four players face each other around the board, each opposite pair forming a team. A team wins by connecting its sides of the board with a chain of its color. Four-Handed Zig-Zag is identical to Criss-Cross Zig-Zag except that four players alternate taking turns rather than two.

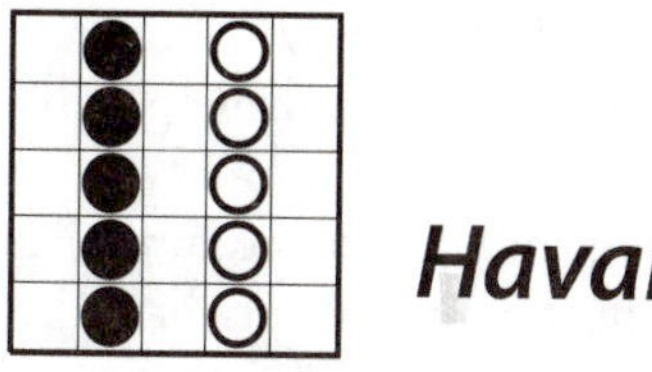

Havannah

Havannah is one of the most complex connection games despite its simple rule set, and takes a good deal of skill to play well.

Rules

Havannah is played on a hex hex board that is initially empty. Two players, Black and White, take turns placing a piece of their color on an empty cell.

The game is won by the first player to complete at least one of the following connections with his pieces:

- a *ring* around at least one cell (either empty or occupied by either player),
- a *bridge* connecting at least two corner cells, or
- a *fork* connecting at least three sides (excluding corners).

Figure 7.143 shows the three winning conditions. A game can be drawn, although this is extremely unlikely.

Notes

Despite the simple rules, Havannah is one of the more demanding connection games. Players must be constantly vigilant of the large number of possible winning conditions (as indicated in Section 4.4.2):

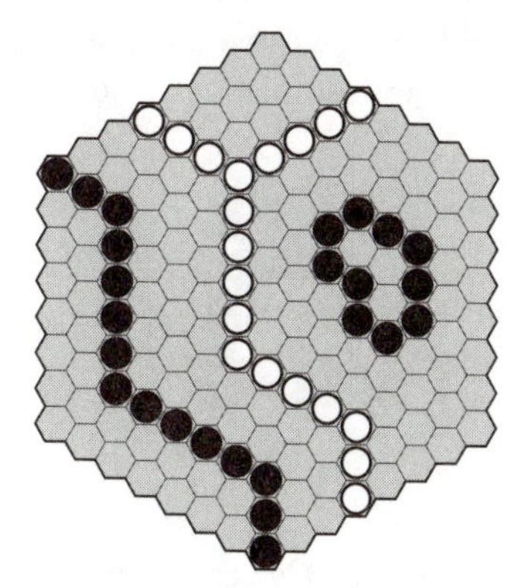

Figure 7.143. A black bridge, a white fork, and a black ring.

ten different combinations of sides that could yield fork wins, 15 different combinations of corners that could yield bridge wins, and many possible ring formations.

This level of complexity can make it hard to look ahead and formulate strategies for all but experienced players, pointing to a possible lack of clarity. It is often hard to define whether a move is good or bad, making Havannah difficult to program. In fact, Christian Freeling acknowledges the elusive nature of Havannah by offering a €1,000 reward for any programmer who can write a program that can beat him in one out of ten games.

The published Havannah board has eight cells per side, however, the inventor's preference appears to be for a board with ten cells per side. This larger board leads to complex games but is large enough to allow a player to recover from an error, a rare feature in connection games.

Strategy and Tactics

The most important strategic concept in Havannah is the *frame* [Freeling 2003], which is a virtual connection that threatens to achieve one of the winning conditions. Tempo comes into play here, as the only way to defeat an opponent's frame is to attack it with a faster frame.

For instance, Figure 7.144 shows a formation of white pieces that are virtually connected and appear to form a ring frame (left). However, this formation is subject to attack by Black, who intrudes with a faster ring frame formed by moves *1* to *9* (middle). In each case White is forced to reply as shown to maintain the ring threat. Move *9* is a killer move that threatens to complete Black's faster inner ring. White is forced to reply ***10*** and Black can then play ***11*** to finally break White's ring (right).

Black won this battle through superior tempo; although both ring threats required six pieces to complete, Black had the move in hand.

Figure 7.144. Not a safe ring frame.

Paradoxically, attacks such as this help solidify the slow frame (due to the opponent's forced replies) but eventually break it.

This example, called the Mill by Freeling, gives a taste of the interplay between tactics and strategy in Havannah. Freeling and other enthusiasts have identified many such plays, and a glossary of Havannah terms has been growing ever since the game's inception.

Although ring threats are useful as leverage for making forcing moves, games are usually won with bridges or forks rather than rings themselves. Havannah is classified as a Path Race game rather than a Cycle Race game for this reason, though it shares characteristics with both categories.

History

Havannah was invented by Christian Freeling in 1976 and published by Ravensburger in 1981.

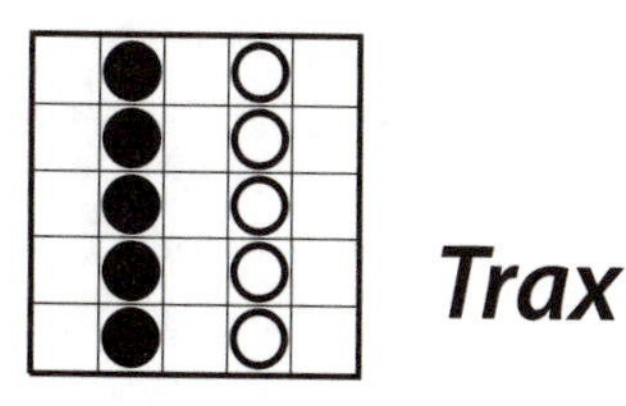

Trax

Trax is an elegant boardless connection game featuring shared tiles but distinct paths.

Rules

Trax is played on a boardless playing surface. Two players, Black and White, share a common pool of tiles, each of which shows a black track and white track crossing on one side and a black track and white track curving away from each other on the other side (Figure 7.145).

Figure 7.145. The two sides of a Trax tile.

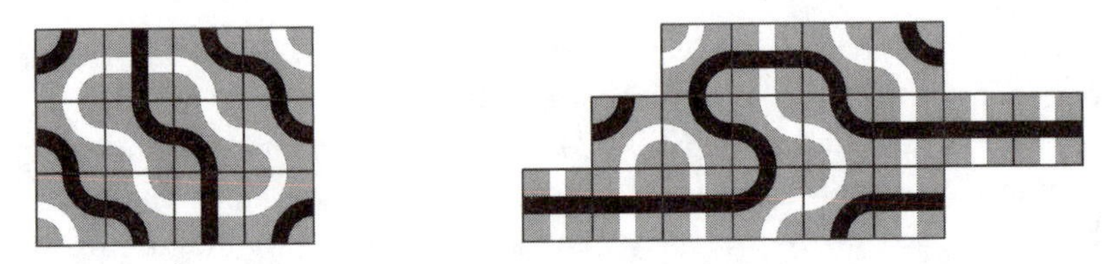

Figure 7.146. A winning white loop and a winning black line.

White starts by placing a tile (either side up) on a flat surface. Thereafter players take turns placing a tile (either side up) alongside at least one existing tile, such that tracks of the same color meet across touching tile edges.

If a tile placement creates any adjacent spaces that can only be filled by one type of tile in one orientation, then the player must also make those plays. Automatic tile placements may lead to further automatic tile placements.

Players are not allowed to make any move that directly or through automatic tile placements would result in any nonplayable spaces, that is, spaces into which three or more tracks of the same color lead.

A player wins by establishing a *loop* or a *line* of his color. A loop is a cycle of any size, and a line is a path that stretches at least eight rows or eight columns (Figure 7.146). If a move creates winning conditions for both players, then the mover wins.

Notes

Trax is unusual in that players share tiles but still have distinct paths. It is one of the most popular connection games, boasting a newsletter, international clubs, and a book on strategy [Bailey 1997].

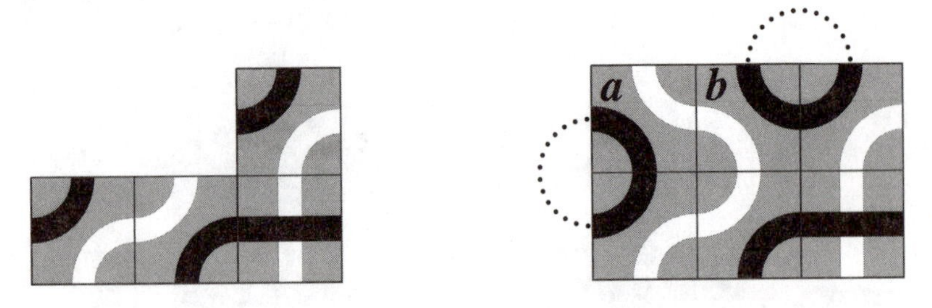

Figure 7.147. An automatic move giving Black a winning fork.

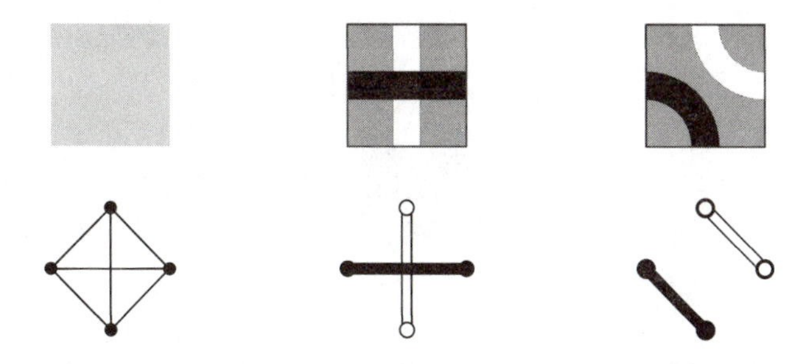

Figure 7.148. Track adjacencies and edge operations.

Figure 7.147 demonstrates the automatic move process in action. Given the formation shown on the left, Black plays tile ***a***. This creates a concave space in which two white tracks converge on adjacent sides, hence tile ***b*** must also be placed as part of the move. Note that the reverse is not true in this case; placing tile ***b*** first does not cause the automatic placement of tile ***a***.

Black now has a winning fork (dotted lines) and can complete a cycle next turn regardless of White's move. This is a common play called the *L threat*.

Figure 7.148 shows the potential adjacencies available to each tile (left) and edge colorings realized by various tile placements. Note that the black edge crossing does not cut the white edge (middle) but passes over it. Both the potential and realized game graphs are nonplanar.

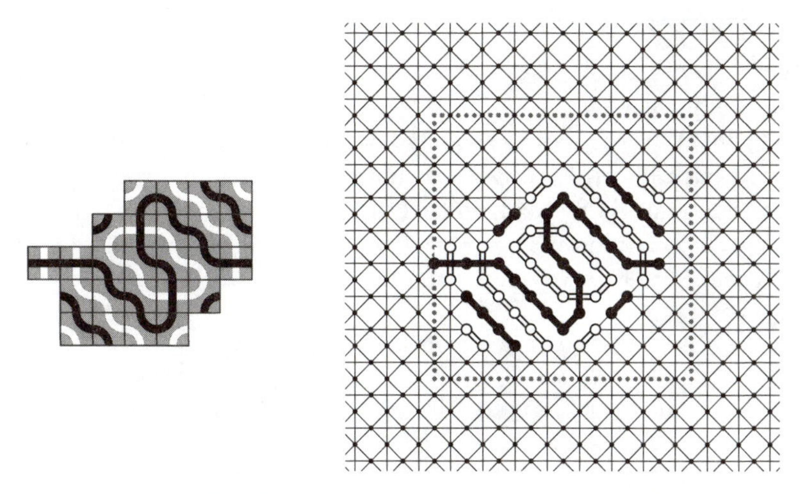

Figure 7.149. A completed game and its game graph.

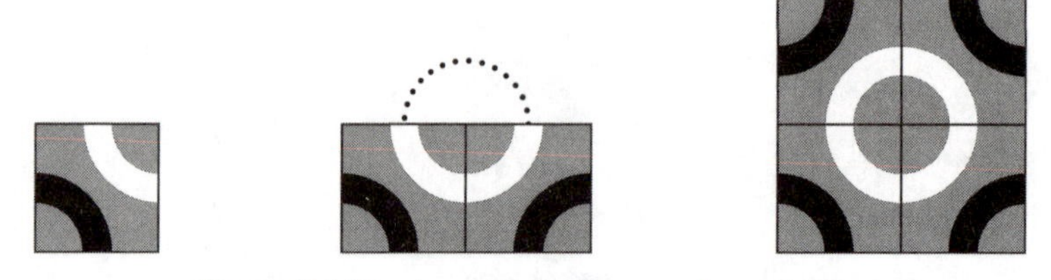

Figure 7.150. A corner leading to a loop threat.

Figure 7.149 shows a game won by White who has completed a loop, and its game graph. Due to the boardless nature of Trax, the game graph is aligned to the opening tile placement and is effectively infinite. The dotted line shows a floating 8 x 8 window used to detect line wins. Black, whose main path spans seven columns, could have won next move with a horizontal line.

This floating window demonstrates that the 8 x 8 constraint does not pose a *maximum* size limit that would eliminate Trax from the Pure Connection category, but rather a *minimum* size limit defined by a virtual board. Almost all connection games feature a minimum size limit defined by their board size, the only difference being that in this case the board boundaries may shift as the game progresses. The key point is that the connection can be of any size and shape provided that it meets the minimum requirement.

Strategy and Tactics

Trax is a very tactical game in which a number of fundamental formations commonly occur.

Figure 7.150 (left) shows a *corner* threat. White can then play an adjacent curved track to form a *loop* threat (dotted line, middle), which is any formation that will form a loop next turn (right). Note that loop threats rely on automatic moves to complete the loop in a single turn. Black must reply immediately to any loop threat.

Figure 7.151. A connectable pair.

Figure 7.151 shows a *connectable pair* formed by adjacent paths, which when linked at either end will result in a loop threat at the other end

Corners and connectable pairs are equally important. Generally, the more corners and loop threats a player has, the better his attacking potential.

Attacking prematurely tends to both expend attacking potential and give the opponent the chance to develop his own attacking potential while defending. Attacks should not be played unless they

- lead to a win,
- foil several independent threats, or
- guarantee an improved position.

As in most connection games, players should endeavor to make efficient moves that achieve more than one purpose simultaneously, attempt to minimize the number of safe plays available to the opponent, and be prepared to sacrifice obvious threats that hide more subtle and dangerous threats. If uncertain, the player should just kill an opponent's corner (stray quarter loop).

History

Trax was invented by David Smith in 1980. Much of the above material was derived from the official Trax web site: http://www.traxgame.com.

Variants

Connexion

Connexion is not so much a variant of Trax as a precursor. It is played on a 5 x 5 square board with tiles similar to the second Trax tile shown in Figure 7.145. A player wins by either completing a path between opposite sides of the board or completing a cycle [Bovasso 1972].

LoopTrax

LoopTrax is identical to Trax except that players win by forming loops only (lines are not counted as a win).

LoopTrax reduces Trax to a more fundamental form that belongs in the Cycle Race category. The removal of the somewhat arbitrary 8 x 8 floating window results in a very elegant game. Unfortunately, it appears that an experienced player can foil an opponent's loop threats indefinitely in the absence of line threats.

The history of LoopTrax is somewhat uncertain. The first known suggestion of the game is from a discussion between Donald Bailey, Don Pless, and Richard Rognlie in July 1996. However, it is hard to believe that David Smith, the inventor of Trax, did not consider this variant at some point and discard it in favor of the official version.

8 x 8 Trax

8 x 8 Trax is played as per Trax except that the board is a fixed 8 x 8 grid outside which players may not place tiles. This was the original form of Trax, which evolved into the boardless game described above.

Chameleon Trax/LoopTrax

Chameleon Trax and Chameleon LoopTrax are played as per Trax and LoopTrax, respectively, except that a player wins by achieving a winning condition in either color. Suggested by Bill Taylor in February 2004.

Tantrix

Tantrix is a score-based game similar in style to Trax, but using hexagonal tiles. Tantrix is not a Pure Connection game. See the Tantrix entry in Section 9.1 for more details.

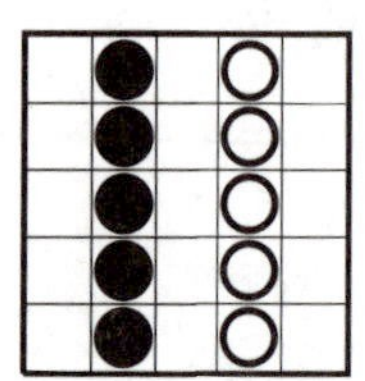

Round the Bend

Round the Bend is a three-dimensional pipe race game with a random element. Primarily a children's game, it is also entertaining for serious gamers.

Rules

Round the Bend is a game for two to four players. Each player owns a set of 20 pipes of his color (Figure 7.152) as follows:

- 2 x Blocker (in),
- 2 x Blocker (out),
- 8 x Elbow,
- 4 x Sprogget, and
- 4 x Klunge.

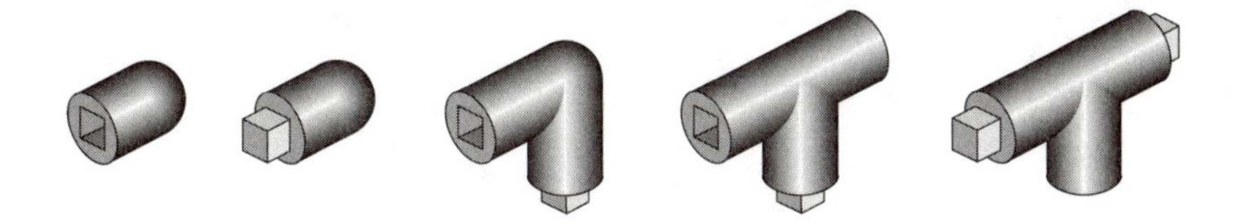

Figure 7.152. The Round the Bend pieces.

A spinner (Figure 7.153, left) is used to decide which piece to play each turn. Some editions of the game provide a die instead of a spinner.

The board (Figure 7.153, right) is a molded plastic 8 x 8 grid featuring alternating pegs and holes that fit the pipe ends. Each side contains a *starting tap* (circled) that matches the color of a *finishing pipe* on the far side of the board. If there are only two players, they must take adjacent sides of the board.

Players alternate taking turns. On their first move players must spin the spinner to decide a pipe (respinning if a Blocker is indicated) and connect that pipe to their starting tap. Pipes may connect in one of four possible rotations (see Figure 7.154).

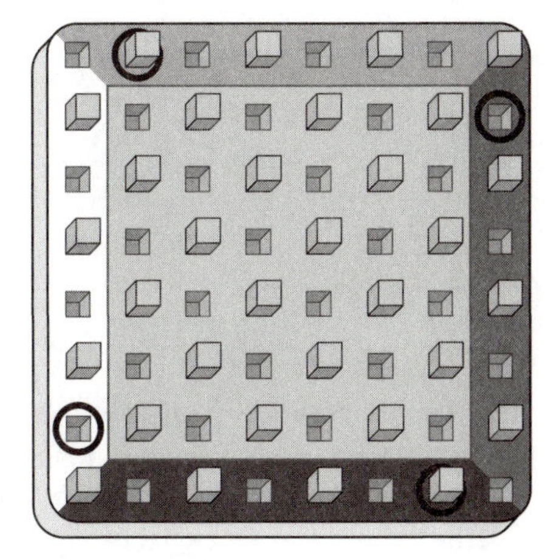

Figure 7.153. The spinner and board.

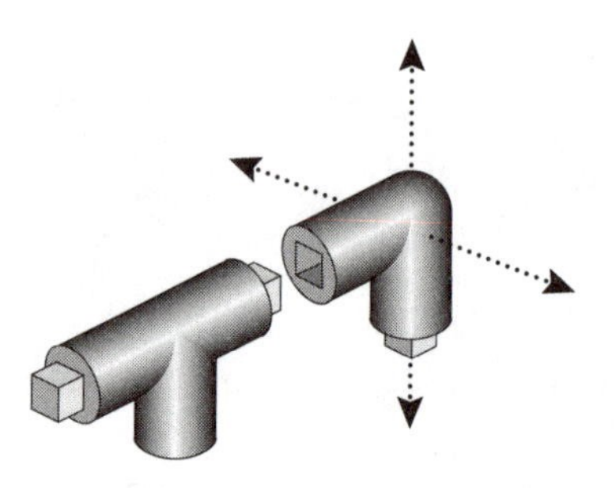

Figure 7.154. Pipes may connect in one of four rotations.

Each subsequent turn players must either

- spin the spinner to decide a piece and connect it to their pipeline,
- remove a piece from their pipeline (if their way is blocked), or
- remove an opponent's Blocker from their pipeline (if it was placed there last turn) and return it to the opponent for reuse.

Players may connect to the board surface at any point that is not a starting tap or finishing pipe of an opponent's color. Players get an extra turn the first time that they connect to the board. Players may build outside the board area. Players miss their turn if they lack the pipe that they spin.

Players may not connect their pipeline to an opponent's pipeline, however they may use their Blockers to plug open ends of opponents' pipelines.

A player wins by connecting a pipeline from his starting tap to his finishing pipe. The winning connection may terminate at a corner point of the finishing pipe.

Notes

The fact that pipes connect in four possible rotations is unique in connection games and allows some interesting tactical play.

The spinner area ratios are similar to the ratios of pipe numbers, except for the Blockers that are relatively harder to spin. These are the only pipes that can be placed directly on an opponent's connection, and are comparatively powerful pieces.

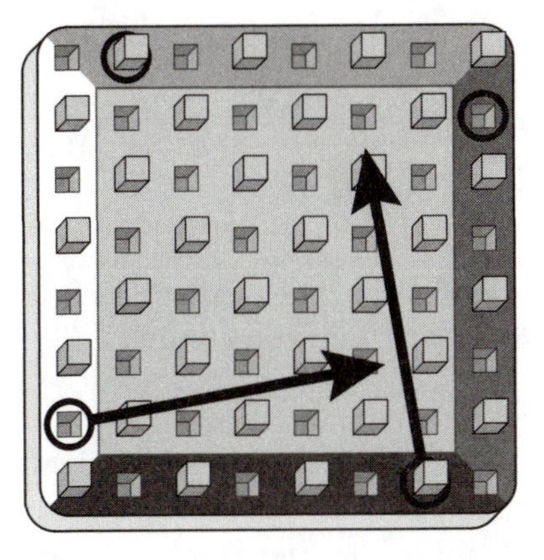

Figure 7.155. Perpendicular directions of play in the two-player game.

Presumably, opponents in the two-player game must take adjacent sides to ensure that they play in perpendicular directions. This tends to increase conflict while minimizing the chance of head-on obstruction.

However, Figure 7.155 shows a problem due to the fact that starting taps are not centered on each side but shifted counterclockwise towards the corners. At some point in the game the horizontal player will be obstructed by the vertical player's pipeline, while the vertical player still has free area in which to build (barring a run of unlucky spins).
This is not a critical problem as it is easy to build over or under existing pipelines, however, it means that the horizontal player is at a one- or two-move disadvantage and should move first (the rules state only that the youngest player should move first). Round the Bend is a rare case of a connection game more suited to three or four players than two.
Round the Bend is one of the few connection games specifically designed for children. It may not satisfy purists due to the random element introduced by the spinner and the resulting lack of strategy, but even serious gamers will admit that it's quite fun to play.

History

Round the Bend was released by Spears Games in 1993. The name of the designer is not known. The rules differ slightly in other editions of the game.

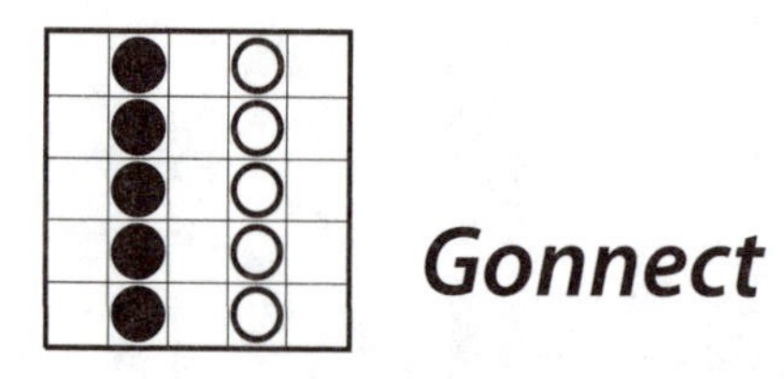

Gonnect

Gonnect is one the best of the recent crop of connection games, successfully marrying the ancient game of Go with a strictly connective goal.

Rules

Gonnect is played on the intersections of a square grid, typically 13 x 13. The board is initially empty.

Two players, Black and White, takes turns placing a piece of the color on an empty intersection, according to modified Go rules as follows

- any chains of stones with no *liberties* (empty adjacent points) are captured and removed,
- a piece cannot be played such that its chain has no liberties, unless that move performs a liberating capture (no-suicide rule),
- board positions cannot be immediately repeated (*ko* rule), and
- players may not pass.

Adjacency is strictly orthogonal for both liberties and pieces (not diagonal). A single-move swap option is recommended.

A player wins if either

- a chain of his color connects either pair of opposite sides of the board, or
- his opponent has no legal move.

Figure 7.156 shows a game won by Black, who has established a black chain between the left and right sides of the board.

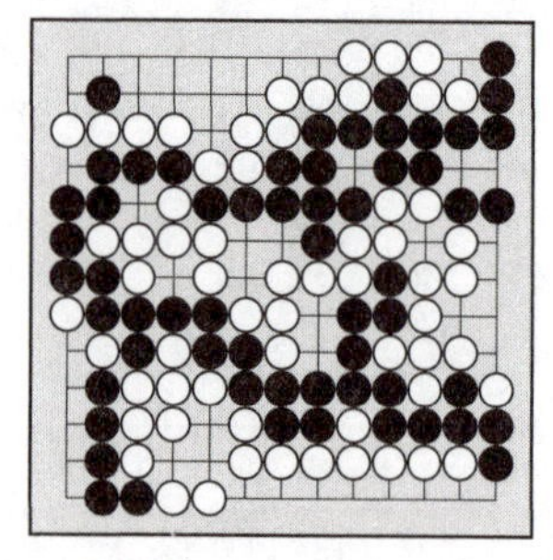

Figure 7.156. A game won by Black.

Figure 7.157. Black move *b* captures white piece *a* by removing its liberties.

Notes

Gonnect is one of the great connection game discoveries. By simplifying the Go rules, it offers more direct conflict and clearer goals than Go, while retaining significant depth. Players may tap into the vast Go literature as background reading but will find that Gonnect offers interesting new tactics and strategies to explore.

Figure 7.157 illustrates the principle of piece-surround capture. This does not introduce a nonconnective element into the game, as captures are connection-based as opposed to pattern-based; the captured chains can be of any size and shape. In fact the surrounding pieces would themselves form connected groups if diagonal adjacencies were counted. Pieces connected to at least two liberties are safe from immediate capture.

The no-suicide rule is illustrated in Figure 7.158, with Black to move. Black has no legal move and loses the game immediately.

Note that both of Black's groups have a single *eye* (enclosed empty point). Normally two eyes are sufficient to guarantee a group's safety, but this is not the case in Gonnect due to an elegant method for resolving deadlocks. This method is a fortuitous conjunction of the no-pass and no-suicide rules, as found in the earlier game of One-Capture Go. For instance, Figure 7.159 (left) shows a game in the middle stages of play. The board is temporarily deadlocked; neither player can win without the removal of some enemy pieces.

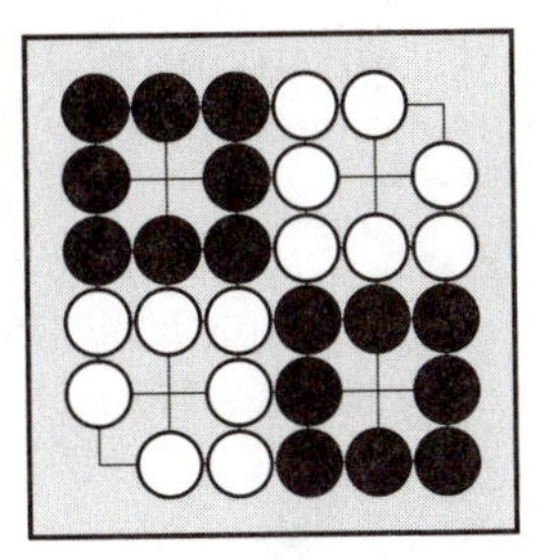

Figure 7.158. Black cannot commit suicide, hence has no legal move and loses.

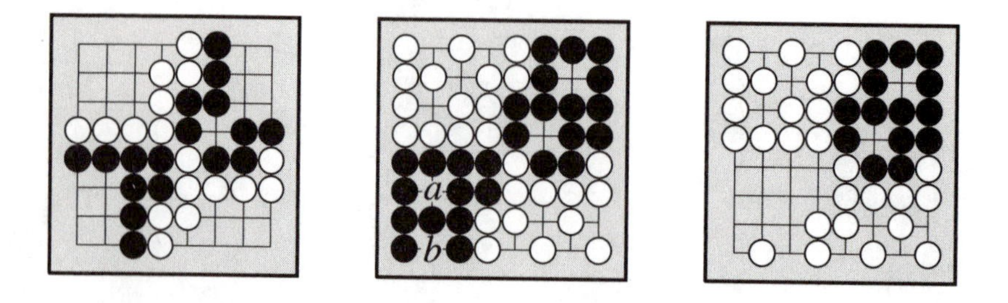

Figure 7.159. The stages of a game.

Figure 7.159 (center) shows the board a few moves later, after players have been forced to fill in some of their territory (Black to move). Both of Black's groups have two eyes, the minimum number that guarantees a group's safety. Unfortunately Black cannot pass and cannot move elsewhere on the board (due to the no-suicide rule) so is forced to put out one of his own eyes. If Black plays at ***a***, then White can play ***b*** next turn to capture the group and break the deadlock (right).

This example demonstrates the two distinct stages of a game of Gonnect: the race stage (hot) and the territory-filling stage (cold). These stages tend to occur in cycles of diminishing length until the game converges to a win. However, subsequent race stages are usually trivial, especially on smaller boards, and it would be prudent for Black to resign in this case.

Larger board sizes favor the territorial aspect, and smaller board sizes favor the connective aspect; the standard size of 13 x 13 achieves a good balance. Go boards and Go stones make ideal playing sets for Gonnect, especially if players seek a deeper 19 x 19 game.

Since Gonnect is played on the intersections, the game graphs (two are required) are taken directly from the board's square grid of intersections. Figure 7.160 shows a temporarily deadlocked game and its game graphs.

Testing for cuts in the graph is problematic due to the possibility of temporary deadlocks; the middle graph is cut, but the game is far from over. A win for each color must be tested in each direction, and the game graphs must be in the Join/Join format.

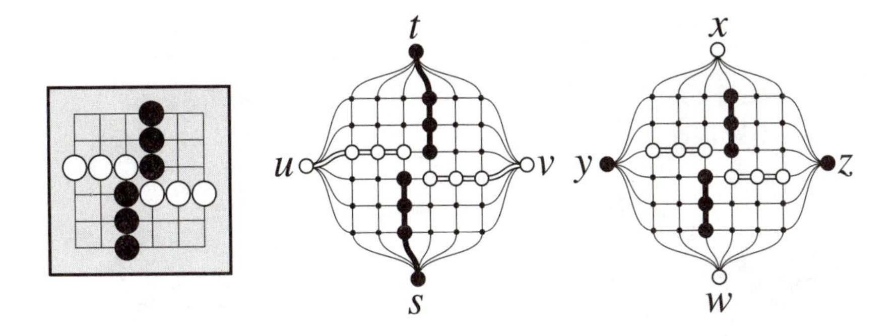

Figure 7.160. A temporarily deadlocked game and its game graphs.

Strategy and Tactics

Players opting for the long game may play defensively to stake out territory and block the opponent, aiming for a deadlock. However, it is equally valid to play aggressively and immediately strive to push through for a connection. While the opening moves in Go tend to be vague as players stake out territory without concern for exact placement, in Gonnect the placement of pieces is critical and a weak move can be attacked immediately [Neto 2004].

Due to the basic similarities between the two games, good Go skills will stand a player in good stead in any tactical Gonnect battles. However, good Go skills can actually be a handicap in terms of strategy; experienced Go players tend to overestimate the importance of territory in their first few games of Gonnect, often to their detriment.

Figure 7.161 (left) shows a case in point with Black to play. Black has the opportunity to capture most of White's pieces with move ***a***, achieving what looks like an overwhelming position. However, White can play move ***b*** next turn and will win the race to connect horizontally by one move. Black should instead have played move ***b*** to neutralize this potential connection before decimating White's main force.

During hot phases *territory is of secondary importance to connection*, but during cold phases the reverse is true. Territory is really just insurance for the end game when the player with the most territory is better placed to weather out cold spells. However, if a player ignores connective threats then the game will probably not last that long.

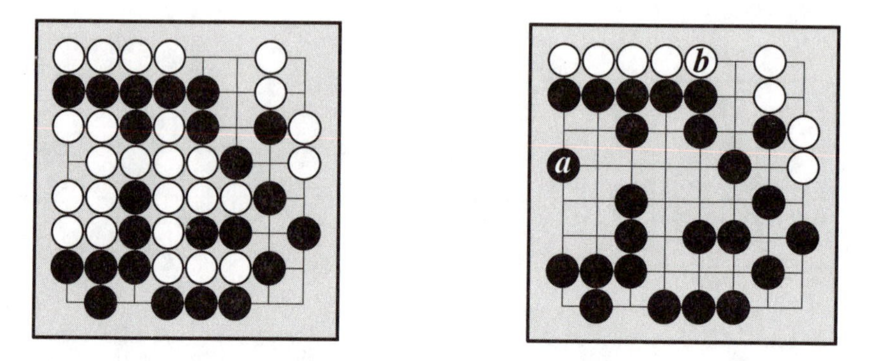

Figure 7.161. Black wins a lot of territory but loses the game.

History

Gonnect was invented in 2000 by João Neto, who describes it as the best of his many game designs. An article describing Gonnect appeared in *Abstract Games* magazine [Browne and Neto, 2001].

Variants

Three-Player Gonnect

Gonnect for three players, Black, Gray, and White, was found to work well with a *mandatory prioritized capture rule* as follows: it is mandatory for a player to capture the next player's pieces if possible, and if not then it is mandatory to capture the following player's pieces if possible. Providing these conditions are met, the player can choose freely among the possible moves.

As in the standard version, three-player capture involves removing all enemy pieces with no liberties after each move. Figure 7.162 illustrates that it is possible to capture pieces of both opponent's colors with one move.

The diplomatic subtleties of three-player games, in conjunction with the mandatory prioritized capture rule, allow some fascinating and amusing situations to develop. For instance, it is possible to initiate forced cycles of moves that gradually develop a player's connection every third turn, which the two opponents can only watch as they trade forced blows. It may be possible to construct infinite mandatory prioritized capture loops, though this has not yet been proven.

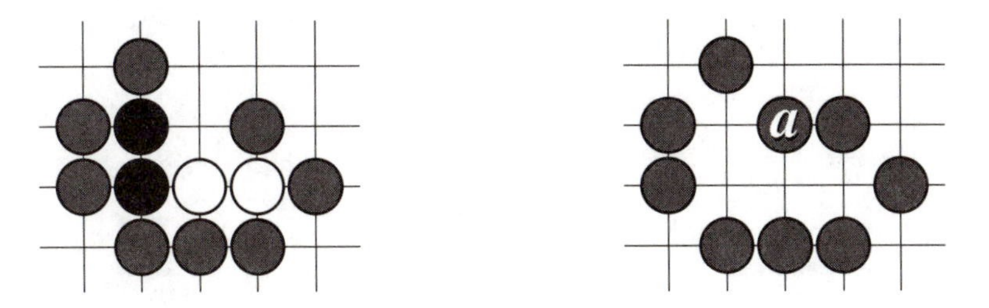

Figure 7.162. Gray can capture both black and white stones with the one move.

Weave

Weave is a Path Race game with alternative (orthogonal and diagonal) winning conditions.

Rules

Weave is played on a square 10 x 10 grid. Two players, Black and White, own alternating sides of the board that bear their color. The board is initially empty. Players take turns placing a piece of their color on the board.

The first player to complete a chain of his pieces between his edges of the board wins. The chain may consist of orthogonally (squarely) adjacent connections or diagonally adjacent connections, but not both.

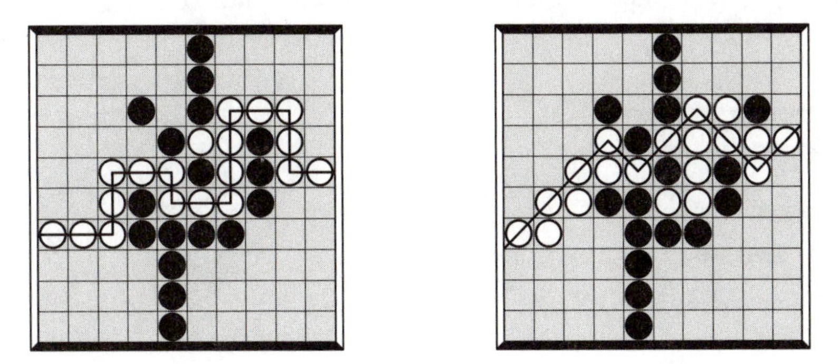

Figure 7.163. Two games won by White. The orthogonal (left) and diagonal (right) winning paths are indicated.

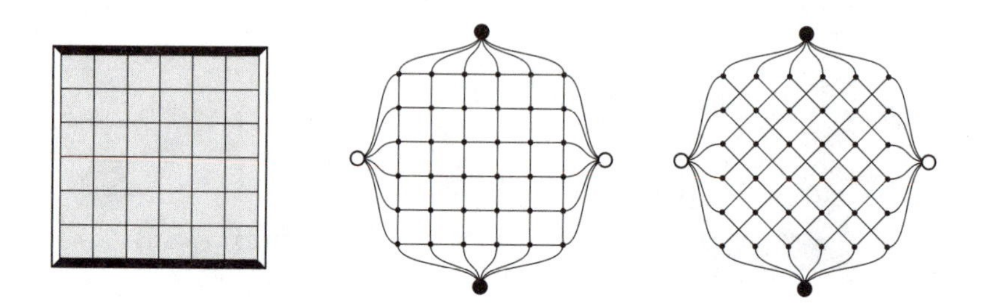

Figure 7.164. A 6 x 6 board and its two game graphs.

Notes

Weave effectively takes place on two independent game graphs, one for the orthogonal component and one for the diagonal component, as shown in Figure 7.164. The winner is the first player to win the game on at least one of these graphs.

The game played on either graph will usually lead to a trivial deadlock, but the superposition of the two graphs may keep the game alive; if one graph deadlocks there is a chance that a win can still be pursued on the other. However, it is possible to deadlock both graphs, and hence the overall game, as shown in Figure 7.165. Neither player can win from this position.

The diagonal component is less likely to yield a win than the orthogonal component due to phase problems. Figure 7.166 (left) shows two diagonal chains that cannot possibly connect. Figure 7.166 (right) shows a similar formation but with the pieces in the second chain shifted up a cell, bringing them in phase with the first chain and connecting with it.

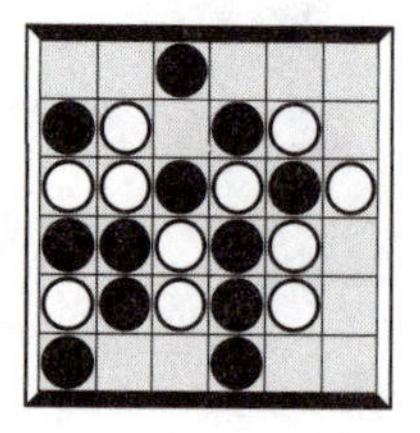

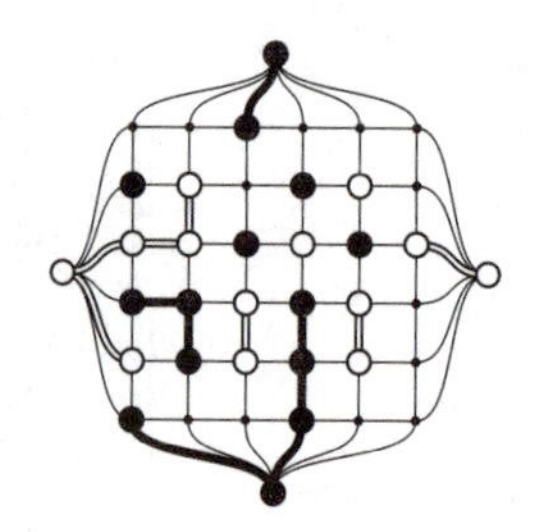

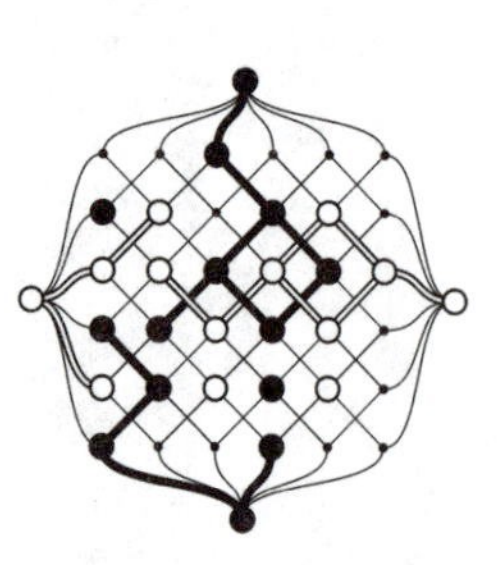

Figure 7.165. A tied game.

Figure 7.166. Phase problems for diagonal connections: out of phase (left) and in phase (right).

History

Weave is Joris game #42 [2002]. Joris describes it as an "old connection game."

7.2.2 Pure Connection > Path Race > Shared Paths

Shared Path games are those in which players may use common paths to achieve their goal, including pieces placed by the opponent. Note that pieces may still be differentiated in some way.

Shared Path games subvert the Cut/Join property of most connection games. Instead, path blocking is achieved in most of these games through the use of directional tiles that forbid path development in certain directions.

The icon for this group shows a path shared by both players stretching in both directions.

Black Path

Black Path is a simple but intriguing tile placement game played on a square grid. It is different from most other connection games in that the first player to complete a connection loses. A winning strategy is known.

Rules

Black Path is played on a square grid, typically 8 x 8. Two players, A and B, share a common pool of the tiles shown in Figure 7.167 (left). Note

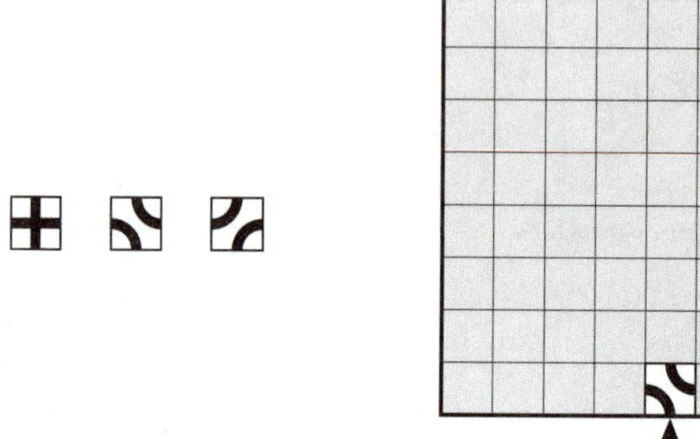

Figure 7.167. The three Black Path tiles, and a board after A's opening move.

that the second and third tiles are rotations of each other. The board is initially empty.

Player A starts the game by placing a tile on any edge cell, and marking a source path leading from the edge. Figure 7.167 (right) shows the start of a game following an initial tile placement.

Players then take turns placing a tile on an empty cell so as to extend the active end of the source path. The first player to run the source path into a board edge loses the game. One player must win.

Notes

Figure 7.168 shows a small 4 x 4 game won by A, who has forced B to play the last tile (bottom left) that runs the source path into a board edge.

If the opening move is made in a corner, then there are two source paths to choose from (Figure 7.169). Only two of the tiles are valid for a corner start: the third constitutes an immediate loss (Figure 7.169, right).

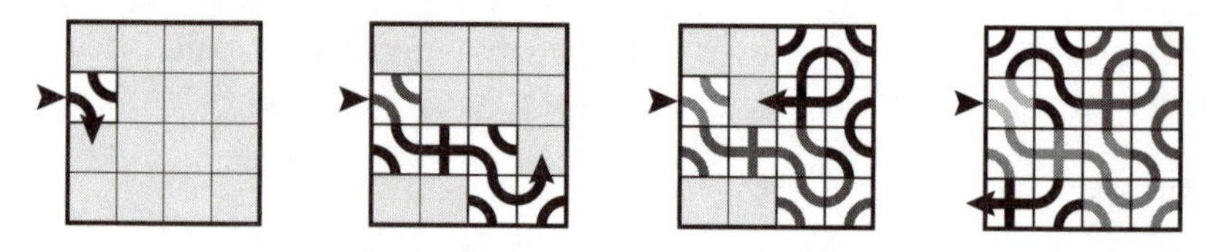

Figure 7.168. A 4 x 4 game won by A.

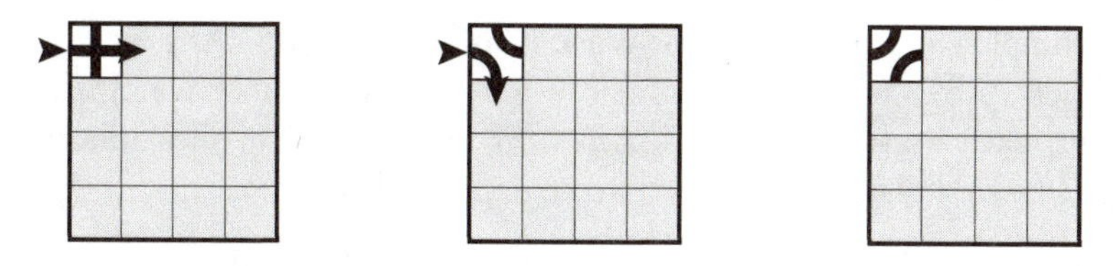

Figure 7.169. Opening in a corner creates two paths to choose from.

The vertices in the game graph of Black Path occur along cell edges rather than at cell interiors (Figure 7.170). This is necessary to accommodate the fact that crossing paths do not actually meet at the cell center to form a junction, but cross in a nonplanar fashion.

Figure 7.171 shows a short game won by B and its game graph. The tiled path completes a connection between terminal vertices *s* and *t*.

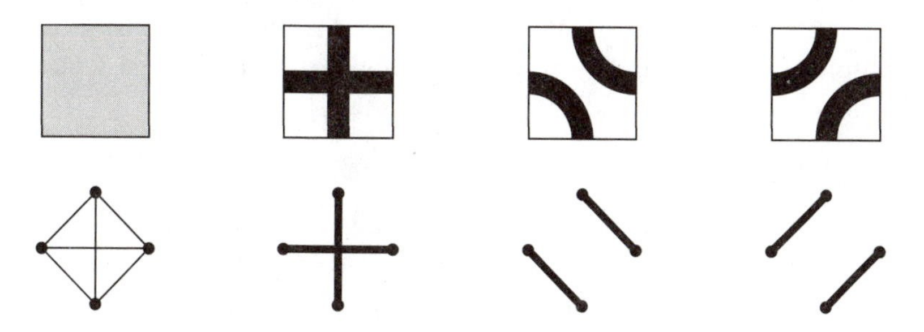

Figure 7.170. Graph operations for each tile.

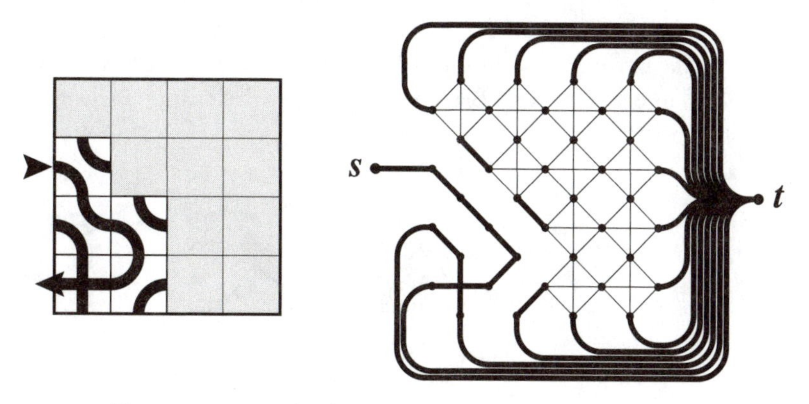

Figure 7.171. A 4 x 4 game won by B and its game graph.

A simple winning strategy for Black Path is described in Appendix E.

Black Path is unusual in that it combines strictly connective play (the source path must be extended each turn) with a strictly anticonnective goal (the first player to connect two edge points loses).

History

Black Path was invented in 1960 by Larry Black after studying Hex and Bridg-It [Gardner 1963].

Variants

A Winding Road

Koch describes the similar game of A Winding Road [1991]. The rules are identical to those given above except that the opening tile *must* be placed in a corner. Koch fails to state that tiles added to the board must extend the source path, though this is implied. This version is also called Black's Road Game, Snake's Road, and Squiggly Road.

Squiggle Game

Squiggle Game is played much like Black Path, except that each tile shows two path segments leading from each edge, and players own tiles with path segments of their color [McMurchie 1979]. Figure 7.172 shows some example tiles.

Players place a marker of their color on their active path end. This is not an independent piece; it merely makes explicit which is the active path end. Players may only place tiles to extend their own active ends. The first player forced to terminate his active path at a board edge loses. Squiggle Game is effectively Black Path played with Separate Paths rather than Shared Paths.

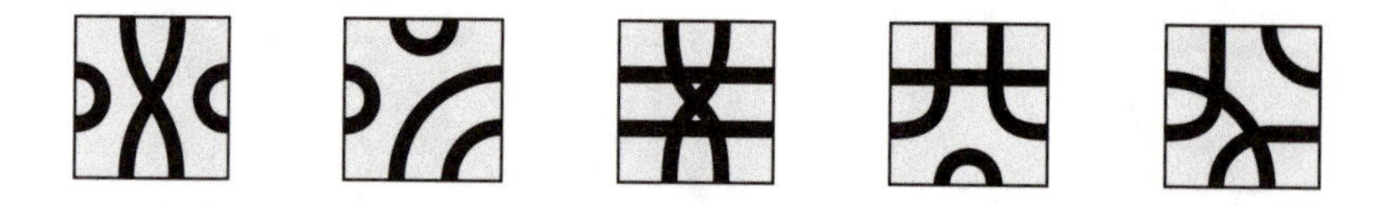

Figure 7.172. Some of the Squiggle Game tiles.

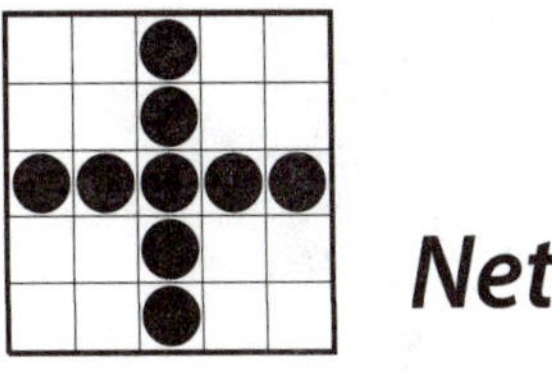

Network

Network is a game by Sid Sackson that incorporates line-of-sight connections.

Rules

Network is played on an 8 x 8 square grid with the corners removed, as shown in Figure 7.173. The top and bottom rows are black goal areas and the left and right columns are white goal areas.

Two players, Black and White, each have ten pieces of their color. Players take turns placing one of their pieces on an empty square. The piece cannot be placed in the opponent's goal area or such that it would create a chain of more than two friendly pieces connected orthogonally or diagonally (Figure 7.173).

Two pieces of the same color are said to *connect* if they are in orthogonal or diagonal line-of-sight. A continuous set of connections is called a *network*.

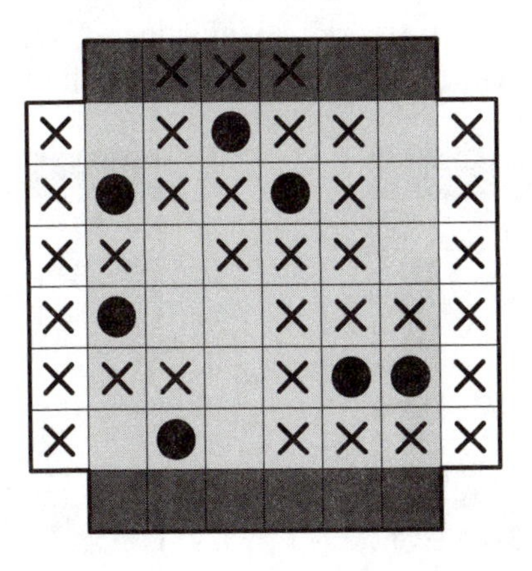

Figure 7.173. Illegal moves for Black are marked X.

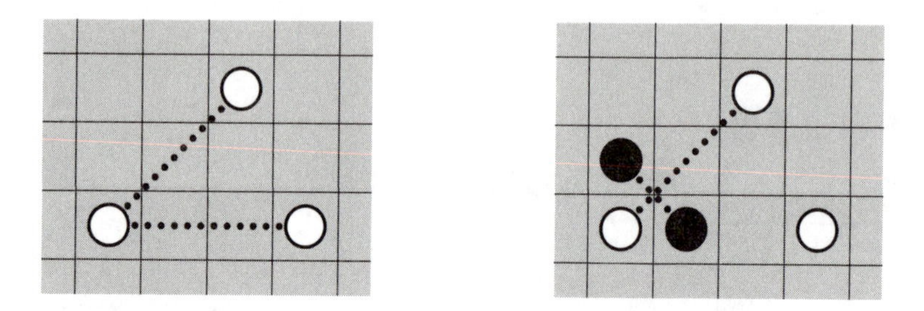

Figure 7.174. Line-of-sight connections between same-colored pieces.

Figure 7.174 shows a network of connected white pieces (left), and interference caused by two black pieces (right). Note that the bottommost black piece cuts the lower white connection, but that connections can cross.

A player wins by establishing a network connecting his goal edges that

- contains at least six pieces,
- does not visit the same piece twice,
- changes direction at each piece, and
- does not visit more than one piece in each goal area.

Figure 7.175 shows two different winning networks for Black on the same board (there are more).

If players run out of pieces before a winning connection is made, then players take turns moving one of their existing pieces to a different square in accordance with the piece placement rules. It is forbidden to make a move that would yield a winning network for both players.

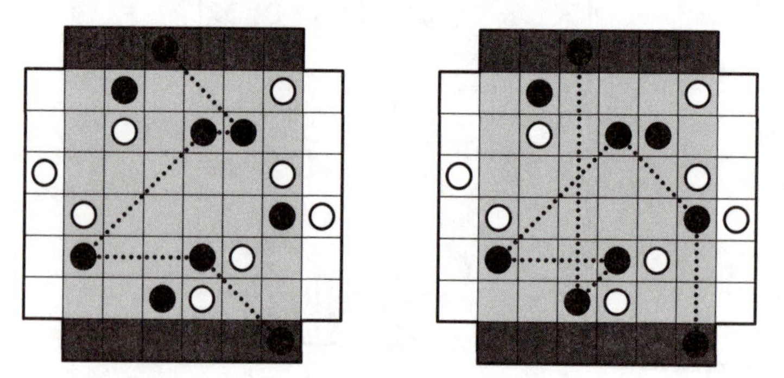

Figure 7.175. Two winning networks for Black.

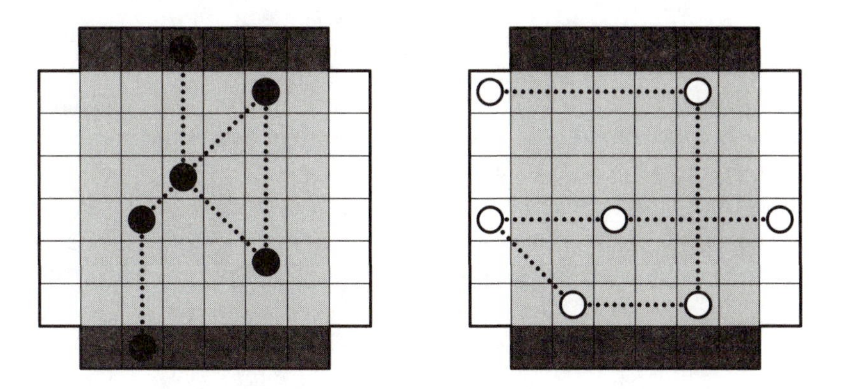

Figure 7.176. Illegal networks: neither player has won.

Notes

Figure 7.176 shows two illegal networks by way of example. The black network in Figure 7.176 (left) contains a cycle that visits the same piece twice. The white network in Figure 7.176 (right) has two problems: it visits White's left goal column more than once, and the connection does not change direction after visiting the central white piece.

Network is intuitively simple and fast to play but is complex in a combinatorial sense, as demonstrated in Figure 7.177.

Figure 7.177 (left) shows the usual adjacency graph for a cell on the square grid. Figure 7.177 (middle) shows the adjacency graph of a single cell on the Network board; note that the cell is potentially adjacent to any cell in orthogonal or diagonal line-of-sight. A friendly piece along a line claims that edge and cuts all prior edges, while an enemy piece along a line cuts that edge and all further edges. Figure 7.177 (right) shows how

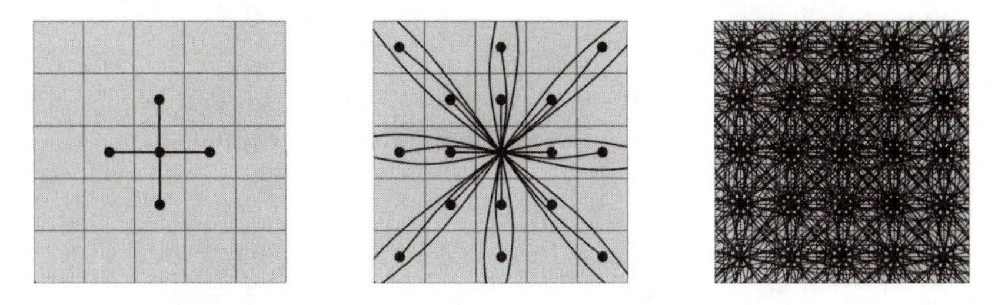

Figure 7.177. Adjacencies on the Network board.

densely nonplanar the game graph for Network becomes when all cells are taken into account.

Sackson recommends that players announce "threat" if they can win next turn. Like announcing "check" in Chess, this reduces the number of games won by error on the opponent's part.

Network on the standard board has a significant first-player advantage. For this reason, Sackson recommends that the second player (White) be allowed to make either two chains of three pieces each, or one chain of four pieces. Other ways to address the first-move advantage in Network include using the swap rule after three moves or playing on a larger board.

History

Network was invented by Sid Sackson and is described in *A Gamut of Games* [1969].

Variants

Estrin's Game

Estrin [1970] describes a game played with flat mirrors that sit upright in slots angled at 45 degrees to the board's edges. Players take turns slotting a mirror onto the board. A player wins if a beam of light projected from one of his edges is deflected to the far edge via a network of mirrors. Points may be scored based on path length, number of deflections, and so on.

Killer Beams

Two players, Black and White, each draw eight circles of their color on an 8 x 8 grid. Players take turns drawing a path (or *beam*) between two of their circles; the beam must take orthogonal steps, must be the shortest path possible, and must not pass through any other of the player's circles. Any opponent's circles that the beam passes through are killed and removed from the game.

The player with the most surviving circles wins the game. Killer Beams is Joris game #25 [2002] and is not a Pure Connection game.

Constellations

Two players, Black and White, take turns claiming squares on an 8 x 8 grid. The first player to make a *constellation* wins. A constellation is similar to a network as described above, except that a constellation

- must pass through at least eight friendly squares,
- may pass through any number of enemy squares,
- must change direction at each friendly square,
- may not pass through the same friendly or enemy square twice,
- must consist of orthogonal or diagonal lines between friendly squares, and
- may not contain consecutive lines of length 1 (that is, it may not contain three or more adjacent friendly squares in a row).

Constellations is Joris game #75 [2002].

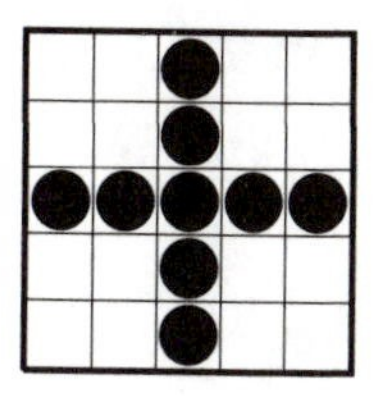

Thoughtwave

Thoughtwave, a popular tile-placement connection game, features distinctly colored tiles but a shared path.

Rules

Thoughtwave is played on a 10 x 10 square grid. Two players, Light and Dark, own alternating sides of the board that bear their color. The board is initially empty.

Each player has the following tiles shown in Figure 7.178:

- 1 x Terminator,
- 10 x Bend,
- 5 x Straight,
- 6 x Tee, and
- 2 x Cross.

Players take turns placing one of their tiles on any empty square, provided that it agrees with all neighboring tiles (if any). No path can be cut off by a blank tile edge. Tiles, once placed, do not belong to either player. Tile coloring is shown merely to indicate which tiles each player has used up.

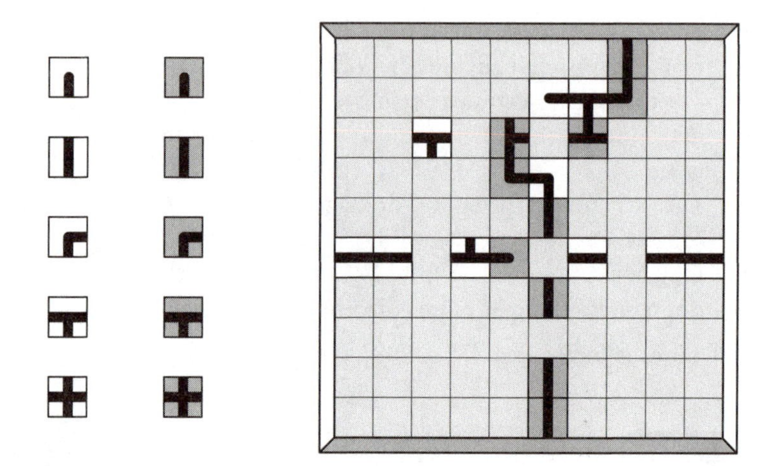

Figure 7.178. The Thoughtwave tiles and a game about to be won by Dark.

The first player to complete a connected path between his sides of the board wins the game. If winning connections are made for both players with a single tile placement, then the mover wins. If all tiles are placed without a winning connection being made then the game is a draw.

Figure 7.178 shows a game about to be won by Dark, regardless of whose move it is. Dark has an unblockable virtual connection between the Dark edges and Light does not have an equal or better virtual connection between the Light edges. Importantly, Dark still has enough tiles in hand to complete the connection.

Notes

Thoughtwave tiles act as edge filters, as shown in Figure 7.179. The game graph consists of a vertex for each board cell with neutral edges to adjacent cells as usual. Tile sides are described as *open* if there is a path drawn to that edge, and *closed* if there is no path drawn to that edge. Each open side colors an edge and each closed side deletes an edge. Figure 7.179 (right) shows that two tiles can be virtually connected without being adjacent.

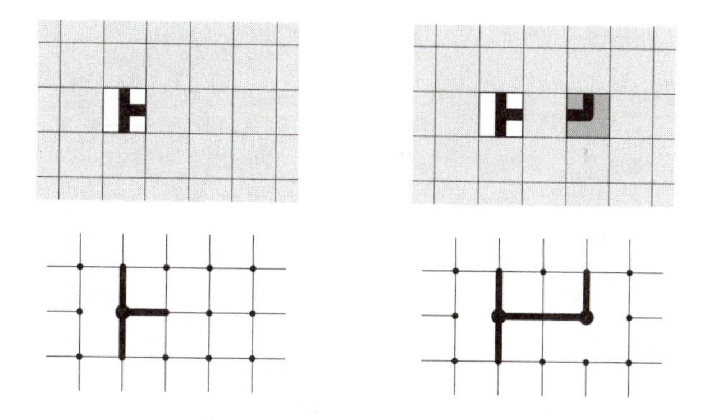

Figure 7.179. Tiles act as edge filters.

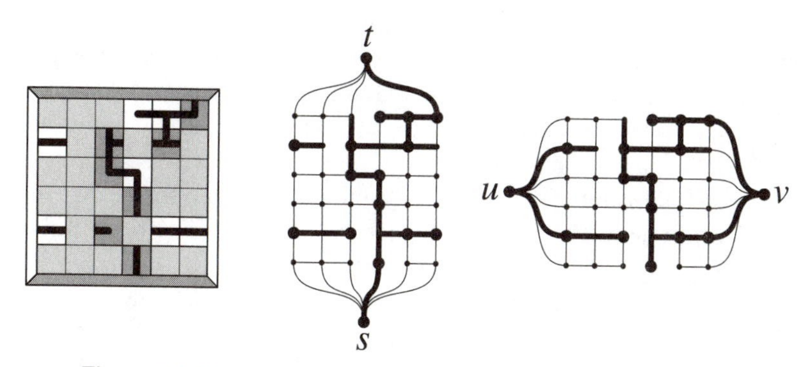

Figure 7.180. A 6 x 6 game won by Dark and its game graph(s).

Figure 7.180 shows a 6 x 6 game won by Dark and its game graphs. Dark wins by connecting terminals ***s*** and ***t***, and Light wins by connecting terminals ***u*** and ***v***. The middle graph explicitly shows Dark's winning path, making the win more obvious than the tiles on the board.

Strategy and Tactics

It is good strategy to lay down alternative routes across the board, well spaced out; a Terminator can swiftly end a single connection threat. Figure 7.181 shows a safe virtual connection that covers more ground than an adjacent chain.

It is important to block from a distance. Figure 7.182 shows a poor defense by Dark, who has tried to plug the Light connection to the right

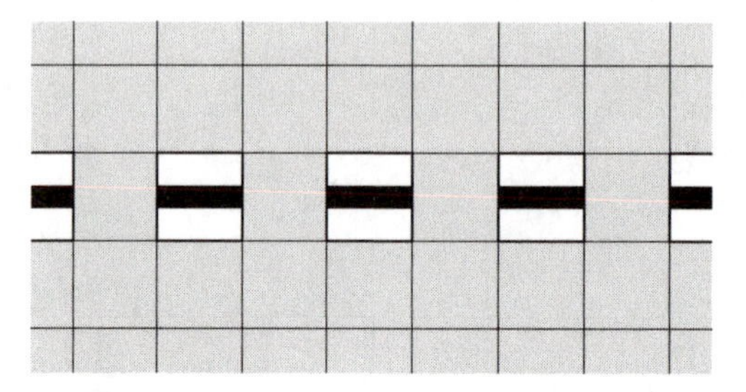

Figure 7.181. A safe progression by Light.

by terminating the path's right-most piece directly. Light is able to step around this block with almost no impediment, and Dark has just wasted their valuable Terminator.

Figure 7.183 shows a better blocking maneuver by Dark, who this time blocks across the connection leaving two spaces between the connection head and the blocking piece. Martin Lester calls this the *leave-two-spaces* rule.

The amount of room needed to block effectively, combined with the relatively small board size means that once a player gets ahead by the slightest margin it is difficult to stop them. As with most connection games, Thoughtwave has a significant first-move advantage, and the second player must take a defensive role initially. The limited board size means little scope for strategy; hence

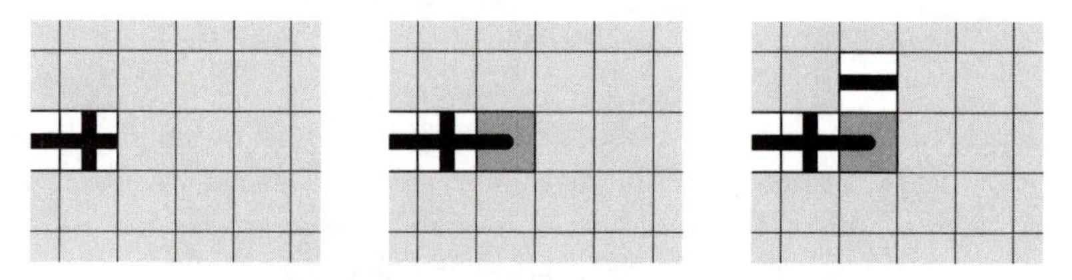

Figure 7.182. A poor defense by Dark.

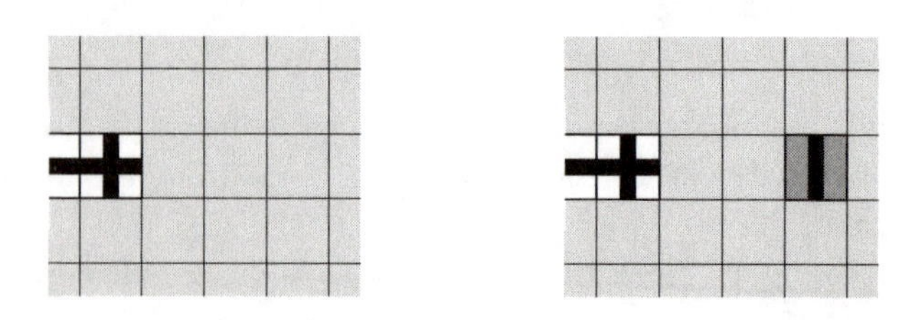

Figure 7.183. A better defense by Dark using the leave-two-spaces rule.

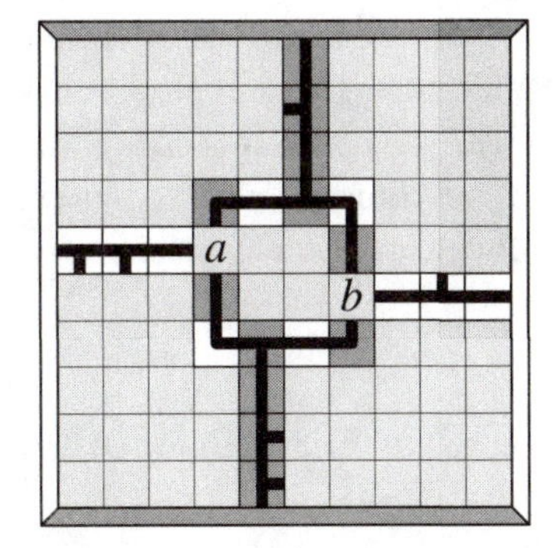

Figure 7.184. A cold war.

Thoughtwave is primarily a tactical game. It is recommended that a first-move equalizer such as the swap rule be used.

Note that virtual paths do not always guarantee a win for the virtually connected player. This can occur if the player does not have the correct tiles to complete his connection, or in the case of a cold war (see Figure 7.184).

Playing at either ***a*** or ***b*** is a losing move for both players, as the opponent can then play at the other square to win the game next turn. As the only possible winning paths from this position pass through ***a*** or ***b***, both players must burn their remaining moves elsewhere on the board until they run out of tiles and the game is drawn. If a player has two Tees or Crosses while his opponent has none, then that player will win.

This impasse is a contrived case but similar situations are not infrequent during play. It demonstrates an important strategy suggested by Martin Lester, that any virtual connection shared by both players is safe if it has at least two gaps common to both players' connection. Generally such a case will not cut all other potential paths, and players can safely ignore the double-gapped threat and attack elsewhere.

The Terminator is a powerful piece and should be used only in emergencies. Straights can be blocked easily, but are useful as filler pieces in building a connection. Bends are also easily blocked, and are not that useful except for cutting connections and forcing the opponent to make detours. Since closed tile sides cut edges in the game graph, they are just as important as open edges in defining connections. Tees are probably the most versatile and useful pieces, and are difficult to block. It is generally good practice for a player to save the bulk of his Tees until needed later in the game. Crosses are difficult to block but also provide a cross-passage for the opponent.

Thoughtwave is not an especially deep connection game, and its complexity drops dramatically with each move. However, there is some art to the optimal deployment of pieces. It is critical that players keep enough of the necessary pieces to complete their connection, while forcing the opponent into using up vital pieces. The scarcity of pieces makes it possible to trap an opponent.

History

Thoughtwave was invented by Eric Solomon in 1973.

Variants

Pista

Pista is Joris game #96 [2002]. It is identical to Thoughtwave except that

- there are no Tees,
- there is an unlimited supply of all tiles except Terminators (each player gets two), and
- tiles must be laid adjacent to at least one existing tile.

Barton's Game

Barton's Game, a precursor to Thoughtwave, is one of the earliest connection games [Barton 1939]. It features tiles identical to the Thoughtwave minus the Terminator. Play starts at a predefined square, then players take turns placing a tile adjacent to at least one existing tile. In one version of the game, the connection must travel around a central obstacle and connect back to the starting edge.

Davies's Game

Davies [2002] describes a set of games based upon the Thoughtwave tiles supplemented with special Poison and Magic tiles. Some games involve the continuous extension of existing paths each turn, and some games involve length-based scoring.

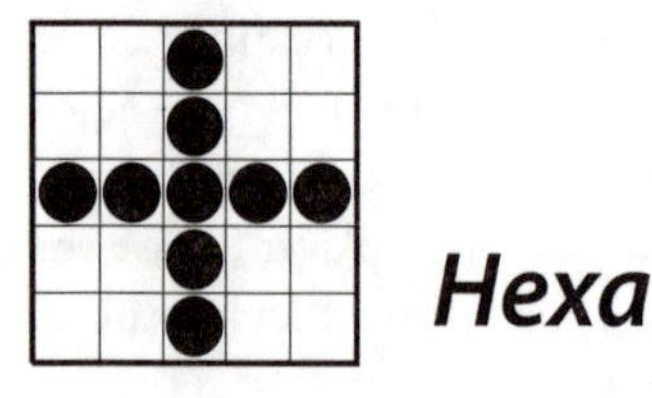

Hexa

Hexa is played with bicolored hexagonal pieces on a six-sided board design.

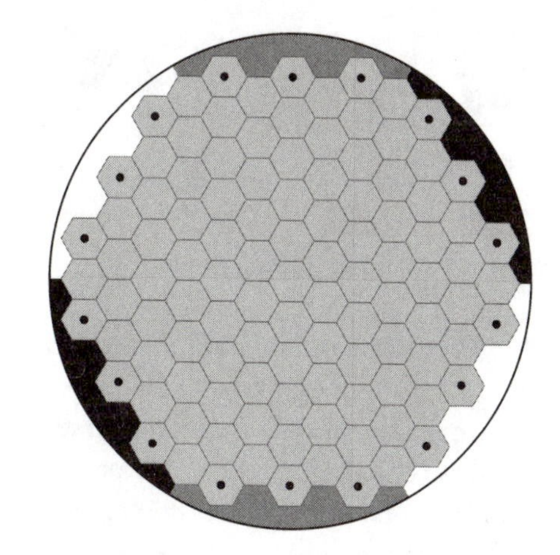

Figure 7.185. The Hexa board.

Rules

Hexa is played on the hexagonally tiled board shown in Figure 7.185. The game may be played by two players, White and Black, or may include a third player, Gray. Each player owns the sides of the board that bear their color. The three outer-most cells along each edge marked with a dot are special goal cells. The board is initially empty.

Each player starts the game with a number of hexagonal pieces marked with path segments (dark) on a neutral background color (light). The pieces are shown in Figure 7.186 and their distribution to each player is as follows: 1, 3, 3, 3, 1, 3 (top row); 3, 1, 2, 2, 2, 1 (bottom row). The piece shown at the top left is a *defensive piece* devoid of path segments. All other starting pieces are *offensive pieces* that consist of various configurations of path segments passing through a colored center.

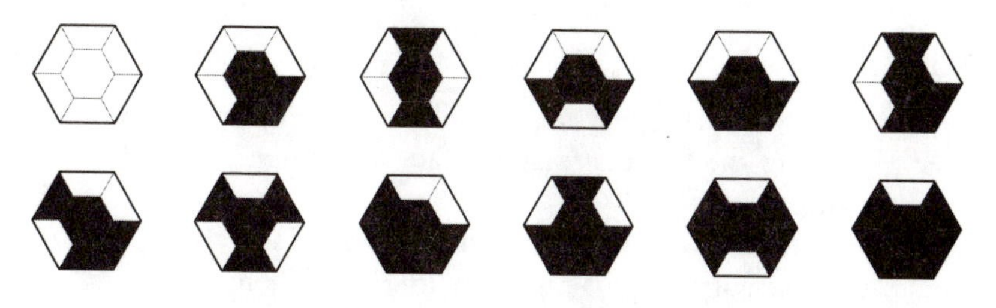

Figure 7.186. Starting pieces for each player.

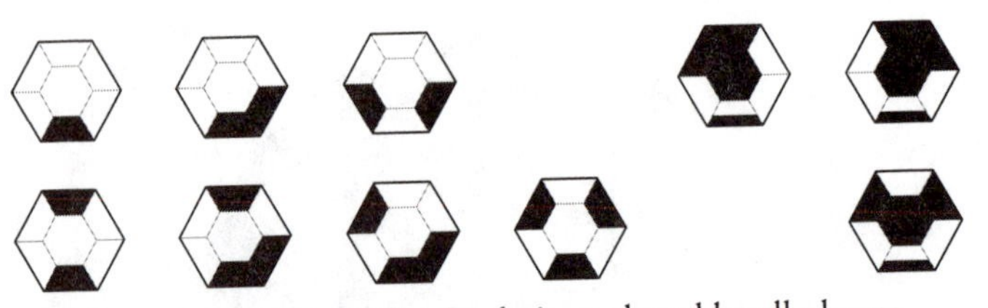

Figure 7.187. Defensive Pool pieces shared by all players.

Players also have access to a common pool of pieces called the Defensive Pool, which contains exactly one of each type of piece shown in Figure 7.187. These include additional defensive pieces with neutral centers (left) and *offensive-defensive pieces* that combine characteristics of both types of pieces (right).

Players take turns placing one of their pieces on an empty board cell. The piece cannot be placed if any path segment edge would adjoin the neutral edge of an existing piece. Players may play in an opponent's goal cell, provided that the piece leaves at least one path edge open to the board. Only two such moves can be made against each opponent.

Prior to his move, the player may purchase a piece from the Defensive Pool with any two of his offensive pieces, which are put aside and play no further part in the game. Only two such purchases may be made by each player over the course of a game.

A player wins by completing a path between at least one goal cell belonging to each of his home areas. A winning path may pass through opponents' goal cells. For instance, Figure 7.188 shows a three-handed game won by White. Note that the aim is not to connect the actual board sides (as Black has done) but to connect the goal cells along opposed sides (as White has done). Gray's connection is broken by an offensive-defensive piece towards the top of the board.

If all possible winning paths are cut off, or there is no winning path after all tiles have been placed, then the game is a draw.

Notes

Hexa is unusual in that it plays as naturally with three players as two. This is largely due to the fact that players share a common path rather than distinct paths.

Purchasing defensive pieces does not exclude Hexa from the Pure Connection category—this is more a matter of piece management than introducing an extraneous trading element.

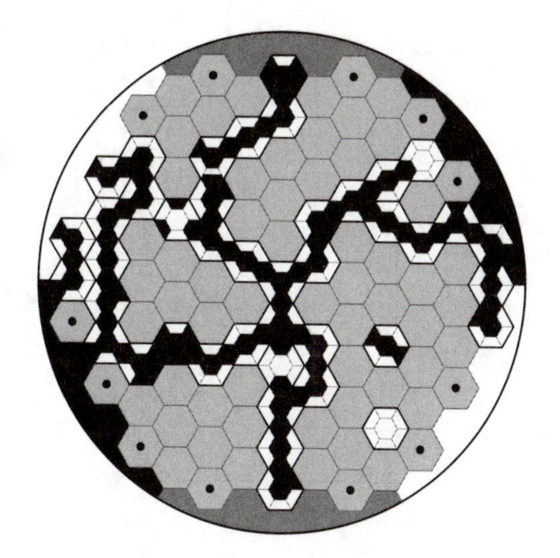

Figure 7.188. A three-handed game won by White.

Defensive pieces are strong as they provide the only means to really impede an opponent. However, players must be careful not to overinvest in defense, and should ensure that they leave themselves enough offensive capability to complete their own connection.

The rules do not state whether it's forbidden to place a piece such that its neutral edges lie adjacent to existing path edges, but this is implied. If this were not the case, then defensive piece placement would be unconstrained and overwhelmingly powerful.

History

Hexa was designed by Thomas Rudden [1980].

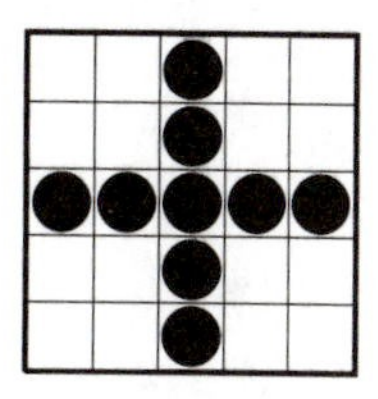

Turnabout

Turnabout is a tile-based path-building game with a somewhat anticlimactic winning condition.

Figure 7.189. The Turnabout tile, and a game won by the Offensive player.

Rules

Turnabout is a game for two played on an 8 x 8 square grid. One player takes the Offensive role and the other player takes the Defensive role. Players share a common pool of 32 tiles, each with a cross on one side and two curves on the other (Figure 7.189, left). Both rotations of the two curves are shown. The board is initially empty.

The Offensive player moves first; thereafter players take turns placing a tile, either side up, on any empty square. The tile does not have to touch existing tiles.

The Offensive player wins if a path is formed between either or both opposite sides of the board. The Defender wins if no such path is made before the tiles run out.

Notes

As with Black Path, the vertices of the game graph occur at the midpoint of cell edges, rather than the midpoints of the cells themselves. This is necessary to accommodate the fact that the crossing paths within the "+" tile do not actually touch, but cross each other in a nonplanar fashion. That is, the path continues straight across the intersection rather than turning at right angles.

Figure 7.190 shows the six potential edge connections per tile (left) and the edges that remain following a tile placement. The path is shared by both players so tile placement does not cut any edges beyond that tile's area.

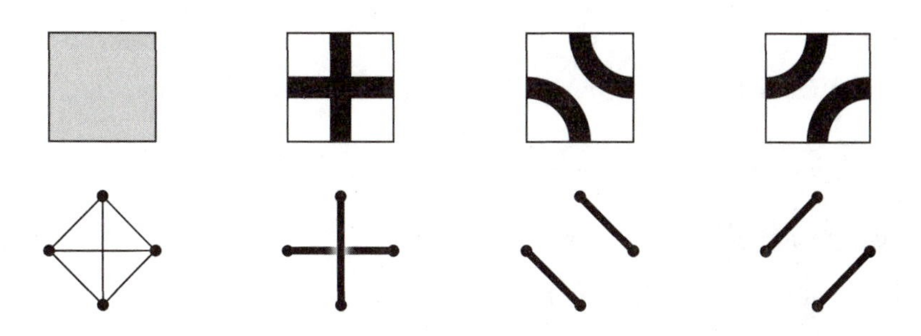

Figure 7.190. Graph operations for each tile.

This underlying structure is suggested by the actual game board itself, which contains protrusions for holding the pieces in place that match the locations of vertices in the game graph.

Figure 7.191 shows a small 4 x 4 game in progress and its game graph. The Offensive player wins by connecting ***s*** with ***t*** and/or ***u*** with ***v***. Note that this game has not been won yet, as the "+" tiles do not connect at right angles. In other words, a winning path must be smoothly continuous and may not follow sharp corners; the game graph is nonplanar and crossing edges do not meet.

It's not possible to know whether the Defender has won until all tiles have been placed. This weak form of win by contradiction (win by the absence of loss) is quite unusual in connection games.

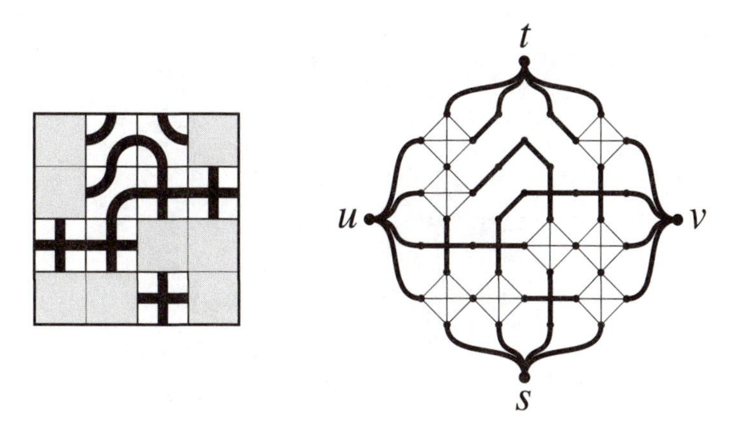

Figure 7.191. A 4 x 4 game in progress and its game graph.

History

Turnabout dates from 1982. Its author and publisher are unknown.

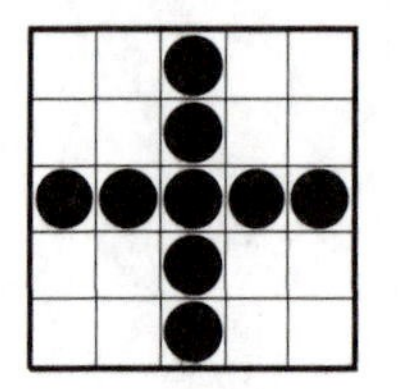

Pipeline

Pipeline has an integrated connection theme and plays equally well for two, three, or four players.

Rules

Pipeline is played on the 15 x 15 square grid shown in Figure 7.192 (right). The board is initially empty and contains the following special squares:

- a central wellhead (shaded),
- three loading docks along each edge (also shaded), and
- eight obstacle squares (marked **X**).

Two to four players each own a board edge and share a common pool of 200 pipeline tiles of the five types shown in Figure 7.192 (left): Plug, Straight pipe, Elbow, T, and Cross. These are identical to the Thoughtwave tiles.

Players draw five tiles at random and place them on a rack hidden from the other players. Each round, players take turns playing one tile from their

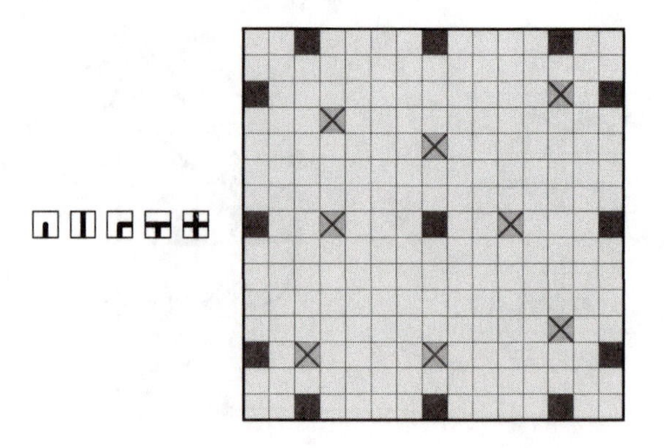

Figure 7.192. The Pipeline tiles and board.

rack onto the board. Players have the option of passing on the last move of each round and keeping their last remaining tile for the next round. Players replenish their racks after each five moves.

The first tile must be placed on the central wellhead. Thereafter tiles must either be placed on empty squares so as to extend the central wellhead connection, or on any of the player's loading docks. Any play that would prevent a branch from eventually joining a loading dock or joining back into the main pipeline is illegal. It is not permissible to place an open end of a pipe against a board edge or obstacle.

A player is removed from the game if all of his loading docks are cut off from the central wellhead or if he has no legal move. The first player to connect a pipeline from the central wellhead to one of his docks wins the game. If this is not possible, then the player to make the last legal move wins. In the rare case that the last legal move connects the central wellhead to an opponent's loading dock, that opponent wins.

Notes

Pipeline is a straightforward game with good clarity, despite the fact that its rules are strongly themed and quite involved for a connection game. It is one of the few games of any type that plays equally well for two, three, or four players.

Figure 7.193 shows the game graph of the Pipeline board. Obstacle squares create holes in the game graph. The central wellhead *s* and loading docks are represented by colored vertices. Each player's loading docks are connected to terminal vertices ***a***, ***b***, ***c***, and ***d***, respectively. The first player aims to connect *s* with ***a***, the second player aims to connect *s* with ***b***, and so on.

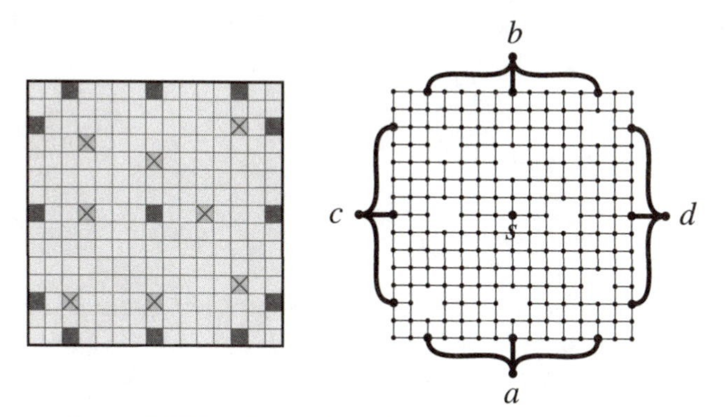

Figure 7.193. The Pipeline board and its game graph.

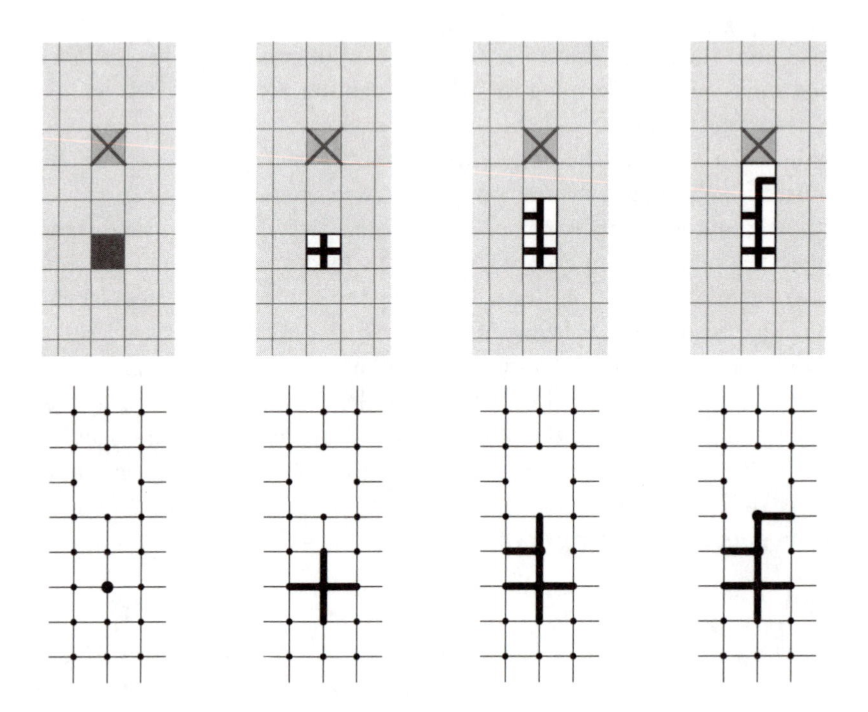

Figure 7.194. Pipeline tiles are edge filters.

Pipeline tiles act as directional path filters, much like Thoughtwave tiles. Each tile placement colors a vertex and all incident neutral edges corresponding to sides of the tile with open ends, and cuts all other incident edges that must be neutral. This is illustrated in Figure 7.194, which shows the first three moves of a game.

Strategy and Tactics

The following points of strategy are drawn from Peter Sarrett's [1993] review of Pipeline.

Save cross pipes for loading docks. The closer a pipeline gets to a player's loading docks, the more important it is for him to play Ts and Crosses to prevent opponents blocking his path.

If a player is about to win, his opponents must cooperate to overrun his dock. This is an example of petty diplomacy in action.

Plugs are strong pieces. It is sometimes wise to pass and hang onto a plug rather than play it ineffectively.

Sarrett recommends using the following rule to avoid the game being spoiled by the luck of the draw: players who draw a handful of straight pipes on their first hand may discard these tiles and draw again. A handful of straight pipes severely limits a player's options.

History

Pipeline was invented by Ed Okimura and published by Playco Hawaii in 1988. Pipeline won the *GAMES* magazine "Game of the Year" award in 1992.

Knots

Knots is a simple, attractive connection game that demonstrates that strategic depth is not essential for a game to be enjoyable.

Rules

Knots is a two-player game played on a 6 x 6 square grid. The board is initially empty.

Players each choose two tiles from a common pool of 40 tiles laid face down on the table. Each tile has two nodes along each edge from which pieces of rope are drawn. Within the tile, each piece of rope starting at a node may

- exit at another node,
- terminate at a frayed end,
- splice into another piece of rope, or
- split into two pieces of rope that exit at two different nodes.

Players take turns placing one of their tiles on any empty board square, then replenishing their hand by drawing another random tile from the pool.

Figure 7.195. A game of Knots in progress.

The first player wins if the top and bottom sides of the board are connected with a continuous path of rope, and the second player wins if the left and right sides of the board are connected with a continuous path of rope. If placing a tile achieves both winning conditions simultaneously, then presumably the mover wins.

The board comes in four pieces that connect jigsaw-style. The dotted lines indicate board joins.

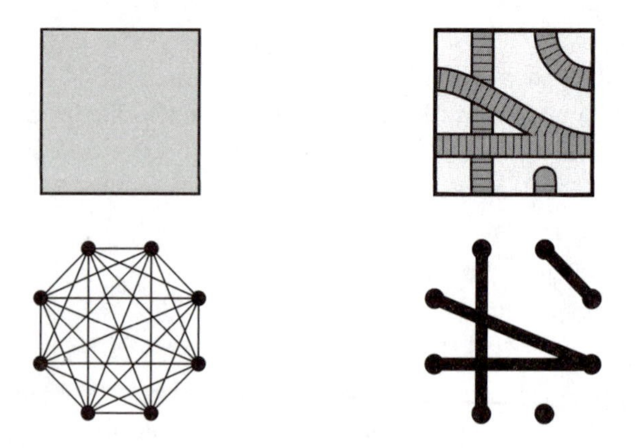

Figure 7.196. Graph operations corresponding to a tile placement.

Notes

Figure 7.196 (left) shows the adjacency graph for a single empty cell. This consists of two vertices along each side, and edges indicating potential connections between every unique pair of vertices.

Figure 7.196 (right) shows edges realized after tile placement. Edges corresponding to rope segments are fixed and all other edges removed from that cell. Unlike almost all other tile-based connection games, each vertex does not necessarily have a unique partner; spliced ropes converge two edges to a single vertex, and frayed ropes create no edge at all.

Figure 7.197 shows the game graph of a small 3 x 3 board after a single tile has been played. Although this graph is quite complex due to the wide variety of connective possibilities, each tile realizes connections in a well-defined way and does not interact with neighboring adjacencies (as does Twixt, for example).

This complexity means that it is hard to plan ahead and form strategies, making Knots a more tactical game. However, almost all recorded comments regarding the game describe it as fun to play despite its lack of depth. This is presumably because the goals of the game are clear and immediately understandable, even if the combinatorial implications of each move are not; it is the clarity of purpose, not method, that is important (see Section 4.3.1).

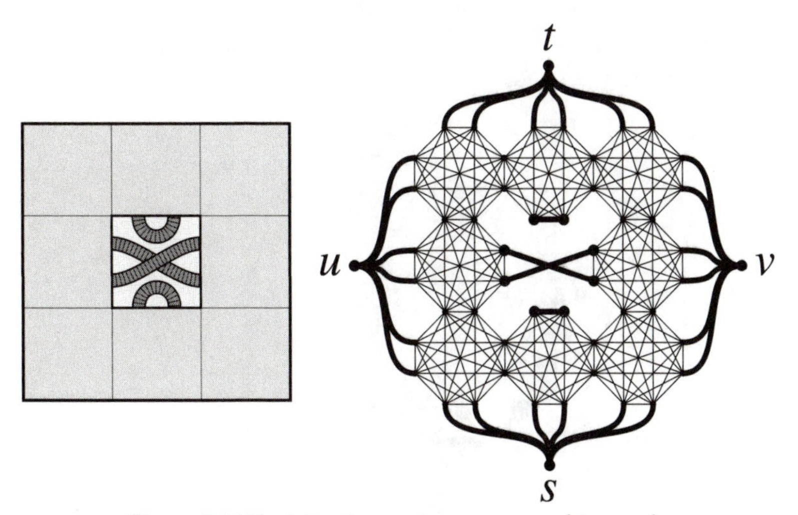

Figure 7.197. A 3 x 3 game in progress and its graph.

History

Knots was invented by Tom Jolly and published by Jolly Games in 1991.

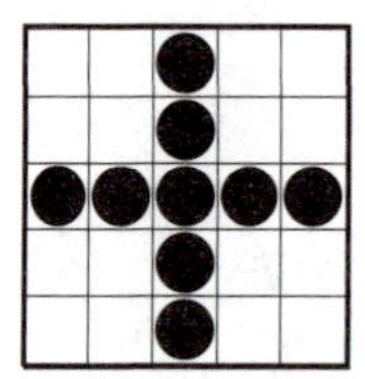

Picture Link

Picture Link is a simple bingo-style connection game designed for children.

Rules

Picture Link is played by two to four players, each controlling a hex hex board with five cells per side. The 61 cells of each board are marked with 61 different pictures (represented by 61 different numbers in Figure 7.198). Each board has the same set of pictures in a different order, and each player can see all boards at all times. All boards are initially empty.

Players take turns calling out a picture that has not already been called. All players then place a token to cover that picture on their board.

A player wins by completing a chain of tokens connecting at least one pair of opposite sides of his board. For instance, Figure 7.199 shows a game won by a player who has completed a chain between the white sides of his board. If more than one player completes a spanning path on the same turn, then those players share the win.

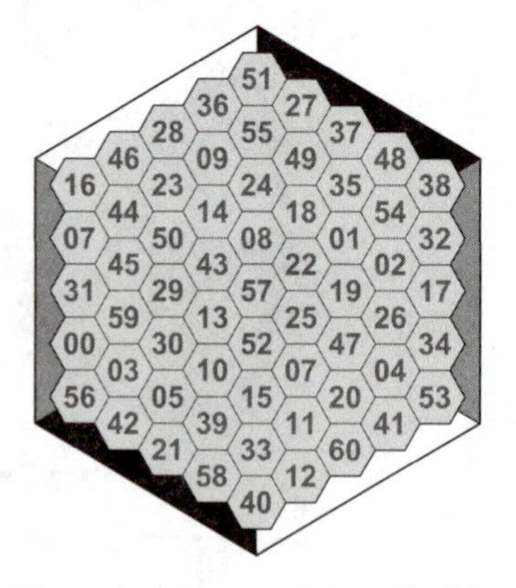

Figure 7.198. Numerical representation of a Picture Link board.

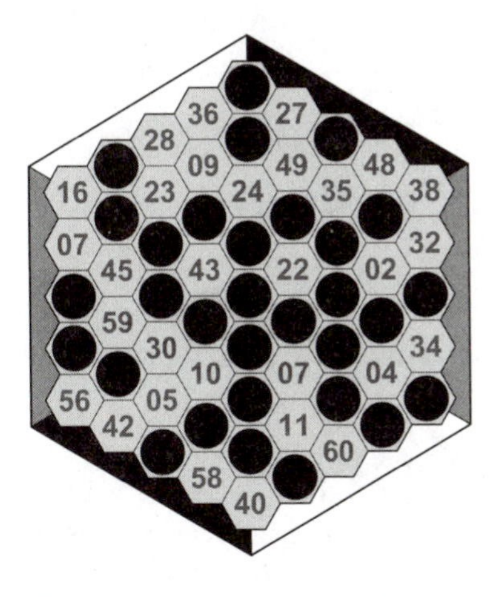

Figure 7.199. A completed game.

Notes

Picture Link is one of the few connection games designed specifically for children. The publisher's recommended age of 4 years and over appears to be well judged: reports from parents and teachers suggest that 4-year-olds may initially require some clarification of the game's objectives, but once the concept of connection is understood children tend to find Picture Link clear and enjoyable.

The fact that players can see other players' boards means that some subtle tactical plays may be formulated. The first player should always win with perfect play, but care must be taken to avoid sharing the win.

History

Picture Link was created by Theora Design and published by Thinkfun in 2002.

Variants

Hidden Picture Link

Hidden Picture Link is played as per Picture Link, except that players cannot see their opponents' boards.

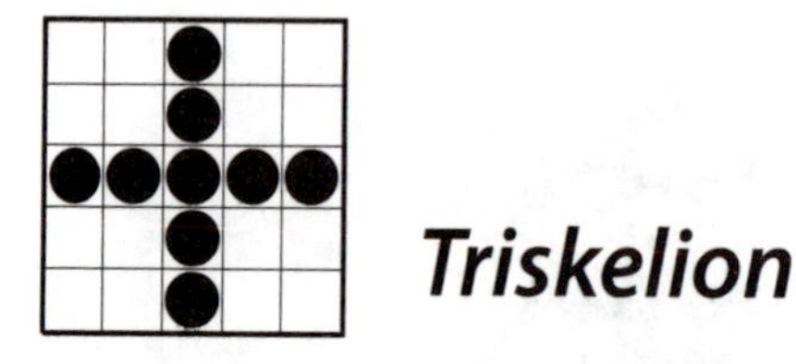

Triskelion

Triskelion is the only known Pure Connection game specifically designed for three players. It introduces the idea of phased piece placement.

Rules

Triskelion is played on a hex hex board with five cells per side. Three players, White, Gray, and Black, each own the opposite sides of the board marked their color (see Figure 7.200). The board is initially empty.

Players take turns placing a white or black piece on an empty cell. If the turn is an odd number then a white piece must be placed, otherwise a black piece must be placed; players therefore place pieces of alternating color each time it is their turn to play.

A player wins by either

- having his sides of the board connected with a chain of either color (a line), or
- forming a Y of either color between three alternating sides of the board.

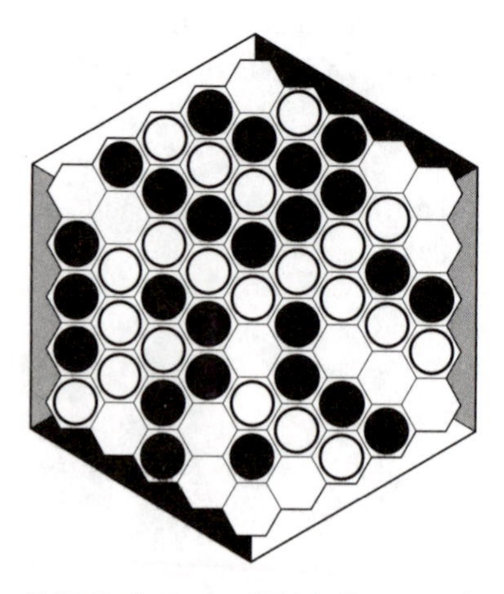

Figure 7.200. A game of Triskelion won by Gray.

If both a line and a Y are formed on the same move, then the line takes precedence. If a line is made between more than one pair of sides on the same move, then the mover wins, else the next player with connected edges in the sequence of play wins. Triskelion uses the Stop-Next rule, that is, no player can make a move that would allow the next player to win immediately if there is another choice. Exactly one player must win.

Figure 7.200 shows a game won by Gray, who has completed a white chain connecting the gray sides of the board. Corner cells belong to both players whose board sides meet there.

Notes

The Y rule avoids draws and ensures victory for exactly one player. It is included more for theoretical completeness and rarely comes into play.

Triskelion has a distinct race feel to it due to the fact that players may share the same path. When developing their strongest connection, players often help their opponents as much as themselves.

The fact that players alternate placing black and white pieces each turn means that planning ahead is difficult, and some very subtle strategies can be developed to exploit the color cycles. Triskelion shares some similarities with games such as Jade and Chameleon in which players choose the color to play each turn. However, the enforced color cycling makes Triskelion somewhat less chaotic than those games.

Frederic Maire suggests that Triskelion may be played for any m players and n piece colors where m and n are relatively prime. The key point is that players should change piece colors every turn, and should eventually play all colors. Note, however, that the game can be tied on a hexagonal tiling if more than two piece colors are used and they occupy different phase spaces (described in Appendix H).

Strategy and Tactics

It is generally better for players to establish pieces near their own edges rather than in the middle of the board. Central pieces tend to become common property, whereas pieces closer to a player's edge will help that player in any race. However, central pieces can still have a dramatic bearing on play as they determine the color and direction of paths crossing the board.

The puzzle shown in Figure 7.201 indicates the depth of Triskelion: Gray to play a white piece and win.

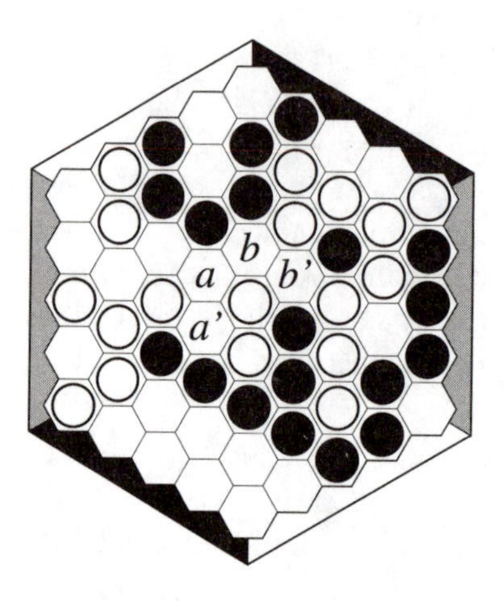

Figure 7.201. Puzzle: Gray to play a white piece and win.

Solution: Gray should burn this move by playing anywhere except at the key points ***a***, ***a'***, ***b***, and ***b'***, then play a black piece at any of these points on their next move. Proof of this solution is subtle and too detailed to include here, but invokes most aspects of the game including the Stop-Next rule.

History

Triskelion was invented by Bill Taylor in 2003.

Variants

Diskelion

Diskelion is played on a standard Hex board. Rules are the same as for standard Hex, except that players alternate piece colors each turn and win by connecting their edges with a chain of either color. Hence, if White opens then the sequence of pieces played becomes: white, black, black, white, white, black, black, etc. Diskelion was devised by Cameron Browne in 2003. The name was coined to describe a "dual Triskelion."

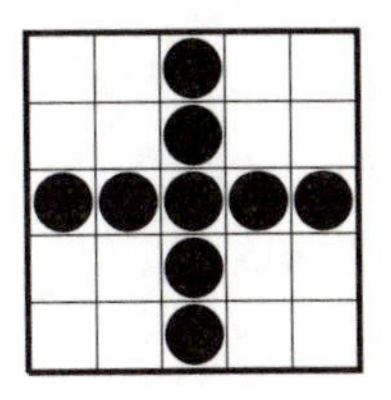

Chameleon

Chameleon is a simple but interesting color-swapping Hex variant. Carelessly placed pieces can be exploited by the opponent to turn a game on its head.

Rules

Chameleon is played on a standard Hex board (usually 11 x 11). The board is initially empty. Two players, Black and White, own alternating sides of the board that bear their color. Players take turns placing a piece of either color on the board.

The first player to connect his two sides with a chain of either color wins the game. If a move achieves winning connections for both players simultaneously, then the mover wins. Exactly one player must win.

For instance, Figure 7.202 (left) shows a game won by White, who has completed a chain of pieces between his edges; the fact that the pieces are black does not matter. Figure 7.202 (right) shows a situation in which the next player to move can win the game by playing a black piece at *x*.

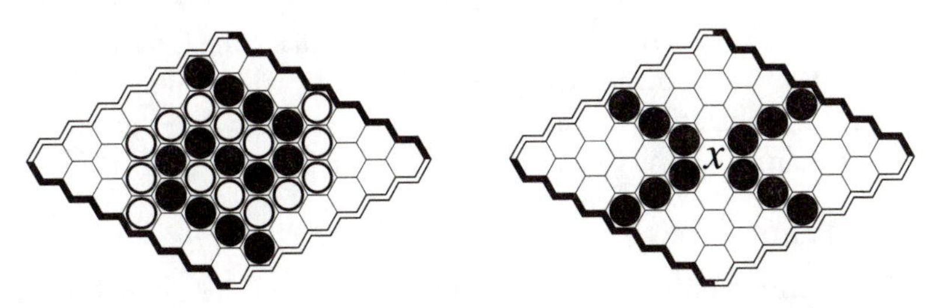

Figure 7.202. A game won by White (left) and a game about to be won by the mover (right).

Notes

Chameleon has a similar feel to Jade but is simpler, is more elegant, and has clearer goals. It also has less depth than Jade and games tend to be much shorter; one slight error and a game can devolve quickly into a race.

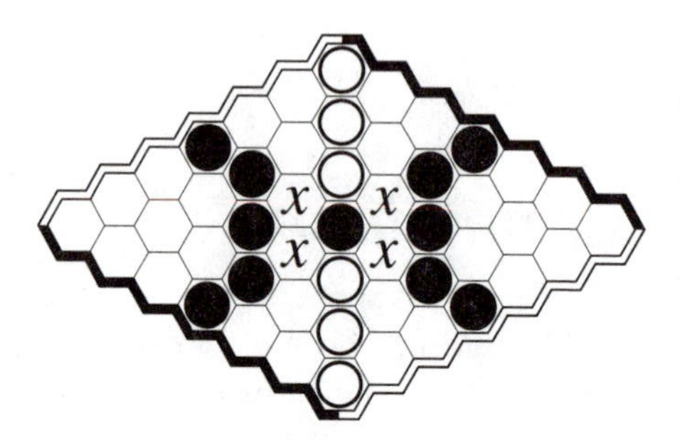

Figure 7.203. A cold war. The first player to make any move *x* loses.

Once a solid group forms in the middle of the board, a player who gets one move ahead is very difficult to stop.

Once each player commits to a particular color for his connection, then the colors effectively become fixed and the rest of the game tends to revert to a standard Hex game. In this sense, Chameleon is like an extended swap option or contract phase where players develop a position until one deems it strong enough to commit to a color. Players do not explicitly indicate the preferred color, but this should become obvious from their moves.

Like Jade, Chameleon is subject to cold wars in which players are forced to waste moves rather than make a move guaranteed to lose the game. For instance, the first player to make move *x* of any color loses the game shown in Figure 7.203. This will be the current mover, as there are an even number of harmless cells to fill before having to make the fatal move.

The Chameleon idea of using either color to achieve either player's goal is a fundamental metarule that may be applied to other games (for instance, Chameleon Y, Chameleon Trax, and so on).

Strategy and Tactics

Playing Chameleon is a constant tightrope act until the colors become fixed. In most connection games, each player can concentrate fully on pushing his connection as hard as possible. However, in Chameleon players must keep their connections strong only in their direction or risk having them stolen by the opponent. Players must think very carefully about the implications of each move.

Chameleon often presents interesting situations during regular play due to the players' ability to swap colors mid-attack, throwing the opponent's

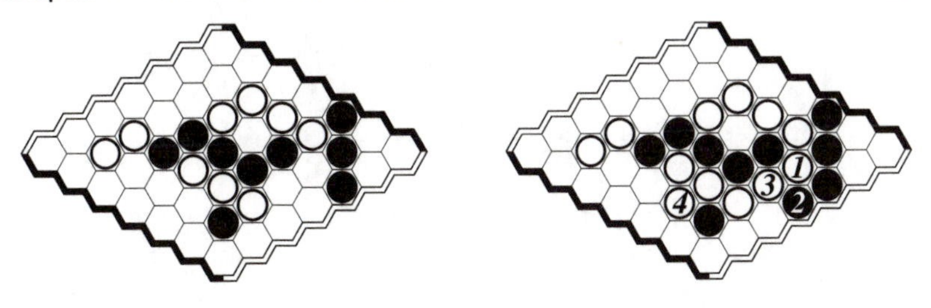

Figure 7.204. White to move and win ... but not like this.

plans into chaos. Figure 7.204 shows a case in point, derived from an actual game.

Figure 7.204 (left) shows a board position with White to move. Figure 7.204 (right) shows the obvious extension of play with White forcing a ladder with moves *1* to *3*. However, black move *4* causes a brilliant paradigm shift. Black has turned White's attack on its head and now threatens to use it to connect the black edges. In fact, Black is now two moves closer to connection than White and is guaranteed to win.

A much better play by White is shown in Figure 7.205. This time White forces a ladder along Black's edge with moves *1* to *4*, blocking this edge from the central group of white pieces. White is then safe to play the previous ladder *5* to *7* for the win.

Unlike most connection games, opening in the center is a disastrous move in Chameleon as that piece may be incorporated into the opponent's connection. The fact that having extra pieces on the board can sometimes harm a player's position means that the usual strategy-stealing argument does not apply (see Appendix D). The best opening in Chameleon appears to be well away from the center and any opponent's edge.

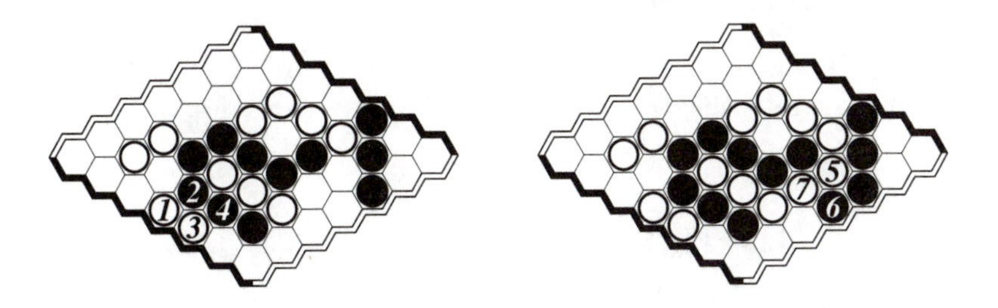

Figure 7.205. A better play by White guarantees the win.

History

Chameleon was first invented by Randy Cox on November 5, 2003. Intended as an entry in that year's Unequal Pieces Game Design Competition hosted by *Abstract Games* magazine, it was originally called Goofy Hex and later Funky Hex. The same game was independently reinvented little more than a week later by Bill Taylor (after an idea by Cameron Browne). Bill named it Chameleon due to the fact that players tend to change colors to suit their environment, and that name has stuck.

Chameleon can be seen as a more fundamental version of Jade. Once again, independent inventors have discovered a fundamental variant at almost the same time. In fact, this form is so fundamental that it is hard to imagine that it has not been implemented previously.

7.3 Pure Connection > Path Majority

Path Majority games are those in which players strive to establish a majority of connections. Such games still fall within the Pure Connection category, as majority counting is a categorical rather than a relative measure; it is still the fact of connection that counts, not the size or shape.

The icon for this group shows Black with a majority of connections.

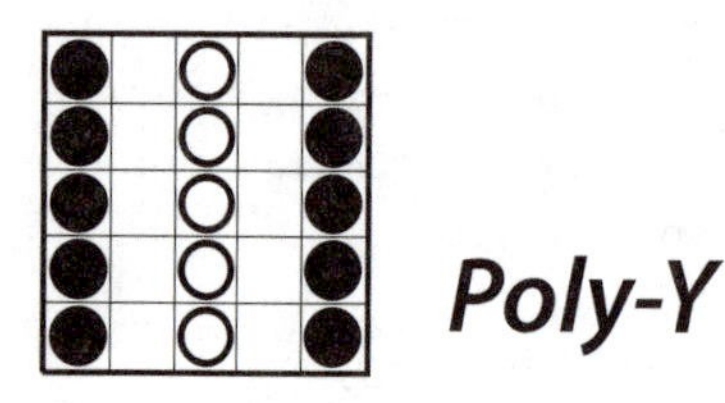

Poly-Y

Poly-Y is a remarkable connection game from the coinventor of Y.

Rules

Poly-Y can be played on a variety of board designs, one of which is shown in Figure 7.206. The board is initially empty. Two players, Black and White, take turns placing a piece of their color on an empty cell.

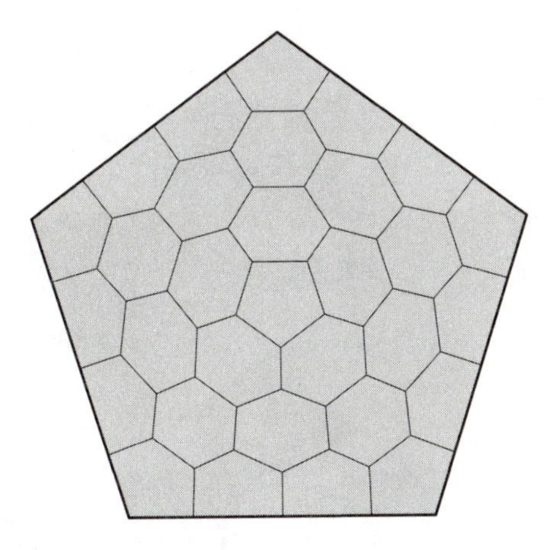

Figure 7.206. A small five-sided Poly-Y board with 31 cells.

A player claims a board corner by forming a Y with his pieces that connects the two sides adjacent to the corner with any other side of the board. A player wins by claiming more than half of the board's corners. Exactly one player must win.

Notes

Poly-Y does not have a standard board size or shape; the small five-sided board is used for illustrative purposes only. Poly-Y boards tend to have five or nine sides and display either five-fold or three-fold rotational symmetry [Schensted and Titus 1975], but Poly-Y can be played successfully on almost any board that

- has an equal number of cells along each side,
- has an odd number of sides (to avoid draws),
- has at least five sides, and
- is tessellated by mostly six-sided cells.

Figure 7.207 illustrates the concept of corner ownership. Black has claimed a majority of three or four corners to win each game. These examples are taken from Schensted and Titus [1975].

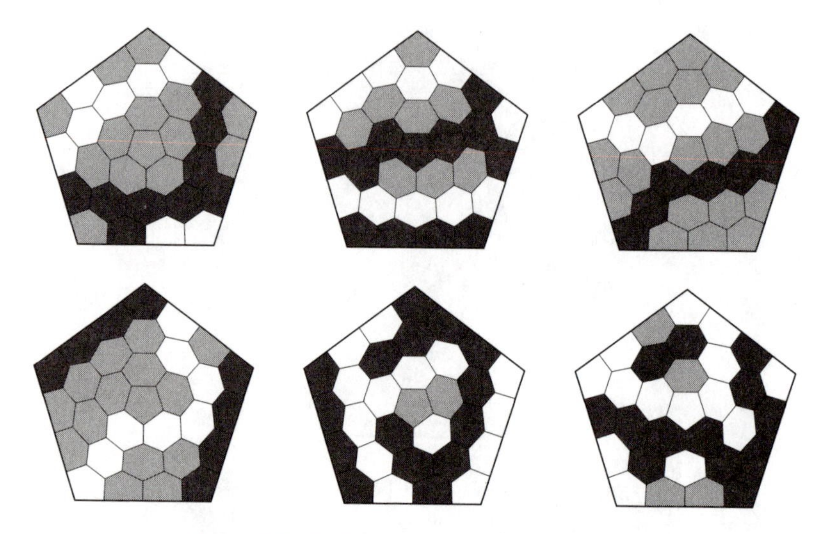

Figure 7.207. Example games won by Black.

Figure 7.208 shows that having to connect to a third side before claiming a corner is not just a nicety of the rules, but is in fact a fundamental part of the game's mechanics. Black (left) has the top corner surrounded, but it is still possible for White to surround the entire black group (middle). Requiring that Black must also connect to a third edge (right) cuts all such white connections; the black group can no longer be surrounded [Schensted and Titus 1975].

Schensted and Titus also offer an informal but elegant proof that at most one player can claim any particular corner. Consider the example shown in Figure 7.208. To claim the top corner, Black must connect the top left edge, the top right edge, and any of the three remaining edges. If

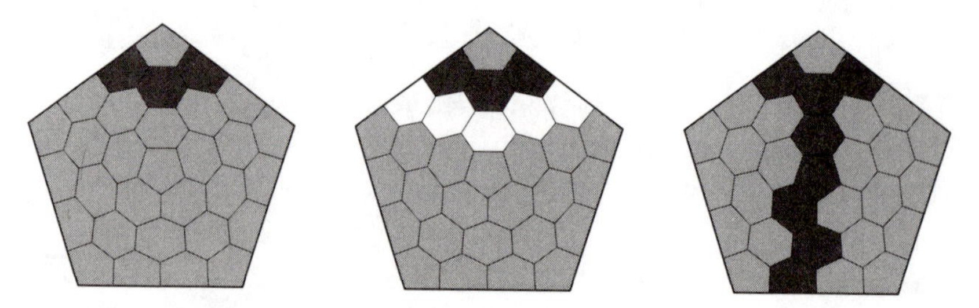

Figure 7.208. Black is vulnerable until connecting to a third edge.

the three remaining edges are considered to represent a single edge, then Black's task is equivalent to connecting the three sides in a game of Y, and only one player can achieve that (see Appendices F and G). Playing Poly-Y on a three-sided board is identical to playing a game of Y; Y is a special case of Poly-Y.

Poly-Y achieves a good balance between tactical and strategic play. It is an elegant, deep, and subtle game that deserves wider exposure than it has received so far.

History

Poly-Y was invented by Craige Schensted in 1970 [Schmittberger 1983].

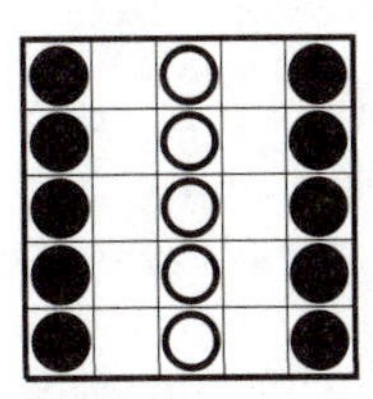

Quadrant Hex

Quadrant Hex is recursive in nature; a standard Hex game is overlaid with miniature games played by the same rules.

Rules

Quadrant Hex is played on a standard Hex board with an even number of cells per side, typically 10 x 10 or 14 x 14. The board is divided into four equal quarters, as shown in Figure 7.209. The board is initially empty.

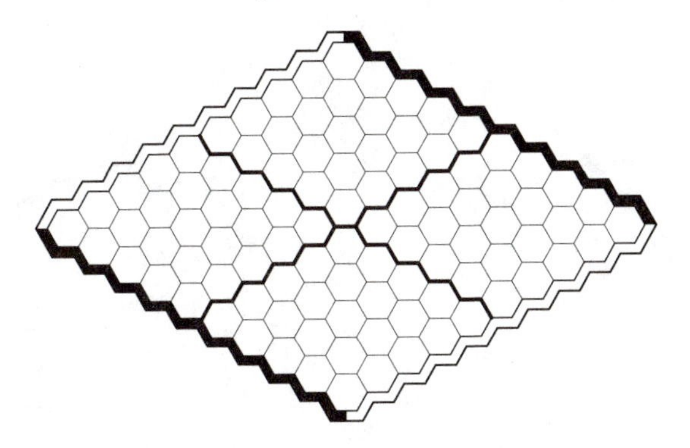

Figure 7.209. A 10 x 10 Quadrant Hex board.

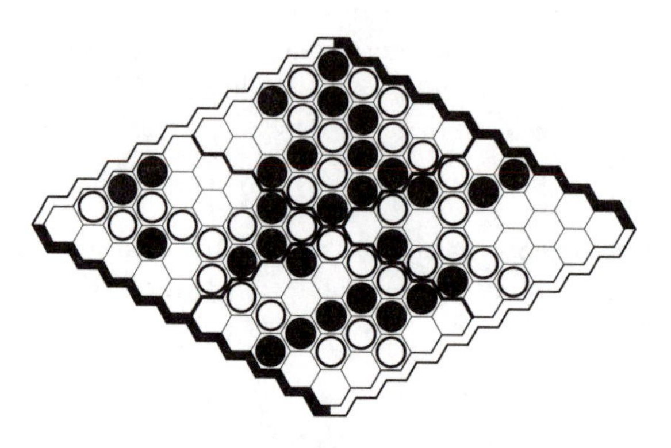

Figure 7.210. A match won by White.

Two players, Black and White, take turns placing a piece of their color on an empty cell. Players may pass. It is not permissible to mirror the opponent's moves for ten or more turns in succession. A three-move swap option is recommended.

In addition to a standard game of Hex played over the entire board, a mini game of Hex is played within each quarter-board, giving a total of five component games per match. The object of Quadrant Hex is to win at least three of these component games.

For instance, Figure 7.210 shows a completed match won by White. White has won the full-board game (Figure 7.210) plus two of the quarter-board games (Figure 7.211) for a total of three out of the five possible wins.

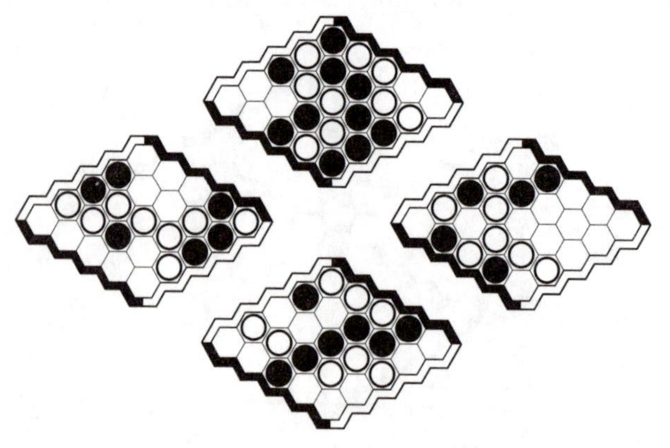

Figure 7.211. The four quarter-board games: two to White and two to Black.

Notes

Exactly one player must win each of the five component games due to the no-tie property of Hex. As there are an odd number of component games, then exactly one player must win the overall match.

While a player can win the match by winning three or more quarter-board games this does not usually occur in practice, as quarter-board connections are relatively easy to defend. More often than not, players will win two quarter-board games each and the overall match will be decided by the winner of the full-board game.

The rule forbidding a player to mirror the opponent's moves for ten or more moves in succession is presumably to avoid any point-pairing strategies (see Appendix E).

Strategy and Tactics

The recursive nature of Quadrant Hex emphasizes the distinction between local and global play more than most other abstract board games. However, these aspects integrate seamlessly and players can develop their connections on both the quarter-board level and the full-board level simultaneously. This allows very efficient and subtle moves that not only threaten multiple connections within levels, but also across levels.

History

Quadrant Hex was invented by Steven Meyers in 2000, and was originally called Quadrant Trellis [2001a].

Phil Bordelon points out that Quadrant Hex can be played on a board with an odd number of sides if the central row and column are simply excluded from the four subgames. Bordelon calls this version Framed Quadrant Hex as the central row and column act as a frame around the subboards.

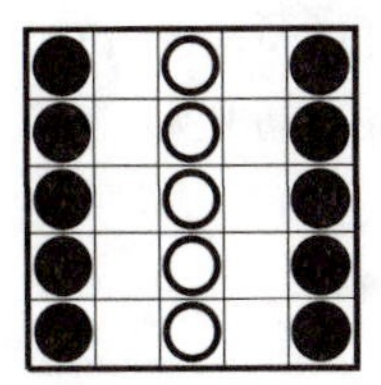

Eight-Sided Hex

Eight-Sided Hex is a Hex variant that shows how a simple change in board design can yield interesting new twists in play.

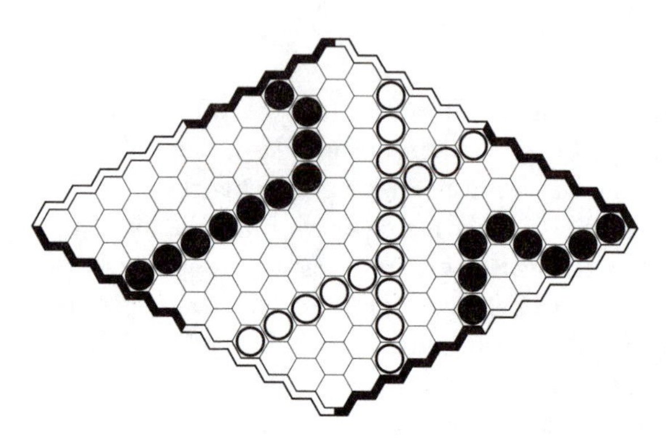

Figure 7.212. A game won by Black.

Rules

Eight-Sided Hex is played on a standard Hex board, except that the board sides are divided into alternating black and white halves (Figure 7.212). Corner cells and cells at the midpoint of odd-sized board edges belong to both players. The board is initially empty.

Players take turns placing a piece of their color on an empty cell. Eight-Sided Hex is won by the first player to connect either:

1. two pairs of his goal areas with pieces of his color, or
2. at least three of his goal areas with a single chain of his color.

For instance, Figure 7.212 shows a game won by Black, who has completed two different black chains between two different black areas. Figure 7.213 shows a game won by White, who has connected three white areas with a chain of white pieces. Exactly one player must win.

Notes

Eight-Sided Hex satisfies the important no-tie property of Hex-type games. Once a player achieves either winning condition, then the opponent can connect at most one pair of his own goal areas.

History

Eight-Sided Hex was invented by Larry Back [2001a].

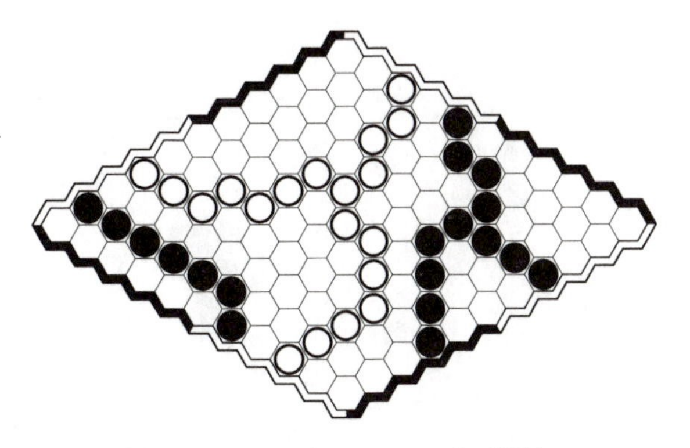

Figure 7.213. A game won by White.

7.4 Pure Connection > Cycle Making

Cycle Making games are those in which players compete to be the first to complete a *cycle* of their pieces or tiles. A cycle is a closed path enclosing some area that may be either empty or contain enemy pieces, depending on the game. Cycles are defined more precisely in Appendix A, Basic Graph Theory.

Although the cycle is a fundamental type of connection, there are few pure Cycle Making games. Most connection games involving cycles tend to include alternative winning conditions, as cycle threats can generally be defended ad infinitum unless some additional mechanism limits an opponent's escape. For example, see Havannah, Trax, Andantino, Beeline, Cylindrical Hex, Antipod, and so on. Cycle Making games are generally race games, with the notable exception of Projex.

LoopTrax, a Cycle Making game that belongs in this category, is described in the Trax variants section.

The icon for this group shows a closed path forming a cycle.

Projex

Projex features the rare case of equivalent complementary goals defined by the playing pieces as the game develops. Its rules are conceptually pure but it is difficult to play well.

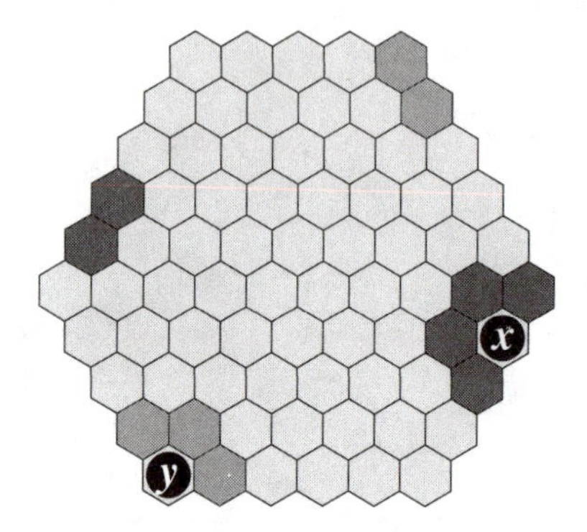

Figure 7.214. Cell x is 6-connected, and cell y is 5-connected.

Rules

Projex is played on a six-sided board tessellated by hexagons. The number of cells along each side alternates between five and six, as shown in Figure 7.214. The board is initially empty.

Each cell is adjacent to its immediate neighbors. In addition, each edge cell is also adjacent to those two edge cells whose distance around the border is the maximum possible; there is a projective connection between each edge cell and those two cells directly opposite on the far side of the board.

Figure 7.214 shows the six cells adjacent to piece x and the five cells adjacent to piece y. Projex is effectively played on a projective plane, on which connections may leave the board from one side and re-enter from the other. Each such connected entry/exit pair is called an *edge crossing*.

Two players, Black and White, take turns placing a piece of their color on an empty cell. The player to complete a *global loop* of his color wins. A global loop is a cycle of pieces with an odd number of edge crossings. Exactly one player must win; a filled board must contain a global loop, and any global loop cuts all possible global loops by the opponent. A three-move swap option is recommended.

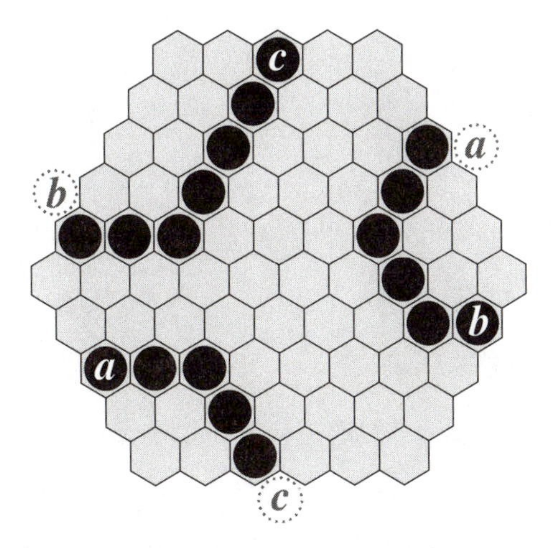

Figure 7.215. A game won by Black.

Figure 7.215 shows a game won by Black, who has completed a cycle of black pieces with three edge crossings ***a***, ***b***, and ***c***. The ghosted symbols indicate projections of edge pieces to the far side of the board.

Notes

Figure 7.216 shows two examples of winning global loops by White. The global loop on the left features one edge crossing, while the global loop on the right features two edge crossings. Black does not have any edge crossings in either case.

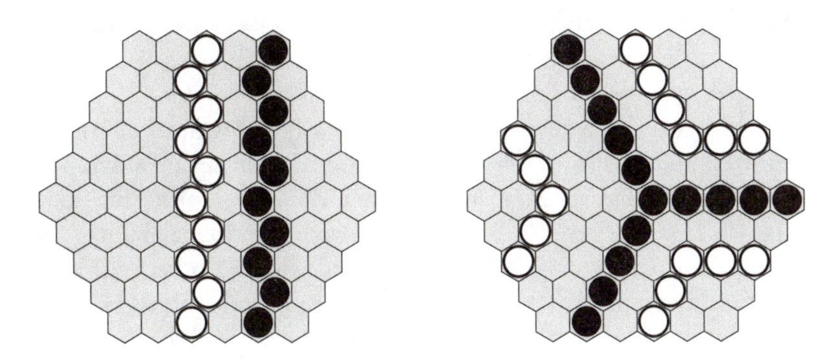

Figure 7.216. Two games won by White.

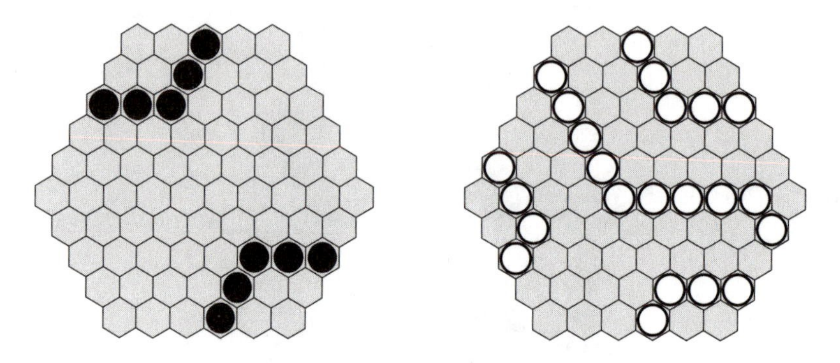

Figure 7.217. Two incomplete games.

Figure 7.217 shows two incomplete games. In each case the pieces form a loop with an even number of edge crossings (two and four, respectively), hence neither forms a global loop and the game continues.

Projex is unique in a number of ways. It is the only known connection game apart from Y and its derivatives to feature equivalent complementary goals. While it shares features common to games in both the Cycle Making and Absolute Path categories, it is the only Cycle Making game that is not a race, and the only Absolute Path game that involves cycles.

Since its rules can be stated simply (make a global loop with your pieces) and its goals are totally self-contained without the need for specific target areas, it may appear that Projex is one of the conceptually purest connection games. In practice, however, Projex is one of the hardest connection games to play well. Many players have difficulty merely visualizing global loops, let alone developing strategies to achieve them, and even experienced players can struggle to follow connections that cross multiple edges. It may help to understand that a global loop leaves the rest of the board connected, whereas a local loop does not (it has two sides).

The Projex board is reminiscent of the Star board with its alternating sides of length n and $(n - 1)$. In Star, this design guarantees an odd number of edge cells so that no game will end in a draw. In Projex, this design guarantees that every edge cell has exactly two projective neighbors.

The corner points, only being 5-connected while all other cells are 6-connected, are equivalent to pinch points in the projective plane. However, cell strengths are still sufficiently uniform across the board that having the first move is a huge advantage no matter where it is placed. A single-move swap option is therefore insufficient and the three-move swap option is recommended.

Strategy and Tactics

Most standard connection tactics appear to apply to Projex. In the absence of board edges, ladders tend to occur relative to walls of pieces or critical lines of the board.

If in doubt where to move, playing to block the opponent's best line is usually a strong play due to the complementary nature of the game's goals. If still in doubt, the most intuitive move is usually a reasonable play; winning a game of Projex without really understanding why has been known to happen.

History

Projex was invented by Bill Taylor in 1994 as a viable solution to Dan Hoey's suggestion of playing connection games on the projective plane.

Variants

Frankel [2000] describes a variety of games that may be played on the lines of a sphere tessellated by triangles. One of these games involves the completion of a cycle upon the sphere, under certain constraints of piece placement.

Ringelspiel

Ringelspiel is a simple Cycle Making game that can be played with a standard Go set.

Rules

Ringelspiel is played on a square grid, typically 19 x 19, which is initially empty. Two players, Black and White, take turns placing a piece of their color on an empty point.

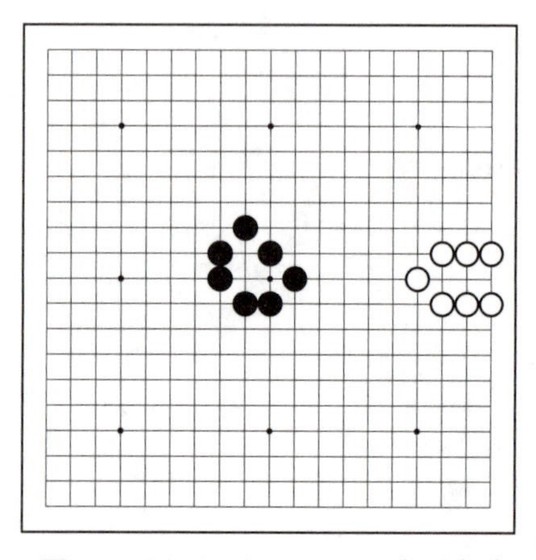

Figure 7.218. A game won by Black.

The first player to complete a cycle enclosing at least two points wins the game. The surrounding path may be composed of orthogonal and diagonal connections, and the enclosed points may be empty or occupied by enemy pieces. Board edges do not count as part of a cycle.

For instance, Figure 7.218 shows a game won by Black, who has enclosed three points with a cycle of black pieces. The white pieces do not form a cycle, but White could have completed a cycle on the next move.

Notes

Figure 7.219 (left) shows how diagonal connections allow black and white paths to cross without interference. This means that the game graph (right) is nonplanar.

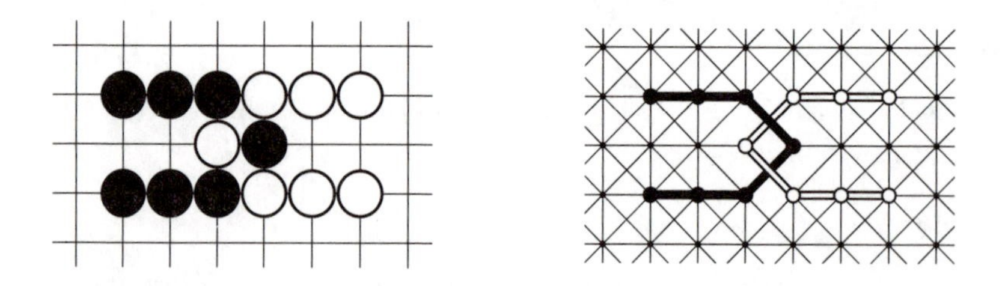

Figure 7.219. Black and white paths crossing, and the corresponding region of the game graph.

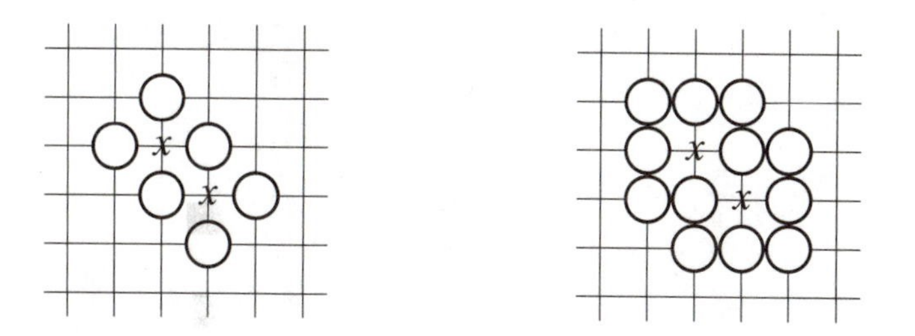

Figure 7.220. An ambiguous case and a suggested resolution.

Peter Danzenglocke points out that several Go tactics are relevant to Ringelspiel. In particular, the concept of *geta* (fencing in the opponent with a clever move) appears to be very important.

History

Ringelspiel is described briefly in Mala's *Das Große Buch der Block- und Bleistift-Spiele* [1995].

Variants

Figure 7.220 (left) shows an ambiguous case: has White enclosed both points marked x in a single winning cycle? Peter Danzenglocke suggests avoiding such ambiguity by playing the game with orthogonal connections only, in which case Figure 7.220 (left) would not be a win for White but Figure 7.220 (right) would be. Phil Bordelon calls this variant Orthospiel.

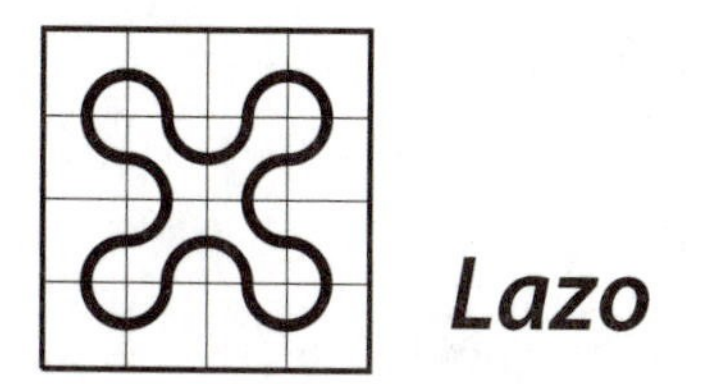

Lazo

Lazo is a tile-stacking race game that introduces the concept of loops in three dimensions. Directional tiles are used to solve phase problems with hexagonal stacking.

Figure 7.221. The Lazo board.

Rules

Lazo is played on a trapezoidal board with 57 triangular holes laid out in a triangular grid, as shown in Figure 7.221. The board is initially empty.

Two players, Black and White, each have 91 nonagonal (nine-sided) tiles of their color, as shown in Figure 7.222, in addition to 40 smaller markers of their color. Each tile has a triangular peg on the underside that fits the board holes, so that when tiles are placed on the board they always face in the same direction.

Note that three tiles placed together as in Figure 7.222 (right) form a triangular gap the same size and orientation as the board holes. Such gaps provide points at which higher-level pieces may be stacked, still facing in the same direction.

Players take turns placing a tile of their color in the playing area. The tile's peg must fit into either a board hole or the gap made by three touching tiles; if the latter, then one of the support tiles must be of the same color. In a slight variation of the usual swap, Black plays first and then White has the option of either making a normal move or swapping the opening black tile to white.

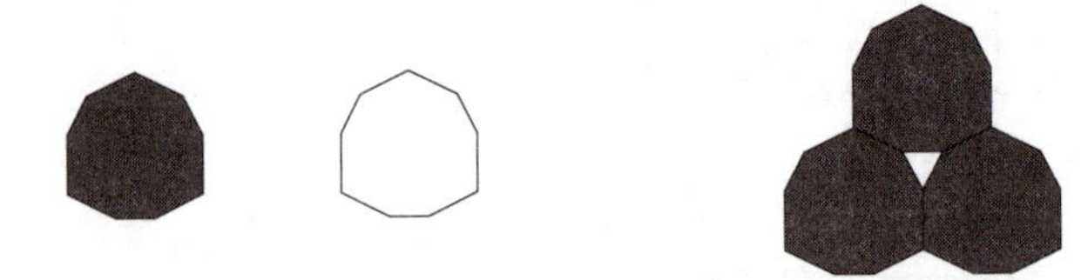

Figure 7.222. The nonagonal Lazo tiles.

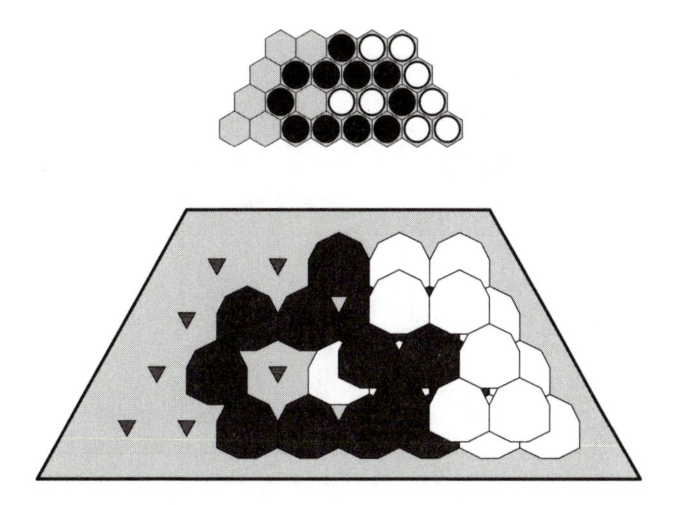

Figure 7.223. A game won by Black.

The first player to form a *loop* wins the game. A loop in this case is defined as either

- a board surface loop surrounding at least one empty space or enemy tile, or
- a three-dimensional loop (explained shortly).

If the board fills up before either player wins, then the game is a draw. A player loses if there is no legal move and the board has not yet filled up.

Notes

Figure 7.223 shows a small 26-hole game won by Black, who has completed a board surface loop. This is not immediately obvious as the winning loop is partially hidden by stacked pieces; hence it is recommended that players place a small marker corresponding to each of their board level moves on a miniature board copy to avoid any confusion (Figure 7.223, top).

A three-dimensional loop is a path of same-colored tiles forming a loop either with itself or the board, through which a path of enemy tiles or vacant spaces passes. This can be visualized as building a bridge of friendly pieces over a river of nonfriendly pieces. For instance, Figure 7.224 (left) shows a black three-dimensional loop; the black bridge connects across a river of white tiles. Figure 7.224 (right) is not a three-dimensional loop as

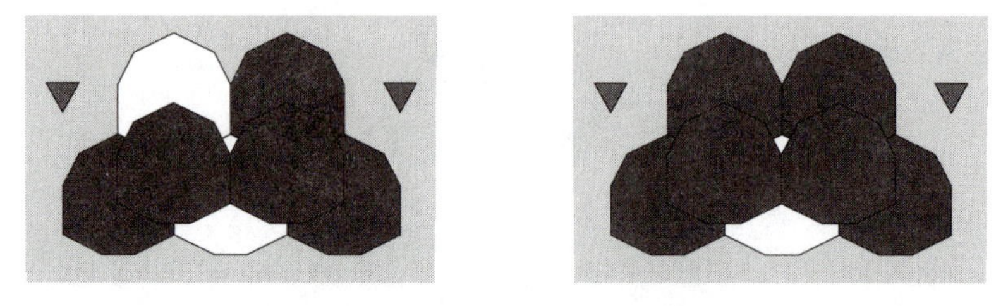

Figure 7.224. A black three-dimensional loop (left) and not a black loop (right).

the white tile forms a dead-end path, rather than a path passing under the black bridge and continuing in both directions.

Due to the connective nature of the nonagonal stacking, it is not necessary to know the color of buried pieces in order to identify a three-dimensional loop; visible pieces provide all the information necessary. David Bush provides a more complete definition of a three-dimensional loop: http://www.geocities.com/twixtplayer/Lazo1.html.

It can be seen that the nonagonal pieces result in a hexagonal packing when stacked. Lazo was developed at the same time as Akron, and employs a different technique to avoid phase problems with hexagonal stacking (described as *dead zones* by David Bush). While Akron sidestepped this problem by moving to the square grid, Lazo introduced directional pieces that ensured all pieces belonged to the one phase. Note that cuboctahedrons would work just as well as the Lazo nonagons in this respect.

Strategy and Tactics

It makes sense to spread out tiles in the early stages of the game, staking out as much influence as possible. As it is not possible to build a loop over a single enemy tile, players must look for ways to attack clumps of enemy tiles. Players must also ensure that a river of enemy tiles has room to extend in both directions when building a bridge over it.

There are essentially two ways a player can force a win:

- by a series of forcing moves that culminates in an unavoidable double threat, or
- by out-waiting the opponent in a cold war.

Games on larger boards provide more scope for connective threats, while games on smaller boards tend to be attritional battles of waiting moves.

In terms of combinatorial game theory, Lazo gets colder the smaller the board size.

History

Lazo was invented by David Bush in January 2003, and was originally called Loop. Much of the above material was derived from David's Zillions of Games help file for Lazo.

8 Connective Goal

Connective Goal games are those that end as soon as some specific connection is achieved. While still connection games, they do not belong to the Pure Connection category due to some non-connective aspect during play. Connective Goal games can be described as being connection-based at the global level rather than the local level.

8.1 Connective Goal > Path Making

Path Making (Connective Goal) games are those in which players strive to complete a path between designated goals, using either their own pieces or a shared track.

The icon for this group shows intersecting Black and White paths, and the winning move for the next player.

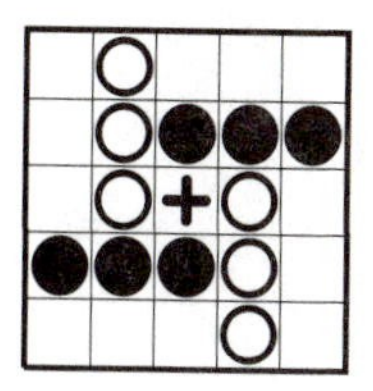

Orion: Hydra

Orion is a gaming system [Hale-Evans 2001] that introduces a mechanical element to abstract board games. The Orion board consists of a 5 x 5 square grid of *rotors*.

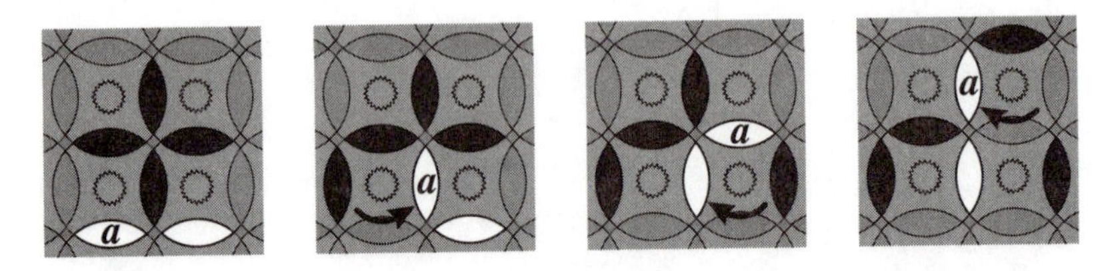

Figure 8.1. White smashes through Black's defense with a three-click move.

Pieces placed in the area of overlap between adjacent rotors are moved whenever those rotors are turned.

The Orion set comes with an official rule book describing more than a dozen different Orion games. Each move in any Orion game consists of one, two, or three *clicks* (quarter turns of a rotor) with which the player may shuffle a piece from one rotor to another, as shown in Figure 8.1.

Figure 8.1 (left) shows a position in which White wants to push piece ***a*** through the apparently solid black defense. This looks difficult, but can in fact be done in a single three-click move. The first click moves piece ***a*** counterclockwise to the rotor housing the other white piece. The second click moves both white pieces clockwise, shuffling piece ***a*** to a third rotor. The third click turns the third rotor clockwise, completing White's breakthrough and leaving Black's defense in disarray.

Note that all clicks within a turn must continue to move the same piece, in this case ***a***. Each move is therefore constrained to shuffle a single piece that may visit at most three rotors, though other pieces may also move as result.

Figure 8.2. A game of Hydra won by White.

Hydra, an Orion game with a strictly connective goal, is described in the official rules. A player wins Hydra by completing a chain of his pieces between the top and bottom sides of the board. For instance, Figure 8.2 shows a win for White. A player loses upon completing a spanning chain of the opponent's pieces, and the game is drawn if spanning chains are created for both players simultaneously.

Orion is a fascinating game system that has a dedicated following amongst abstract board game players. The multiclick moves, in conjunction with the fact that clicking a rotor can move pieces belonging to both players simultaneously, emphasize the combinatorial tactical aspect, and make long-term strategic planning difficult.

Hydra does not appear to suffer unduly from the constrained nature of piece movement in Orion. The dynamic movement caused by shuffling pieces across the board provides sufficient upheaval to keep things interesting.

Orion was published by Parker Brothers in 1971 and patented by Daniel Pierson [1972]. This patent also describes a hexagonal Orion system, a system with differently sized rotors, and a system in which rotors cannot spin unless all neighbors are aligned appropriately.

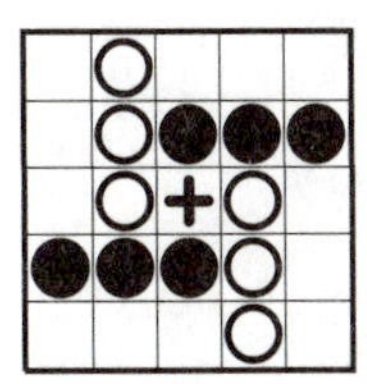

Crosstrack

Crosstrack is an attractive connection game played on the 4.8.8 tiling. Two to four players each own a number of octagonal tiles and share a common pool of square tiles. All tiles are marked with combinations of path segments. Each turn, the current player may either

- add one of his octagonal pieces to an empty octagon,
- rotate one of his octagonal pieces already on the board,
- relocate one of his octagonal pieces already on the board, or
- add one of the stock square pieces to an empty square.

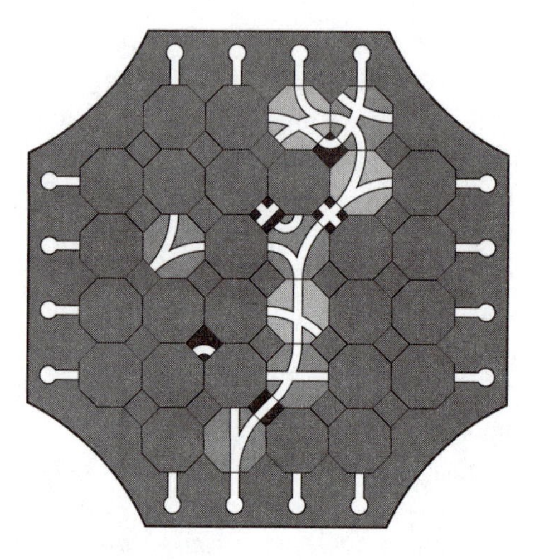

Figure 8.3. A completed game.

The game is won by the first player to complete a path connecting terminals on any two opposed board sides. For example, Figure 8.3 shows a game won by the last player to move.

Crosstrack was patented by Philip Shoptaugh [1972] and published by Shoptaugh Games in 1996. It is not a Pure Connection game as tiles may be arbitrarily moved or rotated. In addition, there is no constraint that tile paths must agree with existing tile paths on the board, further weakening the aspect of connective play.

Turn

Turn is a game of shifting goals that takes place on an 8 x 6 square grid of pegs, on which eight black gears are lined up along the top row and eight white gears are lined up along the bottom row. Each player owns a *Master Gear*, marked with a direction, which starts as the fourth gear from their left.

Players takes turns moving one of their gears to an adjacent empty peg, either orthogonally or diagonally. At the end of his turn, the current player

Figure 8.4. The Master Gears are connected but not in correct synchronization.

may rotate his Master Gear in its specified direction. The player wins if this results in the rotation of the opponent's Master Gear in its specified direction; if not, then the player loses the game.

Figure 8.4 shows a game in progress. The Master Gears are connected by a path of adjacent gears, however they are not in synchronization; turning one of the Master Gears will turn the other in the wrong direction. The current player will lose if he tries turning his Master Gear.

Turn was designed by J. B. McCarthy and published by Bütehorn Spiele Snafoo Games in 1975. Turn and Orion are good examples of the intriguing mechanical board games released in the 1970s. Turn is not a Pure Connection game as piece movement is adjacency-based rather than connection-based.

A precursor to Turn is described by Schaper in the patent "Game Successively Utilizing Selectively Positionable Gear Playing Pieces Varying in Pitch Radii" [1965]. As the rather verbose title suggests, Schaper's Game is played with gears of different sizes that may be selected by die or spinner. Gears, including each player's starting and winning gears, are not moved once placed (unless the last gear played jams the opponent's connection). The object for each player is to build a gear train that drivably connects his starting and winning gears.

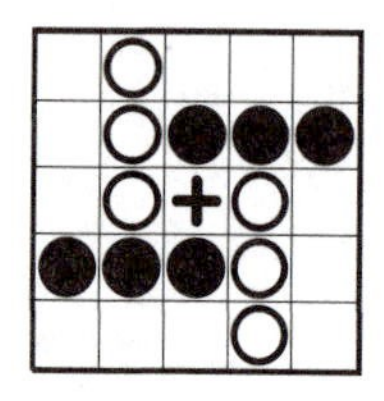

Meander (1982)

Meander is a sliding puzzle game played on a 5 x 5 square grid. All but one space is filled with *truchet tiles* (tiles with diagonally opposed quarter arcs) as shown in Figure 8.5. Players take turns sliding a tile into the empty space, or, in another version, any number of tiles along a row or column towards the empty space.

A player wins by completing a path at least three tiles long that connects the boundary to itself, as demonstrated on the right. This is the inverse of the anticonnection game Black Path, in which the goal is to force the opponent to connect the active path end with the boundary.

The constrained adjacent piece movement disqualifies Meander from the Pure Connection category. The minimum connection size of three tiles does not disqualify it from the Connective Goal category, as this is really much the same thing as the minimum connection size implied by board width, as found in most connection games.

Meander was invented by G. W. Lewthwaite and is described in *Winning Ways for Your Mathematical Plays* [Berlekamp et al. 1982].

Figure 8.5. The Meander tiles, the starting position, and a completed game.

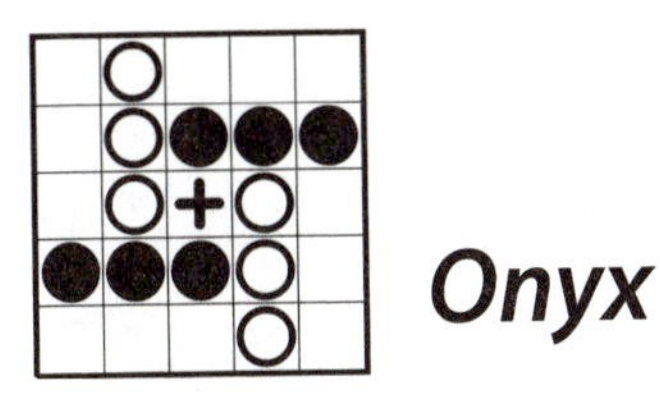

Onyx

Onyx is an innovative connection game that introduces a pattern-based capturing mechanism. The first innovation is the playing grid, which is a modified 3.3.4.3.4 tiling and the dual of the Octagons board. Two players, Black and White, own the alternating board sides that bear their color. The board is initially set up as shown in Figure 8.6.

Players take turns placing a piece of their color on an empty board intersection. Players may not take the midpoint of a square if any of the square's corners are already occupied.

The game's other innovation is that players may capture enemy pieces by completing the pattern shown in Figure 8.7. White piece ***a*** captures the two Black pieces by completing a square with an empty midpoint and all corners occupied by pieces in alternating order. Captures of multiple pairs per move are possible.

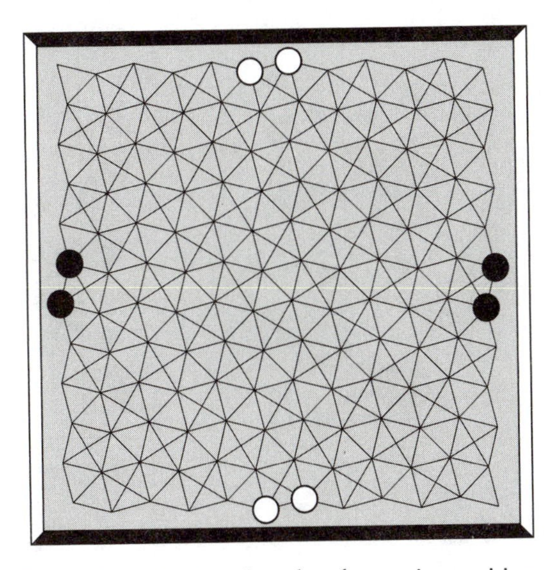

Figure 8.6. An Onyx board at the starting position.

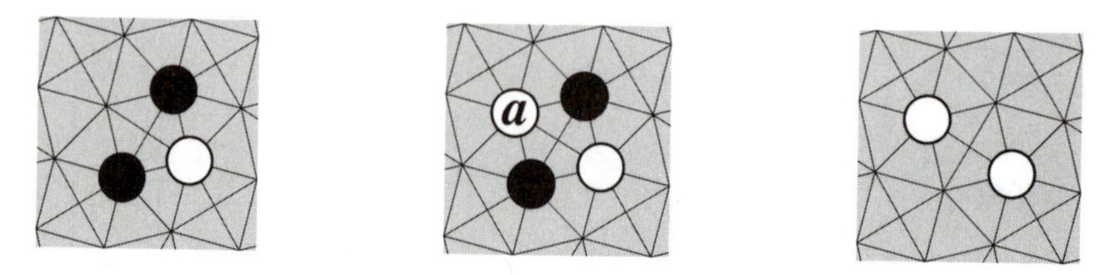

Figure 8.7. White piece *a* captures two black pieces.

The aim for each player is to connect his sides of the board with a chain of his pieces. Figure 8.8 shows an example game won by White, taken from Back [2000].

The idea for Onyx was first conceived by Larry Back in 1984, and was gradually improved over the years to the current rule set outlined above [Back 2000]. Pattern-based piece capture introduces some interesting tactics [Back 2001b], but also disqualifies Onyx from the Pure Connection category.

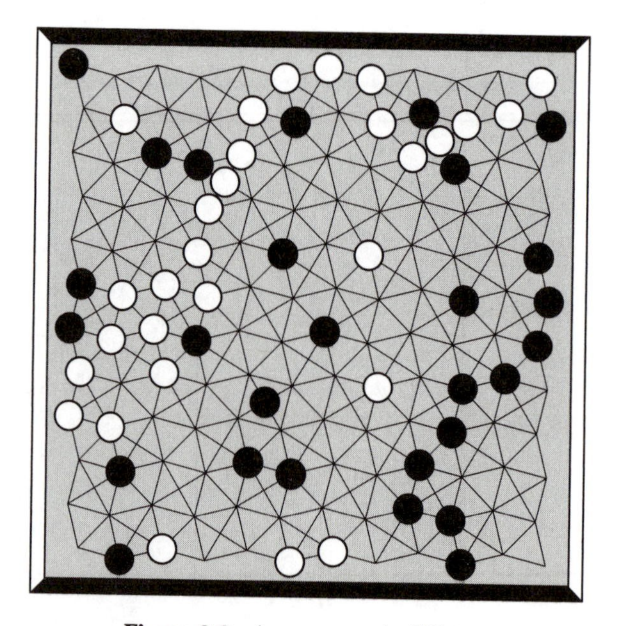

Figure 8.8. A game won by White.

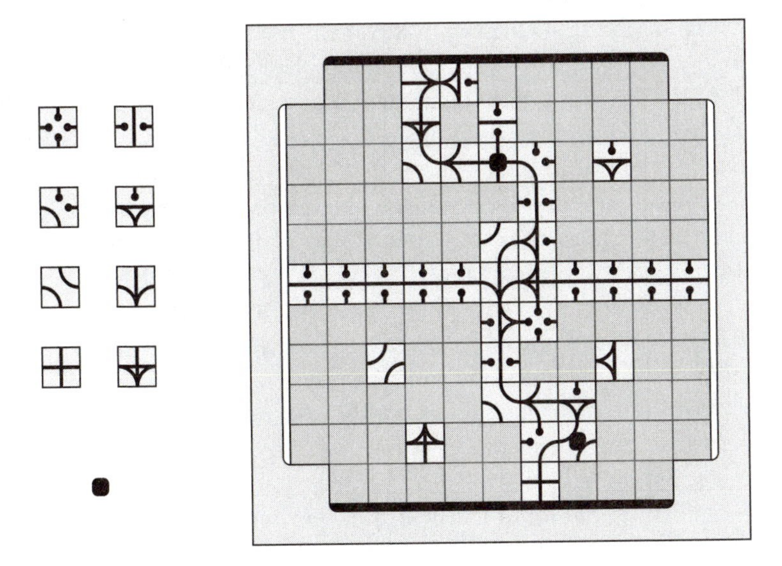

Figure 8.9. The Eynsteyn tiles and board.

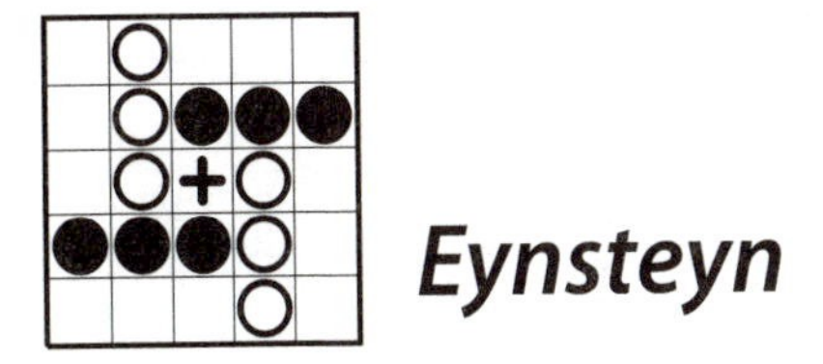

Eynsteyn

Eynsteyn is played on an 11 x 11 square grid with missing corners. Two players, Black and White, own alternating sides of the board that bear their color. The board is initially empty.

Players have access to a common pool of 113 tiles of the following types (reading down the columns of Figure 8.9): Deadend, Bow, DoubleBow, Crossing, Line, Slip, Stem, and BowCrossing. Within each tile, a path segment runs from each edge to either connect to another edge or terminate at a dead end within the tile. Players also get five *marking stones* each (bottom left of Figure 8.9).

In an initial placement phase, players randomly draw ten tiles each and play them at empty squares such that no two tiles lie orthogonally adjacent to each other. The remaining tiles are placed, face up, in five stacks. Players take turns drawing a tile until each player has five tiles in hand.

Players take turns placing a tile on the board. The tile may be placed on any empty square, or stacked on top of an existing tile surrounded on all four sides unless that tile has a marking stone on top of it, is already stacked four tiles high, or is one of the special Deadend tiles. The player may then optionally place one of their marking stones on any board tile. Marking stones may not be moved once placed. The player then chooses a tile from one of the five stacks to replenish his hand (players should always have five tiles in hand).

A player wins by completing a path between his sides of the board. Players must announce "Eynsteyn" if they can win next turn. If a move achieves a spanning path for both players at once, then the mover wins. The winning path must be smoothly continuous, and may not follow sharp corners. For instance, Figure 8.9 (right) shows a game won by Black. White has not won because the white paths from the left and right sides do not join in a smoothly continuous fashion in the middle. Marking stones are used to secure two tiles on Black's winning path.

Players should use marking stones to secure crucial tiles—this is their purpose—but use them wisely. Not only does a marking stone placement use up a valuable resource, but it permanently fixes a tile that may not always be to the player's benefit as the game progresses. Deadend tiles are rare and powerful pieces that allow strong blocking moves.

Eynsteyn was originally designed by Till Meyer as a boardless connection game in 1984. The board was added in 1986. Eynsteyn is not a Pure Connection game as tiles may be stacked and the underlying connections effectively deleted.

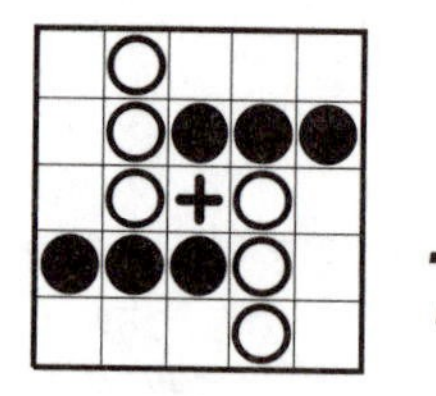

Trinidad

Trinidad is played on a hex hex board, typically with seven cells per side. The perimeter of the board is divided into four alternating black and white sections (see Figure 8.10). Two players, Black and White, own the regions that bear their color. The board is initially empty.

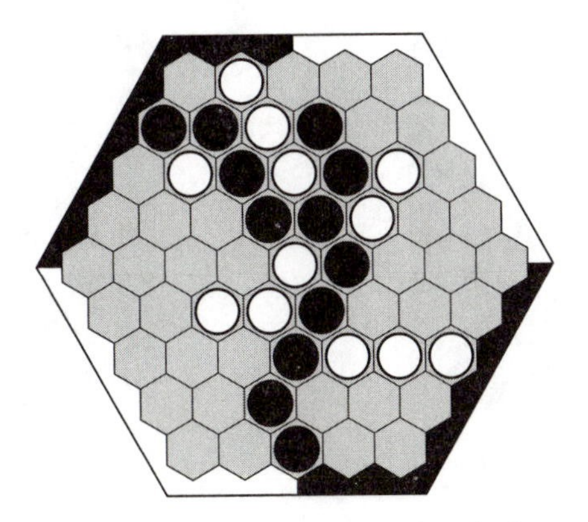

Figure 8.10. A game won by Black.

For the first three rounds, players take turns placing a piece of their color on any empty cell. Thereafter players may either

- place a piece of their color on an empty cell adjacent to an existing piece of their color, or
- move a consecutive line of their pieces a number of spaces up to and including the number of pieces in the group. The line can only move through empty cells.

The player to connect his sides of the board with a chain of his color wins. For example, Figure 8.10 shows a game won by Black. Only one player can win.

Line movement makes Trinidad a tactically rich game; however it also disqualifies Trinidad from the Pure Connection category. Another interesting feature of the game is the division of the six-sided board into four equally sized symmetrical regions. This elegant design demonstrates how a four-goal connection game can be easily adapted to the hex hex board.

Trinidad was invented by Tijs Krammer.

Proton

Proton is a 4 x 4 sliding puzzle game for two players, Light and Dark. Each of the 15 tiles is marked with Light and Dark path segments. The

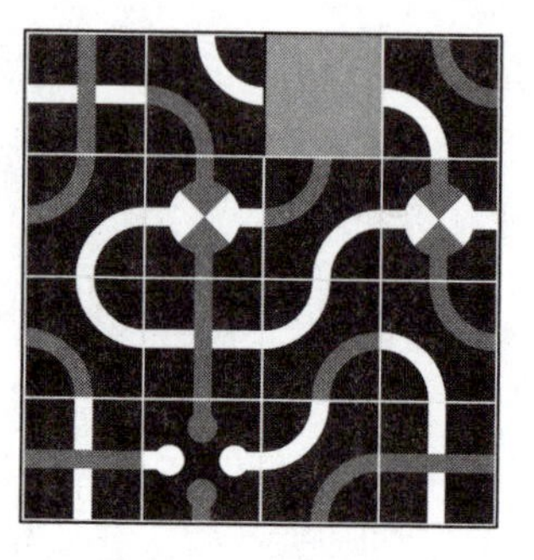

Figure 8.11. The Proton starting position (left) and a game won by Light (right).

game starts as shown in Figure 8.11 (left) with *goal tiles* in the bottom left and top right corners, a *stopper tile* in the top left corner, and the *empty square* in the bottom right corner. The remaining tiles are not in any specific order.

Players take turns sliding one, two, or three tiles along a row or column towards the empty square, effectively moving the empty square each turn. Players cannot undo the opponent's previous move.

A player wins by completing a path of his color between goals of his color on the two goal tiles. For instance, Light has won the game shown on the right. The game is tied if winning paths are completed for both players on the same move.

Proton was designed by Andrew Looney and published by Looney Labs in 1998. It is compact and elegant, and is well suited to solitaire play if no opponent is available.

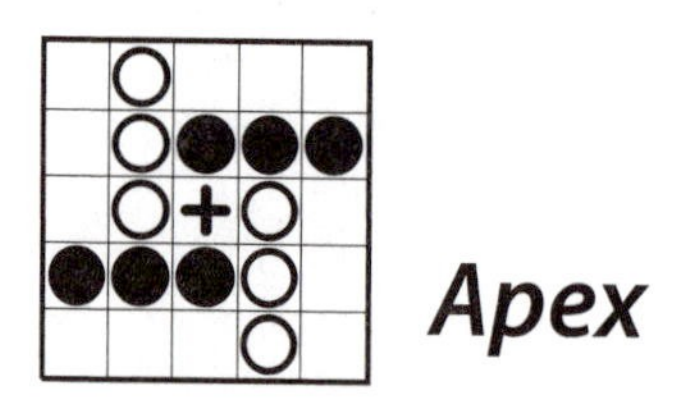

Apex

Apex is played on the rotated square grid with 84 squares shown in Figure 8.12. Two players, Black and White, start with a piece in each of the goal squares along their edges of the board.

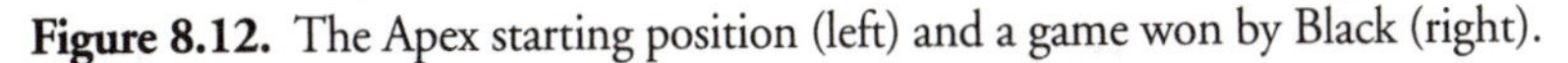

Figure 8.12. The Apex starting position (left) and a game won by Black (right).

Players take turns moving one of their pieces in a straight unobstructed line along a diagonal. The piece may land on either an empty square or, if the move did not start on a goal square and has traveled over at least one empty square, on an enemy piece to capture it. Captured pieces are immediately re-entered onto the board at an empty opponent's goal square of the capturer's choice. In no event may a piece land on an enemy's goal square.

A player wins by connecting his goal areas with a chain of his color, as shown in Figure 8.12 (right). The chain may include both orthogonal and diagonal connections, and does not have to include pieces on the goal squares; merely being adjacent to goal areas is sufficient.

Apex was designed by Jochen Drechsler in 2000.

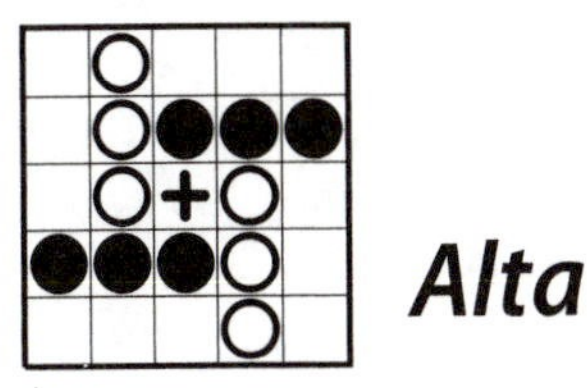

Alta

Alta is a shared path game of toggling switches, played on the board shown in Figure 8.13. Two players, Light and Dark, own the alternating board corners that bear their color. The board is initially empty.

Each player has an unlimited number of tiles called *switches* (Figure 8.13, left) marked in their color. Each switch has a single path segment between diagonally opposed corners, and can be toggled (rotated at right angles) so the path flips to the other two corners.

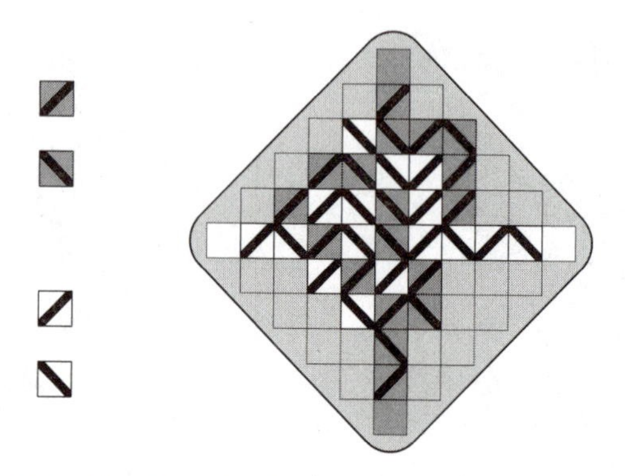

Figure 8.13. The Alta tiles (left) and a game won by Dark (right).

Players take turns either placing a switch of their color on an empty board square, or toggling one of their switches already on the board. Players cannot place a switch in any goal square or adjacent to an enemy goal square; ten squares are therefore forbidden to each player.

A player wins by establishing a path between his goals. The winning path may include tiles belonging to both players; for instance, Figure 8.13 (right) shows a game won by Dark. Draws are possible if a player blocks off an enemy goal with his switches. However, this is rare and should not occur unless the opponent is careless.

Players do not generally win using their own switches exclusively, but by creating double threats and exploiting forced connections through enemy switches. It is generally good to place as many switches as possible and avoid toggling until the end game.

Alta was designed by Dan Troyka in August 2002. It is related to the Shannon game on the edges (see Section 3.2.1). Alta is not a Pure Connection game as players may change connections in an arbitrary way by toggling switches.

Écoute-moi!

Écoute-moi! is played on a 13 x 13 square grid. The top and bottom sides of the board are colored white and the left and right sides colored black.

In addition, the top left and bottom right corners are marked "–" and the bottom left and top right corners are marked "+."

The game starts with a single neutral piece called the *Coureur* (runner) in the center. Players take turns moving the Coureur, which leaves behind it a trail of black pieces. Each move consists of the following actions:

1. **Announce:** The player nominates a number ("un" or "deux") and a direction ("blanc," "noir," "négatif," or "positif").
2. **Move:** The opponent must move the Coureur a number of spaces along the specified axis. If "un" is called then it must move one or three spaces; if "deux" then two spaces. If "blanc" is called then it must move towards either white edge; if "noir" then towards either black edge; or if "positif" or "négatif" then along the corresponding diagonal.
3. **Complement:** The announcing player must then move the Coureur using the complementary instructions. If the opponent moved one space then the player must move three spaces (and vice versa). Movement must be along the axis perpendicular to the axis originally announced.

A black piece is placed on the last square crossed by the Coureur each time it is moved. For instance, Figure 8.14 shows a possible move following the announcement "un positif." The opponent moves the Coureur (shaded gray) three spaces along the positive diagonal and places a black stone on the last square crossed (middle). The announcer then plays the complementary move of a single space in the negative direction, again placing a black stone on the Coureur's last crossed square (right).

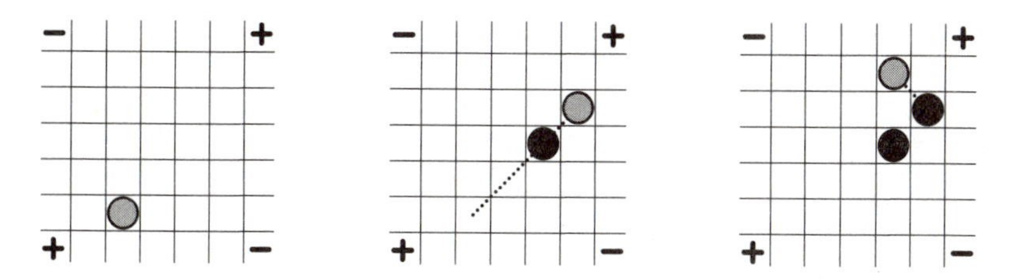

Figure 8.14. A possible move following the announcement "un positive."

Every action in a move must be performed. If any action cannot be made, then the Coureur is moved to another empty square where the action becomes possible. The game is won by the first player to establish a chain of black pieces between his sides, which may include both orthogonal and diagonal connections. If all sides are connected simultaneously then the mover wins.

Écoute-moi! was designed by Ralf Gering in 2002, and was originally called The Bridge. "Écoute-moi" is French for "listen to me."

Quintus

Quintus is played on a hex hex board and involves player-defined goals that may grow during play. The following figures show a hex hex board with six cells per side, but the official size is ten cells per side. The edge and corner cells are called the board's *Ring* (shaded in Figure 8.15) and all other cells are called the board's *Interior*. The board is initially empty.

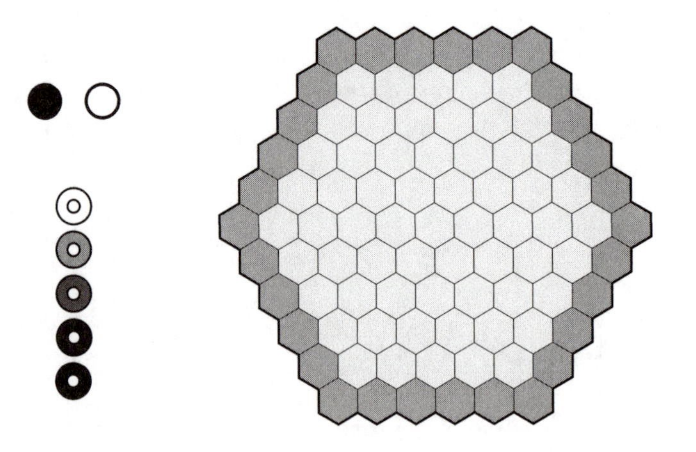

Figure 8.15. The Quintus board and pieces.

Two players, Black and White, own an unlimited supply of pieces of their color. In addition, both players share a common pool of neutral pieces of five different colors (shown with holes). The neutral colors are different than the players' colors.

Players take turns making one of the following types of move:

1. drop one of their pieces on an empty Interior cell,
2. drop a neutral piece (of a neutral color not already on the board) on an empty Ring cell, provided that there are at least five empty cells in each direction around the Ring, or
3. drop two neutral pieces (of neutral color(s) already on the board) on two empty Ring cells, provided that there is at most one chain of each neutral color after the drops.

For instance, Figure 8.16 shows a Type 2 move starting a new neutral chain (left), and a Type 3 move adding pieces to two neutral chains (right). A player wins by either

1. connecting four or five differently colored neutral chains with his pieces, or
2. forming two disjoint chains of his pieces, each connecting three differently colored neutral chains.

Figure 8.17 shows a Type 1 win for White (left), and a Type 2 win for Black (right).

Neutral pieces form player-defined goals along board edges that may grow during the game. This mechanism is reminiscent of Star, in which players must connect their edge pieces to maximize their score, but in Star this is an implied goal rather than an explicit one. Quintus is also

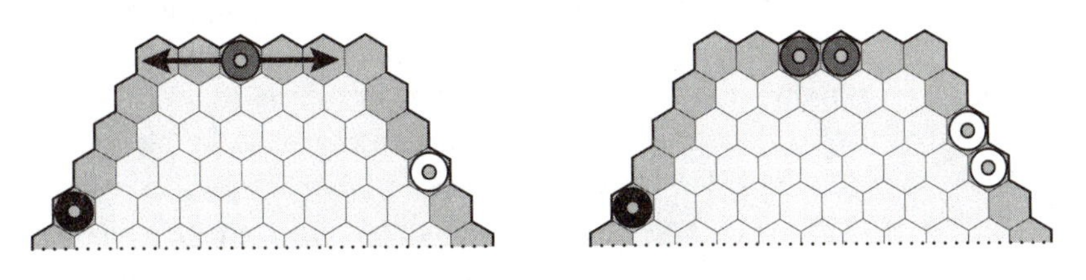

Figure 8.16. Legal goal placements.

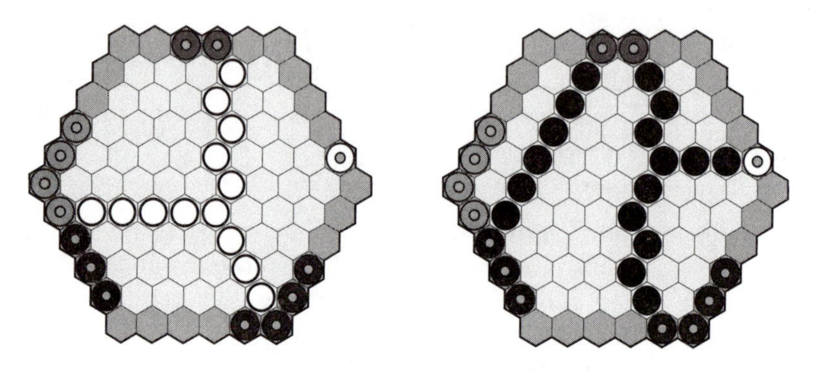

Figure 8.17. Games won by White (left) and Black (right).

reminiscent of Star in that the opposing natures of the two winning conditions lead to a balance between the importance of controlling the center and controlling the sides.

Quintus was designed by Martin Windischer in 2004, who says that the complete game was devised in one hour. Type 2 moves introduce a size constraint that disqualifies Quintus from the Pure Connection category.

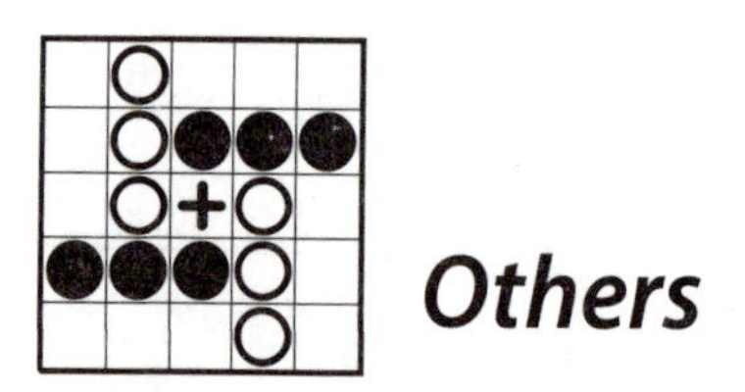

Others

McGaughey's Game

McGaughey's Game is played on an 8 x 8 square grid. Each player owns a deck of cards, each of which shows the shape, name, and number of electoral votes for a particular state in the United States of America. Players take turns randomly drawing a card and placing it on an empty square under certain restrictions. In addition, a player may place his card on top of his opponent's counterpart (showing the same state) if it has already been played. Each player has a wild card that may be placed to cover any opponent's card that has already been played.

A player wins by completing an orthogonally connected path of his cards between the top and bottom sides of the board. The player may score a number of points equal to the total sum of electoral votes along the winning path [McGaughey 1970].

Square Off

Square Off is unusual in that it is a dexterity connection game; quick movement is rewarded as well as quick thinking. It consists of two 5 x 5 sliding tile games placed side by side, each controlled by a player. Each tile is marked with a single path joining two of its sides: either a straight line across the tile joining opposite sides or a right-angled path joining adjacent sides. The sliding puzzles have the numbers 1 to 6 marked at particular points around their edges. In addition, the five cells where the two sliding puzzles meet are randomly marked with the letters A to E.

For each round the tiles are randomly placed within each puzzle, leaving a single empty space in each. One player selects an unclaimed letter, then the opponent rolls a six-sided die to determine a number. Players then modify their connections by sliding tiles towards their empty squares in the usual fashion. The first player to complete a path formed by the markings on his tiles that connects the specified letter with the specified number on his puzzle claims that letter and wins the round. The first player to win three rounds wins the match.

Square Off was designed by Alex Randolph and published by Parker Brothers in 1972. The official rules include an Oriental version, in which players take turns sliding their tiles. This removes the dexterity element from the game.

Morse's Game

Morse's Game is played on a 6 x 6 square grid with a groove along each line. Two players, Black and White, own small lengths of wood of various lengths in their color. Players take turns placing one of their pieces in a groove such that it touches at least one of their existing pieces. Players may also slide any one of their pieces along its groove as part of the move. The first player to establish a path of his color between opposite sides of the board wins [Morse 1976].

Suspension Game

Suspension Game is played by two teams of players, each controlling an empty board and a certain amount of game money. A building element is randomly selected each turn, auctioned off, and awarded to the team willing to pay the most for it. Building elements are abstract shapes with arms that link together, and are color coded for price based upon their usefulness.

The first building element is attached to a starting point on the owner's board, then each subsequent element is attached to an existing element so as to extend the connection. The first team to reach a target point on the far side of the board with its connection wins the game. The target point may be elevated above board level [Richards 1978].

Link, Tack, and Fan

Link, Tack, and Fan are three related metarules specifically designed for connection games (Hex in particular). In each case, a piece is placed followed by an optional move or capture.

Link: After placing a piece of his color on an empty board point, the player may move a consecutive orthogonal line of friendly pieces a number of spaces up to and including its length. The group may push a single smaller line of opponent's pieces, which does not have to be immediately adjacent. Pieces cannot be pushed off the board. Variations on this idea can be found in games such as Trinidad and Abalone.

Tack: After placing a piece of his color on an empty board point, the player then flips any adjacent line of enemy pieces with a friendly piece at each end to his color. This is equivalent to the capturing mechanism in Reversi, and the piece-surround capture of Alak (one-dimensional Go).

Fan: After placing a piece ***a*** of his color on an empty board point, the player may move a different piece ***b*** of his color one adjacent space to capture one or more enemy pieces. All enemy pieces adjacent to the piece ***b*** both before and after the move are captured and replaced with friendly pieces.

Link, Tack, and Fan were proposed by Michael How in the *GAMES* magazine article "Beyond Hex" [1984]. These three metarules do not embody strictly connective play, but may involve strictly connective goals depending on the games to which they are applied.

Tara

Tara is played on a 5 x 5 square grid by two players, Black and White. Black starts with a piece in the top left corner. Each turn, White inserts a piece onto the board from the left edge, and Black inserts a piece onto the board from the top edge. Inserted pieces may push existing pieces one space, and any piece pushed off the far edge of the board is returned to its owner; if this is an opponent's piece, then the opponent cannot use that row or column next turn.

Black wins by establishing a chain of black pieces between the top and bottom sides of the board, and White wins by establishing a chain of white pieces between the left and right sides of the board. Pieces connect both orthogonally and diagonally. Tara was designed by Wil Dijkstra and Ben van Dijk.

Bez's Game

Bez's Game is played on a tapered grid of squares. The grid is 25 units along its base, nine units along its top, and 25 units high. Each player owns a starting square along the bottom of the board, a goal area at the top of the board, and a number of square tiles of his color, each marked with a letter.

Players randomly draw nine of their tiles to start the game, and place them on a rack hidden from any opponent's view. Each turn, the current player must either form a valid English word or pass. Words are formed by placing tiles in a consecutive line on the board. The player's first word must start at his starting point, and all subsequent words must include at least one tile from the player's previous word. The rules do not state whether words formed as by-products of a move must also be valid English words. Players replenish their hand after each move so that nine tiles are always maintained.

The first player to complete a path of tiles connecting his starting point with his goal area wins the game [Bez 1991].

Troll

Troll is played by two players who own alternating sides of an 8 x 8 square grid. Players take turns placing a piece of their color on an empty board point. All opponent pieces in an orthogonal or diagonal line between the placed piece and another friendly piece flip to friendly pieces. A player wins by establishing a chain of his pieces connecting his sides of the board. Troll was designed by Jean-Claude Rosa in 1993.

Panda's Game

Panda's Game, played on a hex hex board with nine cells per side, is like hexagonal Scrabble with a connective goal. Two or three players each own opposite pairs of board sides, and share a common pool of hexagonal tiles, each marked with a letter. The board is initially empty.

Players randomly draw five tiles to start the game. Each turn, the current player must either form a valid English word or pass. Words are formed by placing tiles in a consecutive line on the board. The player's first word must start at his home edge, and all subsequent words must include at least one tile already on the board. Peripheral words formed as by-products of a move must either contain all vowels, all consonants, or be legal English words themselves. Players replenish their hand after each move so that five tiles are always maintained.

The first player to complete a path of tiles connecting his opposed board sides wins the game [Panda 1994].

Sisimizi

Sisimizi is a game in which players strive to connect multiple goals that are defined by playing pieces as the game progresses. Two to four players each own a number of ants and anthills of their color. Each turn, the current player may add to the board one of his anthills and some of his ants (subject to certain restrictions) or move pieces already on the board. Players may perform a limited number of special moves that form a bridge over an opponent's blocking connection.

A player wins by connecting a certain number of his anthills on different regions of the board with a single chain formed by his pieces. The fact that players may move anthills means that Sisimizi is a game of shifting goals. Sisimizi was designed by Alex Randolph and published by Editrice Giochi in 1996.

Kage

Kage is played on an 8 x 8 square grid. Two players each own a piece called the *Bird* and have access to a pool of bars that fit one edge of a board square. Each turn, the current player may either move his Bird to an adjacent empty square or place a bar on an empty line between two board squares. The Bird cannot cross a bar. A player wins by enclosing the opponent's Bird with a cycle of bars.

Kage was designed by Jay Myers and published by DMR Games in 1998. It has a Cycle Making goal.

Split (1999)

Split successfully combines playing cards with a connection theme. It is played on a 10 x 10 square grid, with players owning the alternating sides of the board that bear their color. Players match the cards in their hand with cards placed alongside the board to decide where they may place pieces on the board. The strength of their move is based on the strength of the match. A player wins by completing a chain of his pieces between his sides of the board. Split was published by Hasbro in 1999.

Square Board Connect

Square Board Connect is played on a square grid, typically 8 x 8, 10 x 10, or 12 x 12. Black owns the top and bottom sides, and White owns the left and right sides. The board is initially empty.

Players take turns placing either one, two, or three pieces of their color on empty board cells. All pieces played must be placed in a connected orthogonal line. The first player to connect his sides of the board with a chain of his pieces wins (both orthogonal and diagonal connections are allowed in the winning chain). No more than two pieces can be played on the opening move.

Square Board Connect was designed by Roger Cooper in 2000 [Handscomb 2001b]. The pattern-based piece placement (orthogonal line) disqualifies Square Board Connect from the Pure Connection category. A variation on this idea can be found in the subsequent game Stymie.

Notwos

Notwos is played on a square 8 x 8 grid. Two players, Black and White, own the alternating sides of the board that bear their color. Players take turns either placing a piece, building a stack of their pieces, or distributing a stack of their pieces, according to certain restrictions. A player wins by connecting his sides of the board with a chain of his pieces. A player loses if he connects the opponent's edges with a chain of his pieces, even if he achieves his own connection on the same move (Mover Loses). Notwos was designed by Vincent Everaert in 2000.

Creeper

Creeper is played on a 6 x 6 grid of octagons that form a 4.8.8 tiling. Two players each own alternating corner octagons around the board and control eight pawns. Pawns start on squares near their owner's home corner. Each turn, the current player may move a pawn either

- from square to square,
- to jump over an adjacent enemy pawn to capture it, or
- to jump over an octagon to claim it for the player.

A player wins by establishing a chain of octagons between his diagonally opposed corners.

Creeper was designed by Graham Lipscomb in 1984 and published by Out of the Box Games in 2004. It bears some similarity to Conhex and Caeth in that players must indirectly fight for the key locations that eventually decide the game, but these similarities are only superficial. Octagons may change hands many times per game

8.2 Connective Goal > Convergent

Convergent games are those in which a player wins by converging his pieces into a single connected group. This is a small class of games with one very influential member, Lines of Action.

It is interesting to note that there are no Pure Connection games with a convergent goal, even though this is a fundamental type of connection. This may reflect the fact that Convergent games begin in a minimally connected state (by their very nature) and hence offer little opportunity for connection-based movement until their later stages.

The icon for this group shows a set of pieces that have converged to a single connected group.

Octiles: Team Up

Team Up is an official variant of the game Octiles that fits neatly into the Convergent group. The Octiles board, shown in Figure 8.18, consists of a partial 4.8.8 tiling. Two players, Black and White, share a common pool of octagonal tiles marked with paths. In addition, each player owns five pegs

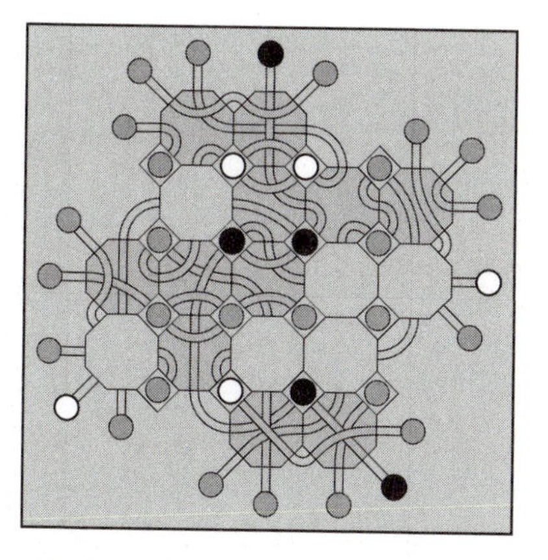

Figure 8.18. A game won by Black.

of his color (called *runners*) that fit into round holes found in each square cell and around the perimeter of the board. The game starts with all ten pegs spaced evenly and alternately around the board in perimeter holes.

Players take turns selecting a random tile from the pool, then either playing that tile on an empty octagonal cell or swapping it for an octagonal tile already on the board. A tile cannot be replaced if that move would change a path connecting two or more of the opponent's runners. After the tile move, the player must then move one of his runners along a connected path. The runner must stop at the first hole it encounters.

The game is won by the first player to establish connected paths between all of his runners. If a tile placement achieves this, then the player can claim the win without having to move a runner.

Octiles was designed by Dale Walton. It was originally published by Kadon in 1984 and later by Out of the Box Publishing.

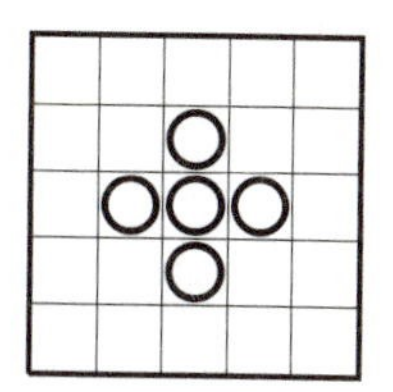

Orion: Banana Boat

Banana Boat is a game for two players using the Orion game system. See the Orion: Hydra section for more information on Orion. Twenty pieces (four differently shaded sets of five) are set up as shown in Figure 8.19.

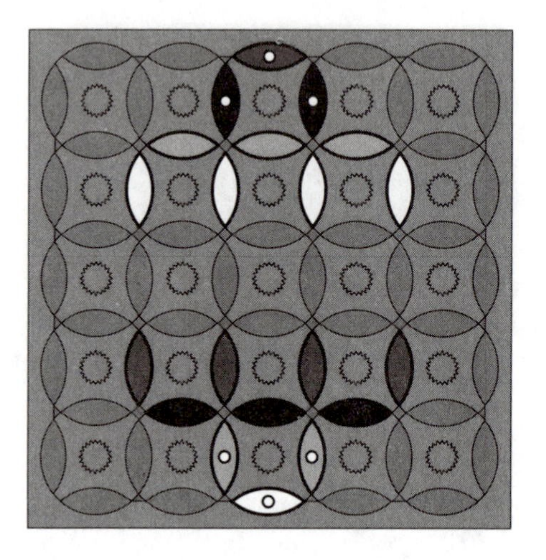

Figure 8.19. The Banana Boat starting position.

The pieces marked with dots in the diagrams are indicated on the actual board by turning them upside down. Each player owns the three marked pieces closest to them.

Players take turns making three-click moves. Players can only directly shuffle their own marked pieces, however, other pieces may be shuffled as a side effect of a move. The first player to create a connected group of five same-colored pieces wins the game.

Banana Boat was invented by Steve Wilson and described by Michael Keller [1984].

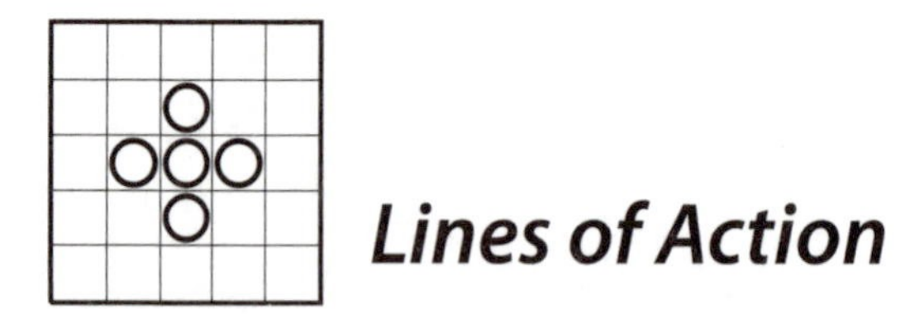

Lines of Action

Lines of Action, the classic game of convergence, is played on an 8 x 8 square grid. Two players, Black and White, start with 12 pieces each, set up as shown in Figure 8.20.

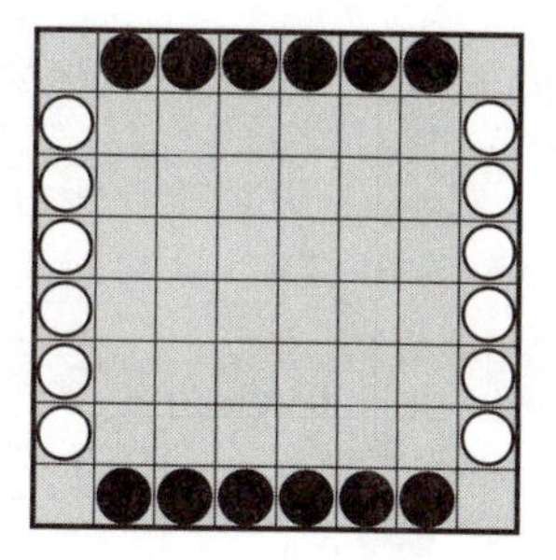

Figure 8.20. The Lines of Action starting position.

Players take turns moving one of their pieces in an orthogonal or diagonal line. The piece must move exactly the same number of squares as there are pieces of either color along that line. The piece may jump over friendly pieces but not enemy pieces, and may not leave the board. The piece may land on an enemy piece to capture it.

The game is won by the first player to move all of his remaining pieces into a single connected group. Connections within the group may be orthogonal or diagonal. If a capturing move creates single connected groups for both players simultaneously, then the mover wins.

For instance, Figure 8.21 (left) shows a winning move by White. The white piece must move three squares as there are three pieces along the line of travel. The white piece jumps another white piece and lands on a black piece to capture it, resulting in a single connected white group that wins the game (right).

Lines of Action is widely regarded as a game of the highest quality, achieving deep tactical and strategic possibilities with a simple and interesting move mechanism. It remained something of a cult game until the 1990s when it started enjoying a wider audience (much like Hex).

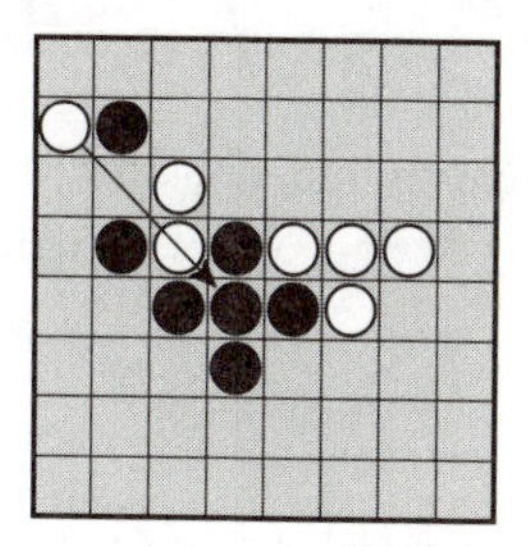

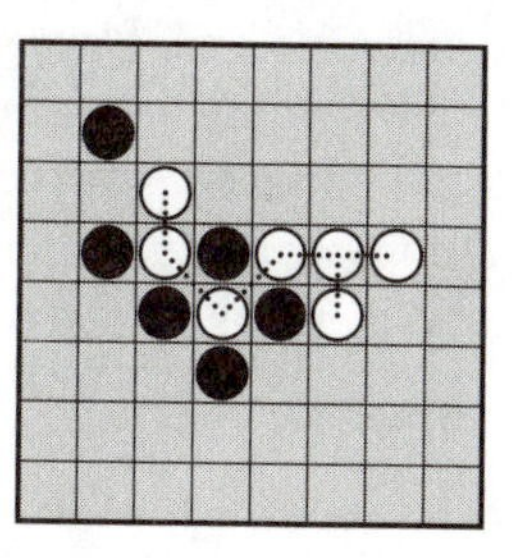

Figure 8.21. A winning move by White.

One of the fascinating features of Lines of Action is that capturing an opponent's piece can harm a player's chances as much as improve them, as the opponent then has one less piece to connect to achieve his goal. This intriguing balancing mechanism, also found in Moloko, is ideally suited to Convergent games.

Lines of Action was invented by Claude Soucie and is described in *A Gamut of Games* [Sackson 1969]. It has inspired a large number of variants, a few of which are listed below.

Neighbors

Neighbors is played as per Lines of Action except that

- players may jump over pieces of either color, and
- move length is based on the number of adjacent neighbors (both orthogonal and diagonal) before the move.

Neighbors was designed by Matthew Montchalin and described in the April 1995 issue of *GAMES & PUZZLES* magazine.

Twirls of Action

Twirls of Action is similar to Lines of Action, except that pieces move in an arc around an enemy pivot stone to another axial cell, such that the distance between the piece and the pivot is maintained. A player wins by forming a single connected group with his stones in which no stone has more than three orthogonal or diagonal neighbors. Designed by Claude Chaunier and João Neto in 2001.

Slides of Action

Slides of Action is a combination of Lines of Action and Proton. It is a 4 x 4 sliding puzzle game with 15 tiles in three sets (five of each color). Each player chooses a color; if there are only two players then one of the tile sets remains neutral. Players takes turns moving the empty space to a new location by sliding lines of one, two, or three tiles into it. Players may not undo the opponent's last move. A player wins by moving all of his tiles into a single connected group, as per Lines of Action. Slides of Action was invented by Clark Rodeffer in 2003.

Zen L'initié

Zen L'initié is identical to Lines of Action except for a different starting position (see http://www.di.fc.ul.pt/~jpn/gv/zen.htm for details) and a special neutral piece called the Zen. The current player may move the Zen instead of one of his own pieces each turn. The Zen moves and captures as per the other pieces, but cannot itself be captured. Zen L'initié was designed by Jean Rousset and E. Deluard in 1997.

Grupo

Grupo is identical to Lines of Action except that the four corner cells are removed from the game. Grupo is Joris game #9 [2002].

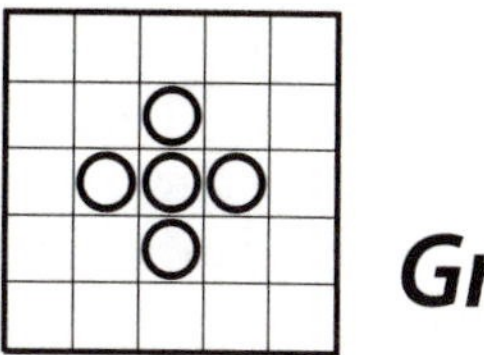

Groups

Groups is based on Lines of Action, but is sufficiently different to merit its own entry. Groups is played on an 8 x 8 square grid. Two players, Black and White, start with six pieces each, set up as shown in Figure 8.22.

Players take turns either

- moving one of their pieces to an adjacent empty square, or
- jumping one of their pieces over an adjacent piece to the empty point immediately beyond.

The game is won by the first player to move all six of his pieces into a single orthogonally connected group.

Groups was designed by Richard Hutnik in 1998.

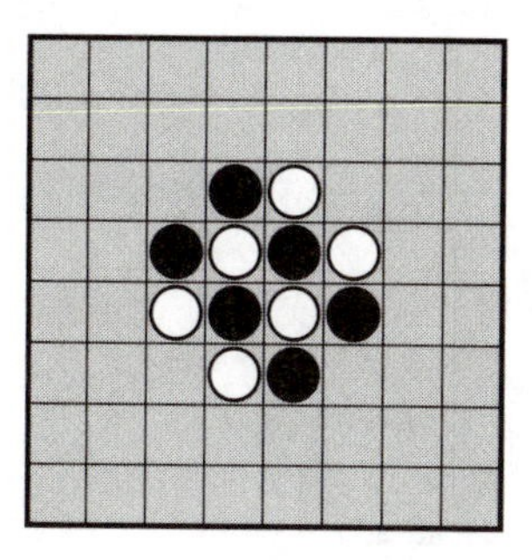

Figure 8.22. The Groups starting position.

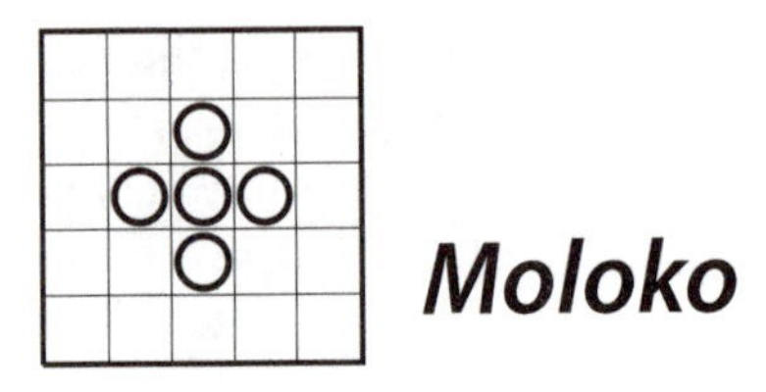

Moloko

Moloko, a connection game that incorporates elements of Backgammon, is played on a 9 x 9 square grid with two walls that form an S-shaped corridor. Players start with their pieces distributed as shown in Figure 8.23 (left).

Figure 8.23 (right) shows White's basic direction of forward movement for each of the nine subareas of the board defined by turns in the corridor. A piece in a particular subarea may only move in that subarea's direction, unless jumping a wall (explained shortly). Rotate this map 180 degrees for Black's basic direction of movement.

White starts with a single four-sided die roll; thereafter players roll two four-sided dice per move. As in Backgammon, players move one or more of their pieces a number of points according to the pips shown on the dice. A double entitles the player to use twice the number of pips shown, and players must use the maximum number of pips each turn.

Pieces may either *step* directly forward or *jump* over a wall. If a piece steps directly forward, then it must step in a straight line a number of points equal to the number of pips shown on one of the dice. If the piece steps onto a single enemy piece, then that enemy piece is pinned until the player moves off it, as in the Backgammon variant Plakoto. Players may stack as many pieces as they like onto any point, but cannot move to any point blocked by an enemy stack of two or more pieces.

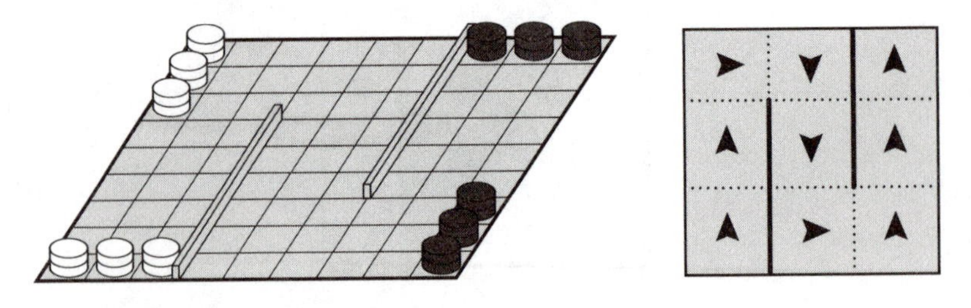

Figure 8.23. Initial board set-up, and White's direction of travel.

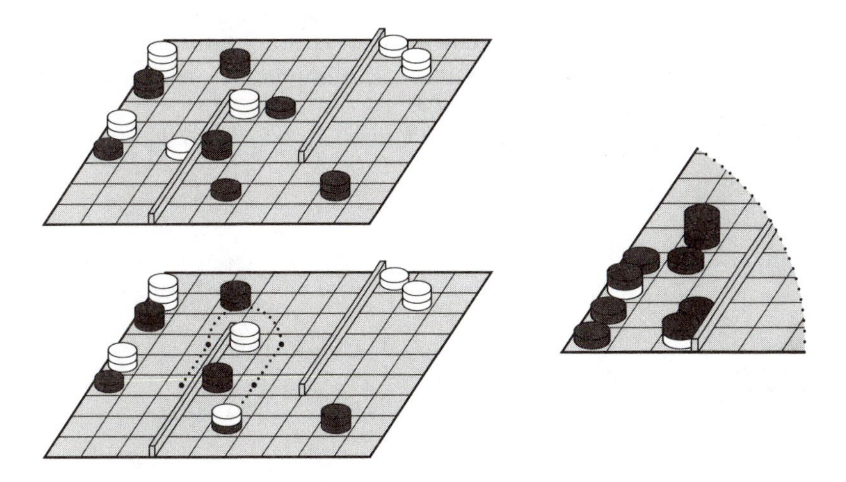

Figure 8.24. White's only move after rolling {2, 2}, and a winning connection for Black.

Alternatively, a piece may jump over a wall if that jump lands on a single enemy piece an appropriate number of points away. The enemy piece is killed and removed from the game. Players cannot jump backwards and cannot cross both walls with the same jump. The same piece may take further steps or jumps, possibly changing direction.

Figure 8.24 (left) indicates White's only possible move after rolling {2, 2} in the board position shown. This move involves one step forward, a jump over a wall to kill a Black piece, then two steps forward to pin another Black piece.

The game is won by the first player to orthogonally connect all of his pieces to the far end of the corridor. For instance, Figure 8.24 (right) shows a winning connection for Black. Pinned pieces, and pieces that pin others, may be included in the winning connection. A player also wins if his last piece is captured, unless the capturing move completes a winning connection for the opponent. If a move creates a winning connection for both players, then the mover wins.

As in Lines of Action, it is not always wise to kill an enemy piece. Jumping over a wall allows a piece to push forward quickly by taking a shortcut, but also means that the opponent has one less piece to connect for victory.

The goal of connecting to the far end of the corridor rather than bearing off pieces as in Backgammon reduces the importance of the random element in the end game.

Moloko was designed by Cameron Browne in August 2004.

9 Connective Play

Connective Play games are those that feature at least some connective aspect and no non-connective aspects during general play. Connection between pieces is paramount during play, and any piece movement must be primarily dictated by connection (see Section 6.1.2 for details). Connective Play games can be described as involving connection at the local rather than the global level.

9.1 Connective Play > Path Making

Path Making (Connective Play) games are those in which players develop paths during the course of the game.

The icon for this group shows potential moves that connect with an existing piece.

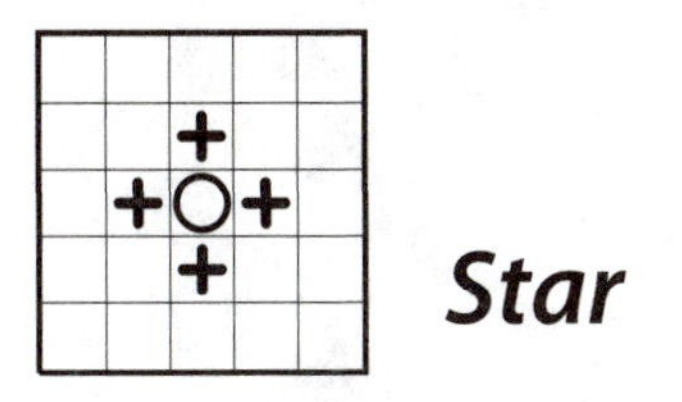

Star

Star is one of the great connection games. It is played on a six-sided board tessellated by hexagons, with side lengths typically alternating between five and six cells. The board is initially empty. Two players, Black and White, take turns either placing a piece of their color on an empty cell or passing.

Figure 9.1 (left) shows external points around the perimeter of the board (marked X) that are adjacent to edge cells. Corner cells have three adjacent external points, and all other edge cells have two external points.

Every chain of a player's pieces adjacent to at least three distinct external points is called a *star*. Each star scores 1 point for each adjacent external point, but also attracts a penalty of 2 points. It is therefore in the player's interest to touch as many external points with as few stars as possible.

For instance, the black corner piece shown in Figure 9.1 (left) is a singleton star that scores 1 point (3 adjacent external points minus 2). Neither of the other two black chains constitutes a star. The white star on the left board scores 11 points (13 adjacent external points minus 2).

The game ends as soon as either the board is full of pieces or both players pass. The player with the highest score wins. For instance, Figure 9.1 (right) shows a game won by Black who has scored 17 points to White's 14.

Due to the odd number of external points there can never be a draw. The final score of a game played to completion will always be the total number of external points minus 2 (31 for the official board shown). It is recommended that players use a single-move swap option to address the first-move advantage.

Star is a game of considerable depth. It involves a subtle balance between establishing edge points, then connecting them across the center of the board to create as few stars as possible in order to minimize the number of 2-point penalties. It is in a sense a game of connecting user-defined goals.

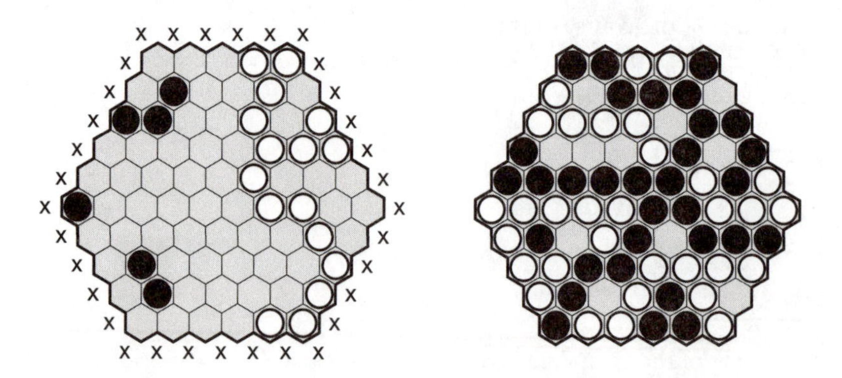

Figure 9.1. Stars and chains (left) and a game won by Black (right).

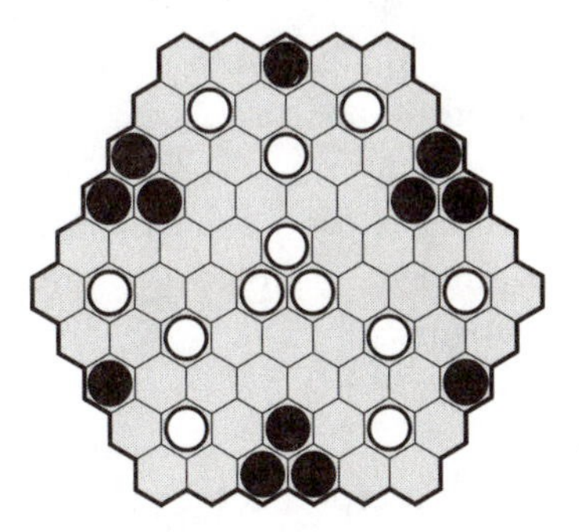

Figure 9.2. White is in a strong position despite having no edge pieces yet.

Star is unusual in that games tend to fill almost the entire board, whereas most connection games are over before even half of the board is filled. This is a function of the scoring system.

The corner cells tend to be taken first, being worth 1 point each in their own right, but then moves tend to be evenly divided between edge and interior cells as the game progresses. Once the edge cells are more or less decided, then Star becomes more like a traditional connection game and the usual connection game techniques apply. A good Hex player, for example, will generally make competent moves at this point.

Figure 9.2 demonstrates the importance of balancing edge and interior play. Black leads this game by 3 points to 0, but has overcommitted to the edges. White is in a strong position to form a single high-scoring star and eventually win the game.

Figure 9.3 shows some tactical end game plays. The situation shown on the left is decided; moving there can score no additional points for either player. The situation shown in the middle offers 1 point for the taking (marked X). White must occupy both empty cells to win the point, but Black only needs to occupy one. The situation shown on the right offers 2 points, and White must take the central empty cell to have any chance of scoring one of these.

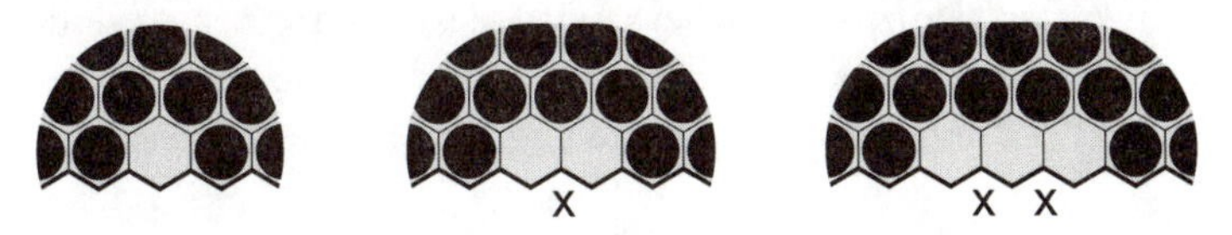

Figure 9.3. Unclaimed points on offer: 0, 1, and 2.

It is somewhat surprising that a game of Star's quality is not more popular. This is possibly due to the scoring mechanism, which may be too abstract and initially daunting for the casual player; a shame, as the game's goals become perfectly clear after only a few moves.

Star was invented by Craige Schensted [Schmittberger 1983]. It has inspired the following variants.

Maxi-Star

Maxi-Star is identical to Star except that the player with the single highest scoring star wins. If both players have equally high-scoring stars, then the next best stars are compared, and so on. Maxi-Star was suggested by R. Wayne Schmittberger.

*Star

*Star is a development of Star played with a more complex scoring system on a more complex five-sided board. See the *Star page for details: http://ea.ea.home.mindspring.com/*Star.html. *Star was designed by Craige Schensted.

SuperStar

SuperStar is played on a larger board in the shape of a six-sided star with truncated points. In addition to stars, players also score points for *superstars* (chains touching three or more sides) and *rings* (cycles around at least one cell). SuperStar was designed by Christian Freeling.

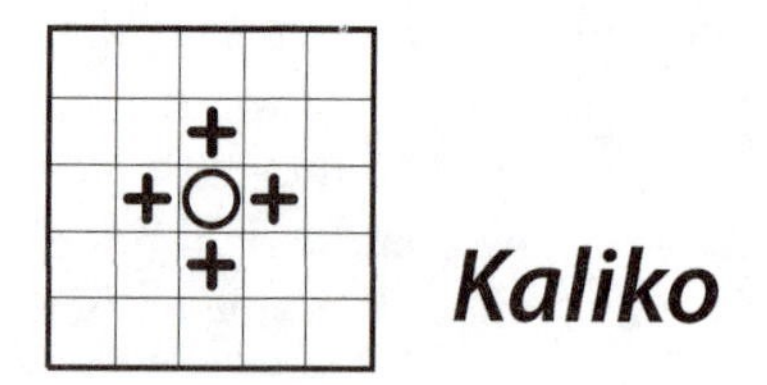

Kaliko

Kaliko is played with a set of hexagonal tiles based on the five designs shown in Figure 9.4. Each tile has three path segments, each of which is one of three possible colors. There are a total of 85 tiles corresponding to the 85 possible combinations.

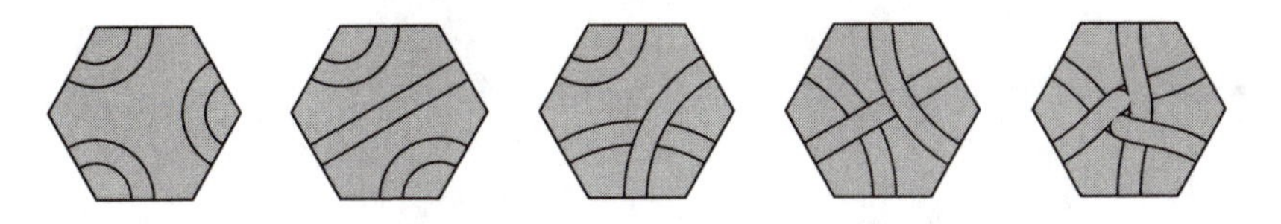

Figure 9.4. The Kaliko tiles.

Each player randomly draws seven tiles, and hides them from the view of other players. An initial tile is placed on the playing surface, then players take turns either passing, playing tiles, or dumping some of their tiles back into the pool. After their turn, players replenish their hand to seven tiles.

Players can only play tiles that extend same-colored paths of existing tiles. Building new paths scores a number of points equal to the path length, with a 3-point bonus for any path that crosses itself. Making a closed path doubles its score.

Kaliko was designed by Charles Titus and Craige Schensted and manufactured by Kadon in 1983. An earlier version of the game was marketed in the 1960s under the name Psyche-Paths, which itself may have been inspired by a previous game by the same authors called Cram [Schmittberger 1983].

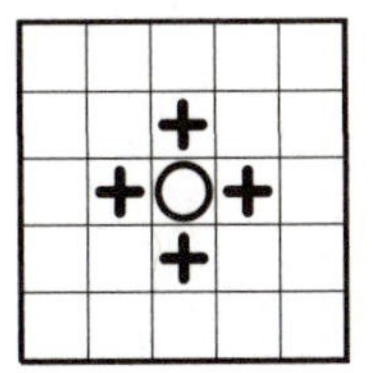

Yaeger's Game

Yaeger's Game is a simple children's game for four players [Yaeger 1984]. It is played with rectangular tiles two square units in size, each of which shows a segment of a snake. There is a neutral tail, a neutral head, and a number of tiles showing a snake segment with three bands of color, some of which are shown on the left of Figure 9.5.

An equal number of the colored tiles are randomly distributed to each player. The tail is placed on the playing surface, then players take turns placing one of their tiles so as to extend the snake, such that colored segments match where they meet. The snake can cross back over itself (Figure 9.5, right).

Figure 9.5. Some of Yaeger's tiles and a completed game.

As soon as all tiles have been played, or there are no more valid moves, then the neutral head tile is added to complete the snake. The game is lost by the player toward whom the head points. Yaeger's Game is unusual in that it does not yield a winner; it only yields a loser.

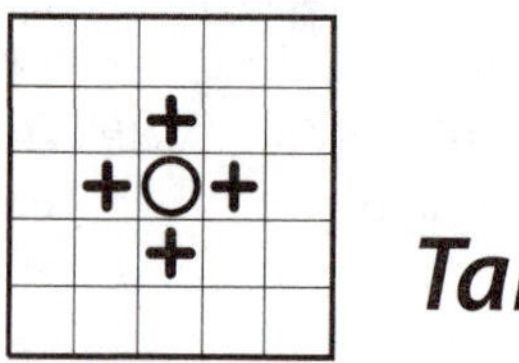

Tantrix

Tantrix is a hexagonal tile-placement game. The tiles are the same as four of the five basic Kaliko tile types (Figure 9.6) but in this case each tile's three path segments are a different color.

Tantrix can be played as a series of solitaire puzzles, or as a two- or three-player game that has as much in common with Trax as with Kaliko. Each player chooses a color and randomly draws six tiles. An initial tile is placed on the playing area, then players take turns placing one of their tiles to touch at least one existing tile, such that all touching tiles match path colors. Players are then forced to fill any spaces surrounded by three or more tiles.

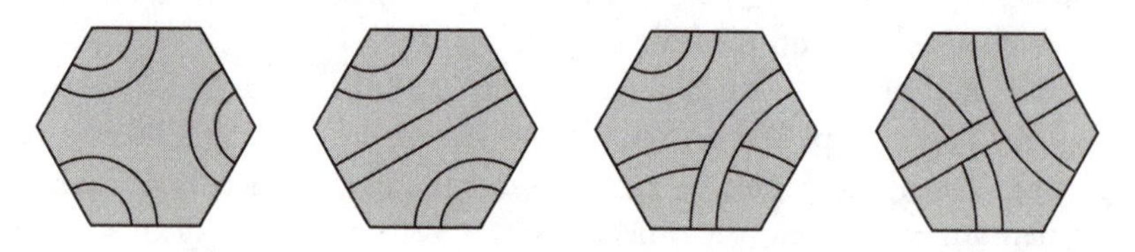

Figure 9.6. The Tantrix tiles.

After all tiles have been placed, each player scores either 1 point per tile in his longest line or 2 points per tile in his longest loop, whichever is higher. The player with the highest score is the winner. Tantrix was designed by Mike McManaway and manufactured by Tantrix Games in 1991.

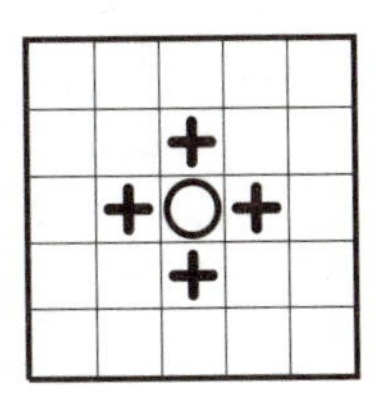

Andantino

Andantino is an elegant minimalist connection game. Game equipment does not come much simpler: it is played with nothing but a number of unmarked black and white hexagonal tiles.

Andantino starts with three tiles placed as shown in Figure 9.7 (left). Two players, Black and White, take turns placing a tile of their color adjacent to at least two existing tiles. Legal moves from the starting position are indicated by dots.

A player wins by either

- surrounding at least one enemy tile with a cycle of his color, or
- completing a line of five tiles of his color.

Figure 9.7 (right) shows a game won by White, who has surrounded a group of black tiles.

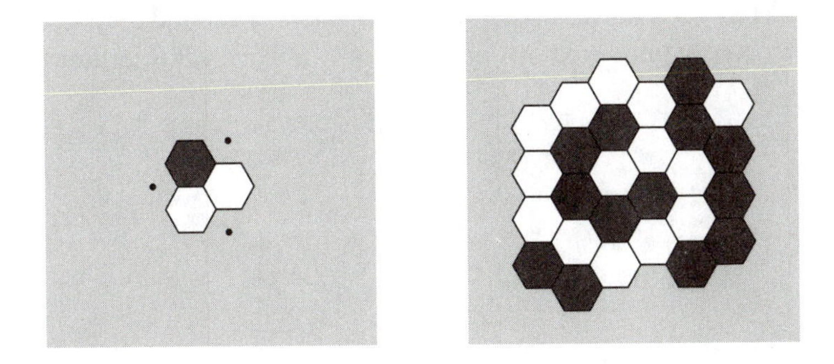

Figure 9.7. The starting position (Black to play) and a game won by White.

Figure 9.8. Three strong formations.

Andantino is a wonderfully elegant game, and would be a prime example of the Pure Connection Cycle Making category if it were not for the pattern-based, five-in-a-row winning condition. This alternative goal is necessary to bring some tension and balance to the game, as without it players could perpetually defend against cycles; line threats allow a player to force the opponent into making disadvantageous moves.

Figure 9.8 shows some basic principles of play. The formation on the left, called the *propeller*, demonstrates that it is good to keep all pieces in a single connected group if possible. White has a strong advantage here and can more or less force play as he wishes, while Black must adopt a defensive role and attempt to protect his vulnerable isolated tiles.

Figure 9.8 (middle) shows a strong 3/3 formation that guarantees a win for Black. Even if this formation is blocked at one end by white pieces, then Black can force a win if it is his turn to move.

Figure 9.8 (right) shows White, whose turn it is to play, in a strong position. The winning sequence of moves is shown in Figure 9.9. White moves *1* and *3* are forcing moves that develop a ladder. White move *5* forces Black reply *6* to avoid an immediate line win, then White move *7* is a killer forking move that sets up both a line threat and a cycle threat simultaneously (indicated by question marks). White will win next turn.

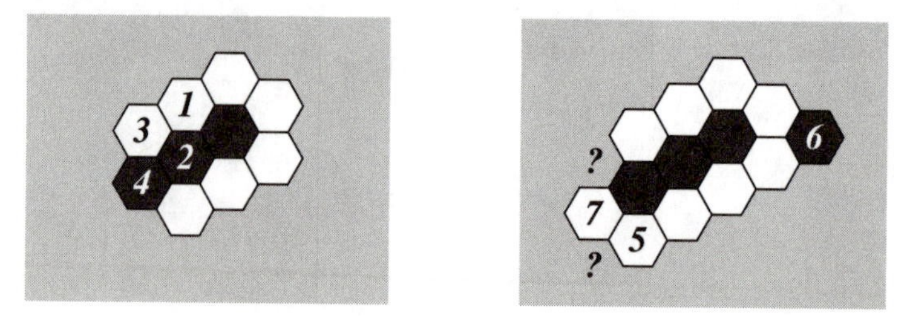

Figure 9.9. White forces a ladder to win.

Andantino was invented by David Smith in 1995. The player that starts with an additional piece (White) appears to have a significant first-move advantage.

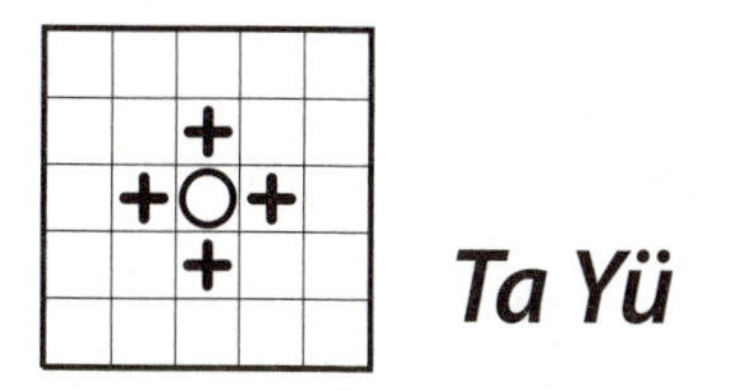

Ta Yü

Ta Yü is a tile-placement game in which players strive to maximize the flow between their edges of the board. The board is a 19 x 19 square grid, which is initially empty.

Two players own the alternating sides of the board that bear their color, and have access to a common pool of 112 tiles, which are placed face down. Each tile is three units long and marked with a path segment that touches the edge at three points, as shown in Figure 9.10 (left). The tile set consists of four copies of the 28 possible permutations of three path exits on a three-unit tile.

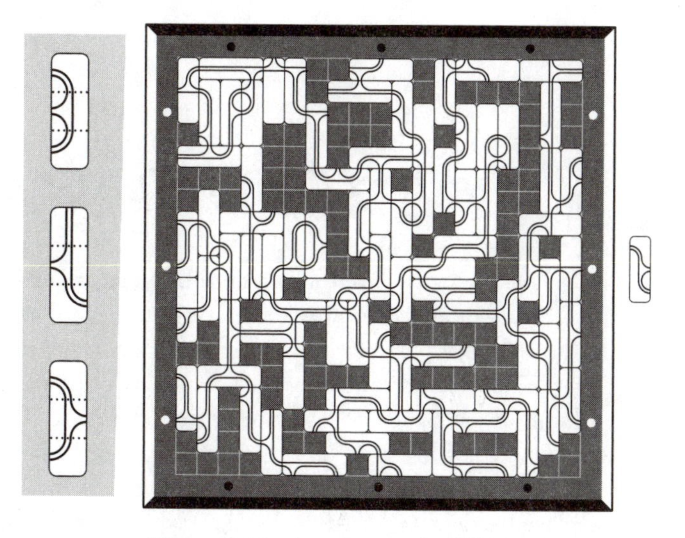

Figure 9.10. A game won by White.

Players take turns randomly drawing a tile from the pool and laying it on the board from the center outwards. The tile must touch at least one existing tile to extend the central path, and may not be placed such that any path meets a blank edge on any adjacent tile. Players may make an educated choice of tile, as tiles whose paths touch three different sides have special markings on their backs.

The game ends as soon as a player is unable to place his selected tile. Players add up the number of path exits terminating at each of their sides of the board (counting 2 points for paths reaching any of the three specially marked squares), then multiply these two tallies together for their final score.

For instance, the game shown in Figure 9.10 has ended as it is not possible to play the tile shown on the right; it does not matter who drew this tile. White wins with 10 x 8 = 80 points, while Black scores 9 x 6 = 54 points. This example is based on an entry on the BoardGameGeek Ta Yü page: http://www.boardgamegeek.com/game/117.

Ta Yü is an enjoyable game with a simple and intuitive method of scoring. A variant including a third player in a saboteur role makes three-handed Ta Yü quite strategic.

Ta Yü was designed by Niek Neuwahl and published by Kosmos in 1999. It was voted one of the top ten German Games of the Year.

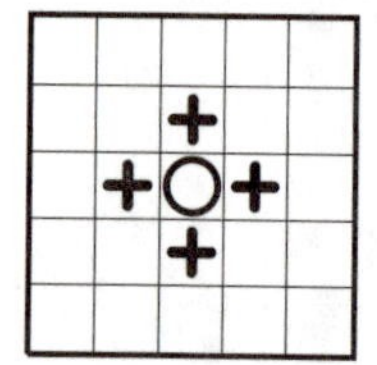

The Very Clever Pipe Game

The Very Clever Pipe Game is played with a deck of 48 rectangular cards, each two square units in size. The cards show various combinations of pipes leading from the six possible midsquare *junctures*, some of which connect to other junctures and some of which terminate in *caps*.

Players randomly draw hands of five cards from the deck. The first player chooses a pipe color (light or dark), then players take turns playing a card from their hand and replenishing the spent card from the deck. Cards may be played adjacent to existing cards, in which case pipes of the same color must match. However, there is no compulsion to play adjacent

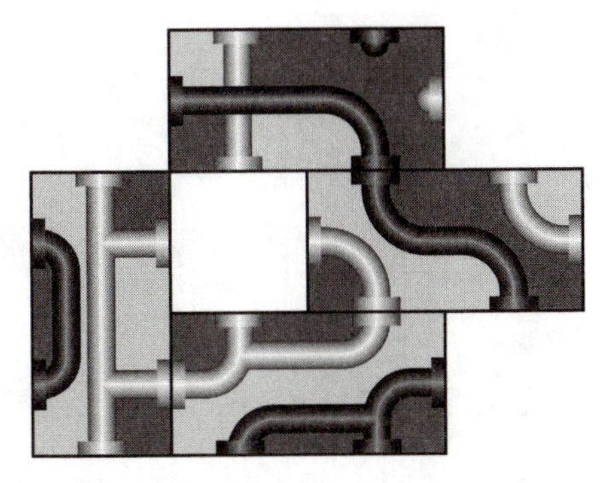

Figure 9.11. Four tiles enclosing a pillar.

to existing cards, and cards may be played some distance away as long as they follow the grid implied by the existing cards.

Any empty square enclosed by four tiles, such as that shown in Figure 9.11, becomes a *pillar* that caps off all entering pipes. Pillars add an element of tactical play as they can be used as additional blockers. Any set of cards showing a connected set of pipes with no freedoms (that is, no open ends) is captured and removed from the board, scoring points for the player who made the capture.

The Very Clever Pipe Game was designed by James Earnest and published by Cheapass Games in 2001. Each card contains alternating light and dark background fields, which come into play in more advanced versions of the game.

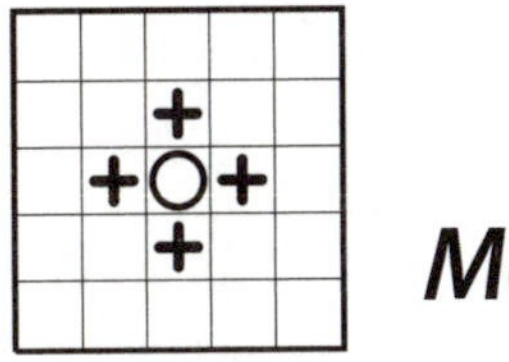

Meander (2001)

Meander combines simple play with a novel method of scoring: players tilt the board and send marbles down their paths to determine their score. Two players, North-South and East-West, share a common pool of square tiles that are played on a 5 x 5 square grid. Each tile has path segments etched into it, as shown in Figure 9.12 (left). The board is initially set up as shown in Figure 9.12 (middle).

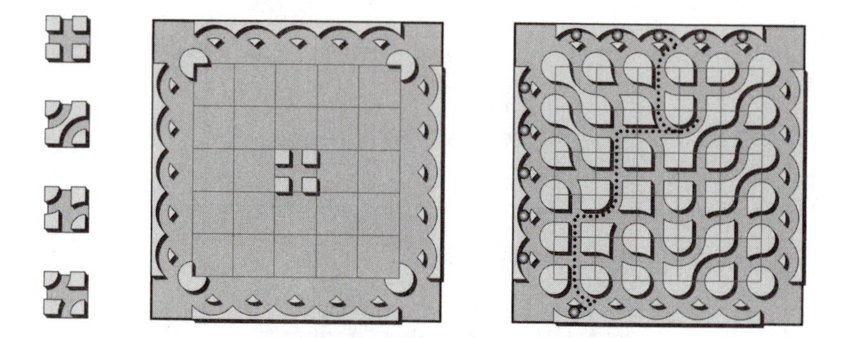

Figure 9.12. The Meander tiles, the starting position, and a game at the ball-rolling stage.

Players take turns randomly selecting a tile from the pool and placing it on a vacant square. After all tiles have been used, each player places five marbles along his home edge as shown in Figure 9.12 (right). The player then tilts the board, and pushes the marbles one by one to follow the network of paths down the board. Each marble will either get caught somewhere in the network, emerge at a corner and drop off the board, or land at the far lip and get caught there.

Players score one point for each marble caught by the far lip. For instance, the dotted line shows the path that North-South's third marble will take through the network. It will eventually land in the far lip and score one point. Gravity has the effect of imposing direction upon connections across the board.

Meander was designed by Justus van Oel and published by De Speelstijl DV in 2001.

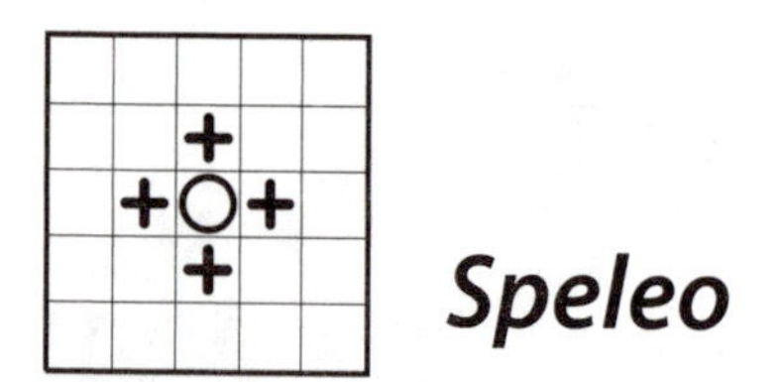

Speleo

Speleo is played on a square grid, typically 5 x 5 or 6 x 6, which is initially empty. Each player is aligned with a direction, the first player with the forward sloping diagonal (/) and the second player with the backward sloping diagonal (\).

Figure 9.13. A completed game of Speleo won by the first player (/).

Players take turns choosing an empty square and drawing a line between the two diagonally opposed corners in the player's direction. Figure 9.13 (left) shows a typical 6 x 6 game after all squares have been claimed.

Players score one point for each tunnel that both starts and ends in their direction. For instance, the game shown in Figure 9.13 is won by the first player (/) who scores three points while the second player (\) scores zero points. Speleo is Joris game #31 [2002].

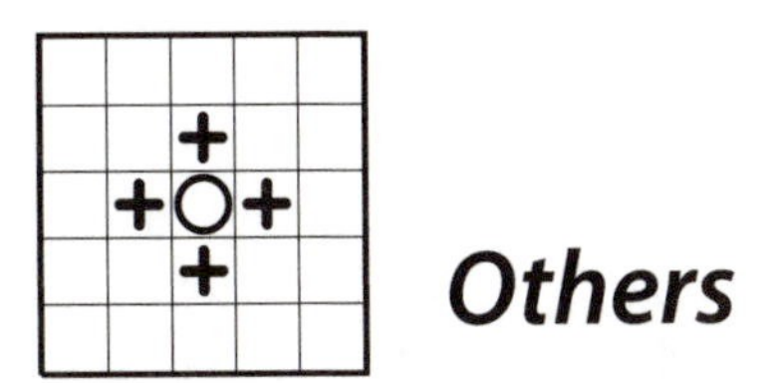

Others

Mongoose Den

Mongoose Den is played on a 21 x 21 square board. The board contains a number of symmetrically placed bonus squares, penalty squares, and a 3 x 3 mongoose den in the center. Two to four players each own a side of the board and share a common pool of tiles, each marked with a score. The tile set includes nine designs showing single path segments in various configurations, and a special power tile containing all valid path segments.

Players takes turns placing a randomly drawn tile on their starting position, which is at the midpoint of their board edge. Thereafter, players take turns drawing a tile and placing it to extend one of the paths on the board under certain restrictions [Dodge 1963]. Once all tiles have been placed, the winner is the player with the highest score determined by

- total score along his path,
- total value of bonus/penalty squares hit,
- whether the path reached a specified target zone, and
- whether the path was forced into the mongoose den.

Nile

Nile is played on a square grid. Players take turns randomly drawing a square tile showing a path segment, and adding it to extend the connection spreading out from a central source tile. Particular tile placements score points for the player. The winner is the player with the highest score when the path connects opposite sides of the board. Nile was published by E. S. Lowe in 1967.

Rivers, Roads & Rails

Rivers, Roads & Rails is a tile-laying game featuring a single linear path that may be expanded at either end. Each tile contains a single path with some combination of river, road, and rail joining one side to exactly one other. There is no branching, and the game can take up a lot of space as the single path grows and meanders.

Players share a common pool of over 100 tiles. Each player randomly draws ten tiles and a starting tile is placed on the playing surface. Players take turns drawing a random tile and placing it to match either end of the existing path, or adding it to their hand if placement is not possible. Once the pool runs out, then players use a tile from their hand each turn. The first player to run out of tiles wins the game.

Rivers, Roads & Rails is credited to Ken Garland and Associates, and was published by Ravensburger in 1969. It is a children's game with limited strategic scope.

The Great Downhill Ski Game

The Great Downhill Ski Game is played on a 26 x 21 square grid representing a ski slope. Slightly more than a quarter of the squares are taken up by trees 1 x 1 or 2 x 2 squares in size. Players share a common pool of 256 tiles each showing a parallel set of ski tracks, including straight lines, diagonals, turns, sharp turns, and crossing tracks. Players randomly draw ten tiles each and reveal them to their opponents.

Players take turns playing their ten tiles in a connected path, which must avoid all trees, downwards from the top of the board. The player then replenishes his hand from the tile pool. Players may swap tiles in lieu of making their turn. If a player's trail becomes blocked he may remove the tiles played that turn and try again.

The game ends when a player's track reaches the bottom of the board. Each player scores a number of points for each tile played, and the finisher scores an additional bonus. The player with the highest number of points wins.

The Great Downhill Ski Game is credited to Terry Rose Martin and was published by Waddingtons Games in 1970. It was later renamed Nancy Green's Great Downhill Ski Game.

Trails to Tremble By: The Scary Crossroads Game

Trails to Tremble By is played with hexagonal tiles, some showing straight path segments, some showing curved path segments, and some showing crossing path segments. In addition, each player has a number of hexagonal *marker tiles* showing appropriately spooky figures such as witches and tombstones.

Players take turns placing a tile to extend the existing path network. If a tile placement creates an empty space surrounded by at least three tiles, then the player may place one of his marker tiles in that space provided that it does not block a path. If that marker tile creates a further three-sided space then another marker tile can be laid, and so on. The first player to play all of his marker tiles wins. Trails to Tremble By was published by Whitman in 1971, as described by Richard Reilly.

Waterworks

Waterworks is a boardless tile-laying game played with 100 cards showing segments of plumbing pipe. The pipes are of different kinds, which may be leaky, vulnerable to leaks, or impervious to leaks. Players take turns placing cards to extend the plumbing network, with the aim of completing a connected pipeline capable of carrying water over a specified distance from their faucet card to their spout card. Players are also allowed special moves that repair leaky pipe segments. Repaired pipes are replaced by cards showing the same connection in better condition. Waterworks was published by Parker Brothers in 1972.

Weber's Game

Weber's Game is played on a 10 x 10 square grid of holes. Each hole has eight radiating grooves leading to orthogonal and diagonal neighbors, which may include terminal markers around the edge of the board. Two players each have a number of balls and links of their color, and control a number of terminal markers of their color.

Players take turns placing one of their links on the board. Links cannot cross existing links, and must lie in a channel such that either a ball or a terminal marker rests at each end. If no such move is possible, then the player may also add a ball to a board hole as part of his turn. Once all links have been placed, players determine their score based on the number of friendly terminals connected, the number of friendly terminals cut off, the total score of links, and other criteria [Weber 1972].

Le Camino

Le Camino is played on a 10 x 14 square grid. Players share a common pool of 144 tiles that show path segments of road, rail, and river, as well as prairie and forest tiles with no paths. Some tiles contain bridges and railroad crossings. Players own alternating board corners and randomly draw six tiles each (redrawing if they have no road tiles). To begin, each player places a road tile in one of his corners and picks a new tile.

Players then take turns placing a tile on the board and drawing a replacement from the pool. Tiles must be placed to connect to existing tiles and match any paths upon them (some restrictions to improve game play exist). The game ends when either player completes a road between their corners. Players score one point per unit distance along that road, plus an additional penalty point for each bridge and railroad crossing. The player with the fewest points wins.

Le Camino was designed by Jean Naudet and published by Atelier des Jeux d'Agremént in 1973. It is unusual in that players do not compete to make the first or longest connection, but the shortest.

Galdal's Game

Galdal's Game consists of a somewhat bizarre combination of familiar connection game elements. The board consists of a 5 x 5 square grid of holes (the starting grid), a 7 x 7 square grid of holes (the playing grid), and a 4 x 4 grid of holes (the scoring grid). The playing grid overlaps the other two grids by one hole each, at diagonally opposed corners.

Each player owns a number of pieces that look like resistors; long cylindrical handles with thin legs protruding down at each end. Some pieces are long enough to fit each leg into orthogonally adjacent holes, some long enough to fit each leg into diagonally adjacent holes, and the remainder long enough to fit each leg into holes placed a knight's move apart. Pieces can also be short or tall.

Players take turns placing one of their pieces on the board such that each leg fits into a hole (each hole is wide enough to accommodate several legs). A player's first move must have at least one leg in a starting grid hole, but thereafter the only restriction is that each subsequent piece must have at least one leg in a hole already occupied by the piece's owner. Taller pieces may be used to bridge over shorter pieces. Players score points for each leg that occupies a scoring grid hole [Galdal 1977].

Deaton's Game

Deaton's Game is a boardless game played with square tiles, each showing some combination of towns, roads, rivers, railways, and topographic features. Players take turns randomly drawing tiles and placing them to extend tracks between towns. Players score points for completing connections between towns [Deaton 1978].

Hextension

Hextension is the Connective Play precursor to Take It Easy (see Chapter 10, Connection-Related Games). Players take turns placing a hexagonal tile on a hex hex board, such that it lies adjacent to at least one existing tile. Each tile is marked with three straight paths of different colors between opposite edges. Once the board is full, players score points by completing paths of pure color in a straight line between opposite sides of the board. Published by Spears Games in 1983, Hextension was most likely designed by Peter Burley (as reported by Mike Hutton).

Würmeln

Würmeln is a boardless game. Each player has seven worm segments of their color that remain connected as whole worms throughout the game. Two posts mark a starting line and two posts mark a finishing line. The game begins with all worms placed behind a starting line, then, each turn, players bid for the right to move. Based on the bid, players may move a certain number of their worm's segments from its tail to its head, effectively moving the worm by that number of segments. The player whose worm first crosses the finish line wins; however, it can be good play to divert a worm's progress to block opponents. Würmeln was designed by Alex Randolph and published by Blatz in 1994.

Schlangennest

Schlangennest is a tile-placement game for two to four players. Each player owns a number of tiles showing both a neutral snake segment and a snake segment of his color. Tiles are of three types: crossing, divergent curves, and head/tail. Players takes turns extending the colored section of their existing snake with one of their tiles. Play continues until no more moves are possible. The player with the longest snake wins.

Schlangennest was designed by Günter Cornett and published by Bambus Spieleverlag in 1995. The rules also describe a variant called Gobi in which enemy snake segments may be captured. Schlangennest is German for "snake's nest."

Puddles

Puddles is played on a Hex board, typically 6 x 6. Two players, Black and White, take turns placing a piece of their color on the board. When the board is filled, players count the size of their chains; the player with the smallest maximal chain wins. If the players' maximal chains are equal in size, then the next largest chains are compared, and so on.

Puddles, devised by Alp Lo, is an anticonnection game in the sense that players strive to minimize their connections. Like Black Path, it is a game with connective play but an anticonnective goal.

Metro

Metro is played on an 8 x 8 square grid. The board shows eight railway stations along each side, and four in the central 2 x 2 region. Players share a common pool of single unit square tiles, each of which show four path segments such that two path segments run to each tile edge.

Players start the game with a number of trains of their color distributed in the stations around the board perimeter, and a randomly drawn tile.

Players take turns either playing their tile on an empty square or drawing a new tile from the pool and playing that one instead. If placed, the tile must connect with either a board edge or an existing tile.

Any move that completes a connection between a station occupied by a train and another station scores a number of points equal to the path's length in tile units, and the train is then removed from the board. Double points are scored for connecting to any of the central stations.

Metro was designed by Dirk Henn and published by Queen in 1997. It was previously called Iron Horse.

Nexus

Nexus is played with a deck of rectangular cards, each showing one to four circles from which one to four paths emanate. Each circle's value is equal to the number of paths emanating from it. A card is placed on the playing surface to start the game. Players then take turns randomly drawing a card from the deck and placing it to connect with at least one existing card, building up a network of circles with connecting paths.

A *nexus* is a system of circles, which is either *open* if it contains open-ended paths or *closed* if it does not. After playing his card, the player places a piece of his color on any circle belonging to an open nexus. If the card just played forms a closed nexus, then the player with the highest value of pieces within that nexus scores the value of unoccupied circles it contains.

Nexus was designed by James Earnest and published by Cheapass Games in 2001.

TransAmerica

TransAmerica is a game of shared paths played on the edges of a hexagonal grid, which is superimposed on a map of North America. The grid is divided into five general regions, and about 50 vertices are marked as cities. Players randomly draw five cards to designate a city from each of the five regions. Cards are hidden from opponents.

Each player starts with an initial hub on the board, then each turn players must extend their connection from the hub by claiming one or two edges depending upon the terrain shown on the map. The game ends as soon as a player connects his five cities. Players score points based on how many moves they are from completing their connection. Players may use other players' edges in their connection; in fact, exploiting enemy connections to advantage appears to be the key to success in this game.

TransAmerica was designed by Franz-Benno Delonge and published by Winsome Games in 2001.

Snap

Snap is a colorful tile-laying game consisting of 36 jigsaw pieces, each with intertwining segments of dragons of three different colors. The official rules describe four different games aimed at different age groups, including one solitaire game. In general, players take turns placing a tile to connect with at least one existing tile such that dragon segments match. When a player completes a dragon of any color with a head at one end and a tail at the other, a number of points are awarded based on the dragon's length.

Snap, designed by P. Joseph Shumaker and published by Gamewright in 2002, is similar to Schlangennest. A special tile showing a tiger may be included in some versions of the game.

The Legend of Landlock

The Legend of Landlock is an attractively decorated tile-laying game. Two players, Land and Water, share a common pool of 40 tiles showing various combinations of water, land, and pathways. Some tiles also include special characters.

After an initial tile is placed, players take turns drawing a random tile and placing it to match at least one existing tile. The layout may not exceed six tiles in any direction, making this one of the few tile-placement games that may enter a cool phase as players' options run out and they are forced to place tiles in disadvantageous positions.

When 36 tiles have been placed, Land scores points for connecting paths between all four sides of the completed 6 x 6 square, and Water scores points for connecting waterways between all four sides. Additional points are awarded for closing off path loops or creating islands, respectively, and for matching up pairs of special characters on a player's terrain.

The Legend of Landlock was designed by Edith Schlichting and published by Gamewright in 2002.

Rumis

Rumis is a block-building game with connective play. Each player owns a set of wooden blocks of his color, consisting of wooden cubes joined together in different configurations. Players must place their blocks to fit within one of the predefined game maps to connect with at least one existing block of their color. Players score points based upon how many of their cubes are visible from above the board at the end of the game. Rumis was designed by Stefan Kögl and manufactured by Murmel Games in 2003.

9.2 Connective Play > Territorial

Territorial games are those in which players strive to enclose territory or divide the play area into connected subregions during the course of the game.

The icon for this group shows territory surrounded by a connected set of pieces.

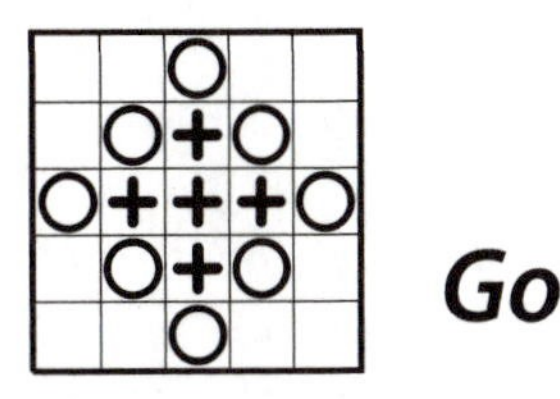

Go

Go is one of the most influential of all abstract board games. It is played on the intersections of a square grid, typically 19 x 19, which is initially empty. Two players, Black and White, take turns placing a piece of their color on an empty intersection. All enemy groups with no remaining *liberties* (orthogonally connected empty points) are then captured and removed from the board.

Figure 9.14 shows a group of black pieces with one remaining liberty (left), and its capture by white piece ***a***, which removes that liberty (middle). To be guaranteed life, a group must surround two or more empty regions, called the *eyes* of the group. The black group on the right, really two mutually supporting black groups, is safe as it contains two eyes.

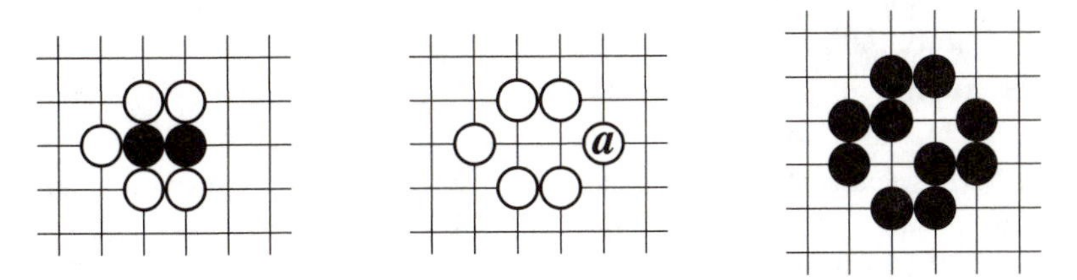

Figure 9.14. Piece-surround capture and a safe black group.

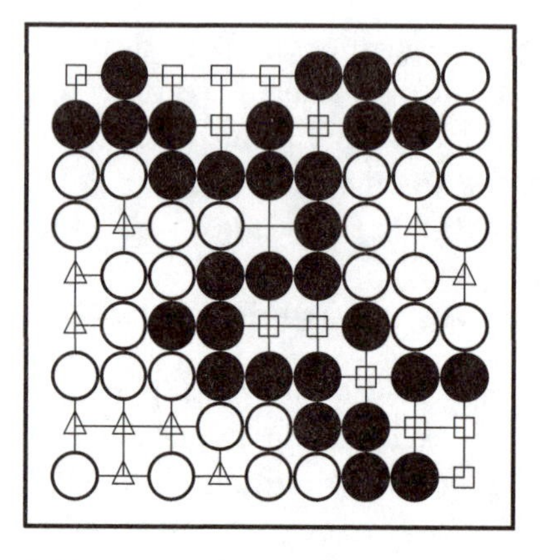

Figure 9.15. A 9 x 9 game with territories marked.

Players may not place a piece that would commit suicide, that is, any piece placed on the board must have at least one liberty or become part of a group that has at least one liberty. In addition, players cannot make a move that would result in a repeated board position (*ko* rule).

Players may pass in lieu of making a move. If both players pass in succession then the game ends, and players' scores are calculated based on the amount of territory under or surrounded by their pieces. Captured stones may also contribute to players' scores, depending upon which version of the rules is used. For example, Black has 12 points of territory (marked with squares) and White has ten points of territory (marked with triangles) in the 9 x 9 game shown in Figure 9.15.

The unmarked empty cell near the board center is not owned by either player, and there is an ongoing world-wide debate over how such cases should be scored. Proponents of *territory* rules argue that such neutral points should not affect either player's score, while proponents of *area* rules argue that the game shown in Figure 9.15 is not over, and that the next player should be able to improve his score by claiming the empty point.

Elwyn Berlekamp points out that at least ten dialects of the Go rules existed in the twentieth century, and that even today Japan, China, Taiwan, North America, and New Zealand each have their own distinctive version of the "official" rules. Fortunately, the subtleties arise so rarely that most players remain blissfully unaware of the potential for disputes [Bozulich 1992].

Go is widely regarded as one of the deepest games in existence [Berlekamp and Wolfe 1994]. It is estimated to be 3,000 years old and boasts many textbooks, clubs, and professional players who spend their lifetimes studying the game. In this brief summary it is only possible to hint at the possibilities that Go has to offer.

Although strongly based on the concepts of war and territory, Go has a substantial connective aspect. In fact, capturing moves can be seen as the formation of orthogonally and diagonally connected cycles around maximal orthogonally connected enemy groups.

Variants

Go has spawned a large number of variants over the years. The following games are just a few of particular interest that emphasize the connection aspect.

Gonnect

Gonnect is one of the best of the recent crop of connection games, replacing the territorial goal of Go with a strictly connection-based one: the first player to achieve an orthogonally connected path of his pieces between any opposite pair of sides wins. Gonnect simplifies the rules of Go, avoiding any possible conflict over scoring. Gonnect is described in detail in Section 7.2.1.

One-Capture Go

One-Capture Go is played according to the Go rules except that passing is not allowed, and the first player to make a capture wins. As a result, a single eye with two empty points is as good as two eyes in One-Capture Go, as demonstrated in Figure 9.16.

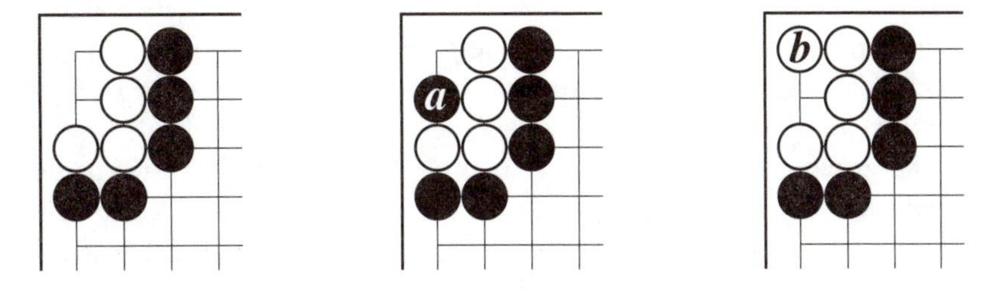

Figure 9.16. A double eye is as good as two single eyes in One-Capture Go.

This example, derived from Schmittberger [1992], shows a white group with a single eye consisting of two empty points (left) and an intruding move ***a*** by Black (middle). White can capture Black's intrusion with capturing move ***b*** to win the game; however, if this were standard Go, then Black could capture the entire white group next turn to gain the advantage.

One-Capture Go still has a strong territorial aspect, but this has become a strategy rather than a winning condition. Since players cannot pass, they must prepare for the end game by claiming as much territory as possible in which to safely waste moves, rather than be forced to play at a disadvantageous (capturable) position. Games between experienced players should reach this cold territory-filling stage unless either player makes a blunder.

One-Capture Go is based on a fundamental variation of the Go rules, and it is not known when this game was first suggested. It is described in Schmittberger's *New Rules for Classic Games* [1992] and is Joris game #53, where it is called Single Take Go [2002].

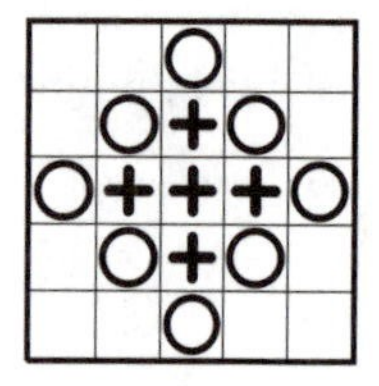

Worm

Worm is not a single game but a family of related games in which each player controls one or more *worms* (orthogonally connected sets of pieces) that grow but do not otherwise move. The aim is to block off the opponent's worm so that it has no more room to grow.

Worms start the game as a single piece, and iteratively grow each turn by the placement of an additional piece adjacent to the head. Worms may not intersect paths left by themselves or other worms, hence worm trails partition the board into decreasing areas until a player runs out of moves and loses the game. Games tend to become increasingly colder as the players run out of places to move.

The game shown on the left of Figure 9.17 is a generic Worm game in which two players, Black and White, grow their worm each turn. The worms started in opposite corners of the board; worm tails are marked with concentric circles and worm heads with dots. Legal moves for each player are indicated by question marks.

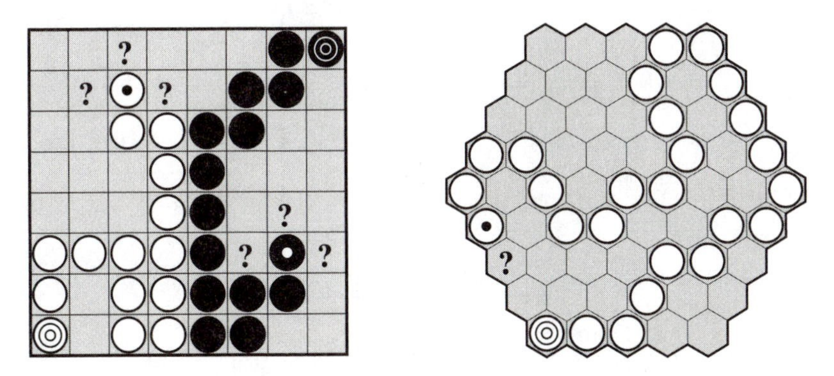

Figure 9.17. A generic Worm game (left) and Blue Nile (right).

The game shown on the right is a variant called Blue Nile in which two players share the same worm on a hex hex board. Again, the worm's tail is marked with concentric circles and its head with a dot. Blue Nile is unusual in that players are not even allowed to move adjacent to the existing worm path (reminiscent of Tanbo). The current player has only one legal move, indicated by the question mark. After this move is made the opponent has no legal moves and loses the game. Blue Nile was designed by Dan Troyka in March 2002.

Another example is Goldfarb's Game [1981], which is played on a 21 x 21 square grid. Players take turns placing a 2 x 1 rectangular piece of their color to cover two empty board squares, such that it extends the player's existing path orthogonally. A player wins if the opponent has no legal move.

There are a large number of other Worm-style games including Snail Trail (Don Green), Cascades (William Chang), Slime Trail (Bill Taylor), Snake Pit (a three-player variant by Bill Taylor), Un (João Neto), Bunch of Grapes (Joris game #21), Erring Line (Joris game #57), and Snake Fight (Joris game #83).

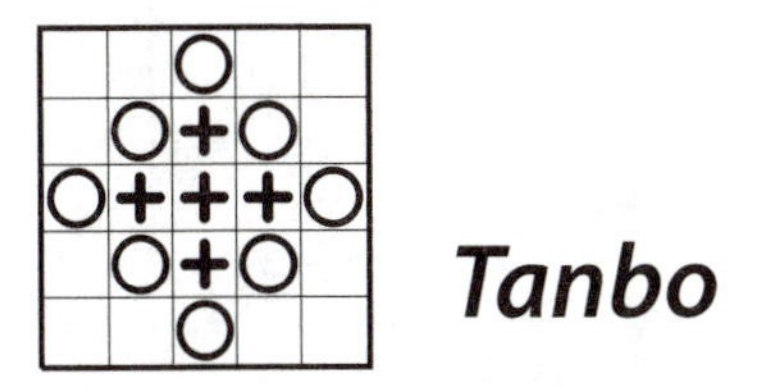

Tanbo

Tanbo is a territorial game played on the intersections of a square grid, typically 9 x 9 in size. Two players, Black and White, start with pieces set

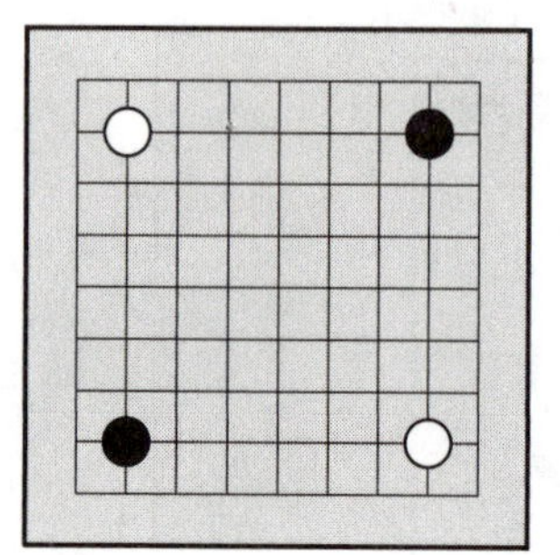

Figure 9.18. The Tanbo starting position.

out as shown in Figure 9.18. Each starting piece is the beginning of a *root*, which is an orthogonally connected set of same-colored pieces.

Black starts, then players take turns placing a piece of their color on an empty point orthogonally adjacent to exactly one existing piece of their color. In this way roots grow incrementally with each move. The *freedoms* of a root are the empty adjacent spaces into which it can legally grow.

If, after placing a piece, the player's root has no more freedoms, then the root dies and all of its pieces are removed from the board. If, however, the root survives and removes the last freedom from any opponent's root, then that opponent's root dies and is removed from the board. The last remaining player with roots on the board wins the game.

For instance, Figure 9.19 (left) shows a board position with Black to play. Black only has two possible moves labeled ***a*** and ***b***. Move ***a*** would be disastrous for Black; the black root would then have no more freedoms, die, and White would win the game.

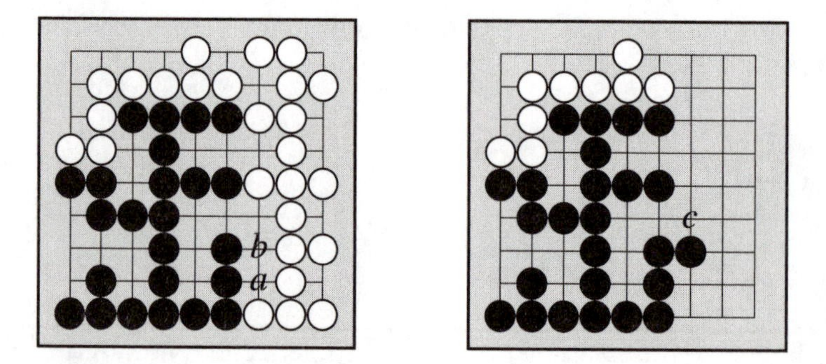

Figure 9.19. Black has two possible moves (left) and chooses the correct one (right).

Move *b* is a much better move for Black. It gives the black root one extra freedom *c*, allowing it to survive for another round; even better, it removes the last remaining freedom from White's right-most root and kills it (Figure 9.19, right), opening up the board and putting Black in a winning position.

Additional starting stones should be used for games on boards larger than 9 x 9; however, they should always be well spaced out, and separated by an odd number of points to avoid possible symmetry strategies.

An earlier version of Tanbo, called Rootbound, was released by Mark Steere in 1984. The current version was created by Steere in 1995. Both were inspired by Go.

Three-Dimensional Tanbo

Three-Dimensional Tanbo is played as per standard Tanbo on a three-dimensional grid. This is an extremely difficult game, and smaller board sizes such as 5 x 5 x 5 are recommended.

Anchor

Anchor features Go-like territorial battles, and is played on a hex hex board typically with eight cells per side. Two players, Black and White, own the alternating corners that bear their color, called their *home corners*. Corners belonging to the opponent are called *away corners*. The board is initially empty.

Players take turns either placing a piece of their color on an empty cell or passing. A three-move swap option is recommended. Players may not mirror their opponent's moves ten or more turns in succession, to stop any possible symmetry strategies.

An *anchor* is a connected set of same-colored stones that touches at least two board sides that do not share an away corner. Corner cells belong to both adjacent sides. Figure 9.20 shows some examples of anchors (left) and some examples of connected sets of pieces that are not anchors (right).

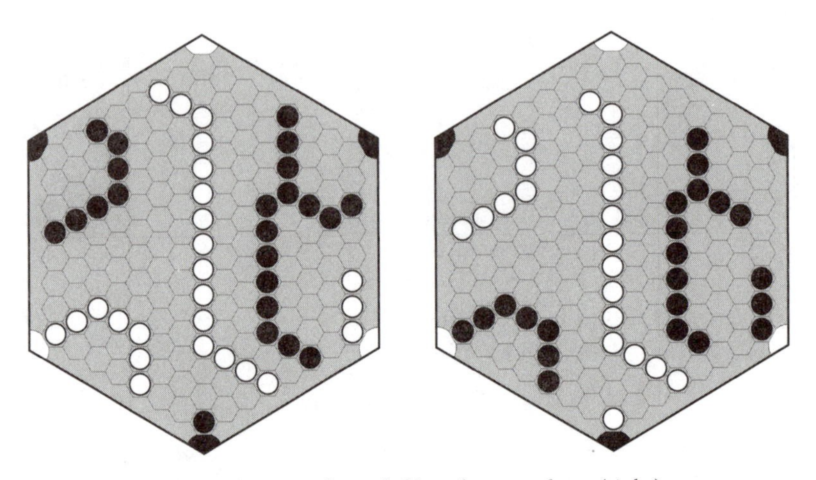

Figure 9.20. Anchors (left) and nonanchors (right).

The game ends when either the board fills up or both players pass in succession. Any piece connected to a friendly anchor is then deemed to be alive, and all other pieces are deemed to be dead. Players calculate their scores based on the amount of territory controlled, including dead enemy pieces.

Anchor was designed in 2000 by Steven Meyers, who describes it as the favorite of his games. See Handscomb [2001a] for further details. Anchor shares many similarities with HexGo, which was invented independently at around the same time.

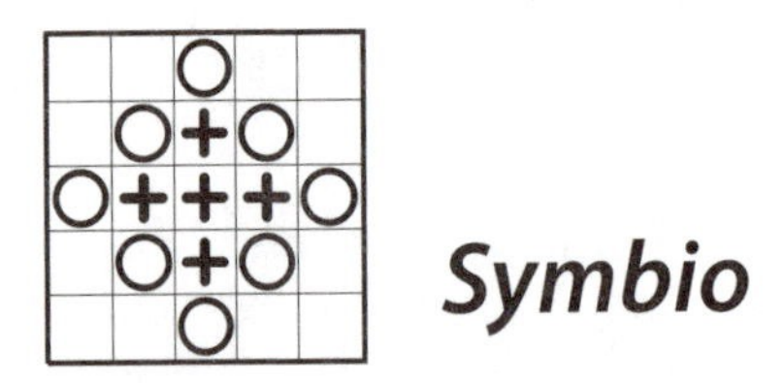

Symbio

Symbio is played on an 8 x 8 square grid. Two players, Black and White, start with six pieces of their color set out as shown in Figure 9.21 (left). An orthogonally connected set of same-colored pieces is called a *body*. A single piece counts as a body. A *free body* is a body with at least one freedom (empty orthogonally adjacent square).

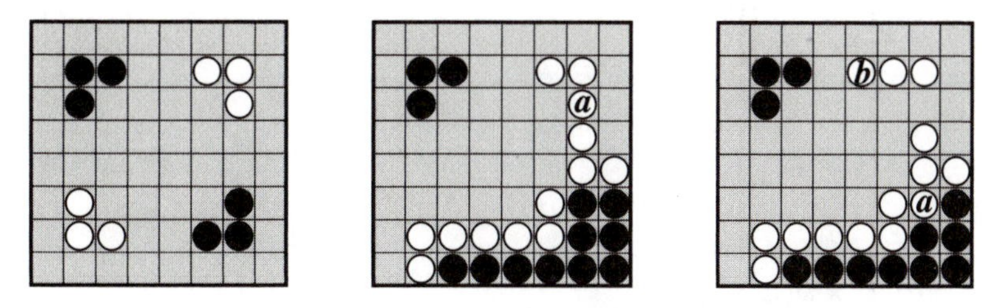

Figure 9.21. The Symbio starting position and a move by White.

Black starts, then players alternate taking turns. Each turn the current player must

1. move one of his free bodies, then
2. add one piece adjacent to at least one of his bodies.

Moving a body involves taking a piece from the body and moving it to any empty square connected to the body, including squares adjacent to the piece itself. The move may land on a foreign body piece if that foreign body has no freedoms. Capture is by replacement.

Figure 9.21 (middle) shows an example board position with White to move. Observe that the two white bodies combine to surround the lower black body. Figure 9.21 (right) shows the result after white piece ***a*** moves to capture one of the surrounded foreign pieces. This move joins the two white bodies but also splits off a new body at the top, to which piece ***b*** is added to complete the move.

Symbio, invented by Cameron Browne in 2000, has a biological theme. Bodies move in a pseudopod-like manner, grow and split to create child bodies, and attempt to surround and consume foreign bodies much like an immune system would. The connection-based piece movement is similar to that of Akron.

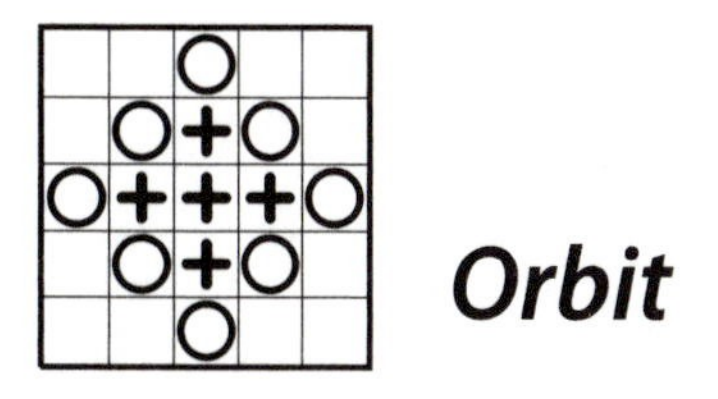

Orbit

Orbit is a Go variant played on the intersections of a 16 x 16 square grid, which is initially empty. Two players, Black and White, take turns either placing a piece of their color on an empty intersection or passing. A three-move swap option is recommended.

An *orbit* is a cycle of same-colored pieces around one or more points (the board edge does not count as part of a cycle). A *half-orbit* is a connected set of pieces that, together with one side of the board, encloses one or more points. Orbits and half-orbits may contain both orthogonal and diagonal connections.

Enemy pieces within an orbit are captured and removed from the board. It is forbidden to play within enemy orbits or half-orbits.

For example, Figure 9.22 shows a white half-orbit in the top left corner and a white orbit immediately below it. A black half-orbit is shown in the bottom left corner surrounding two pairs of white pieces. If Black moves at point ***a*** then an orbit will be completed, capturing the enclosed pair of white pieces; the other pair is not captured.

In the formation shown in the top right corner, White can capture a black piece by playing at ***b***, but then all four white pieces can be captured next turn if Black plays at ***c***. The point marked ***x*** is *shared territory* owned by both players. The formation shown in the bottom right is not yet decided. Black has a half-orbit, but if White plays at ***d*** then the four enclosed black pieces would be captured and the black half-orbit broken.

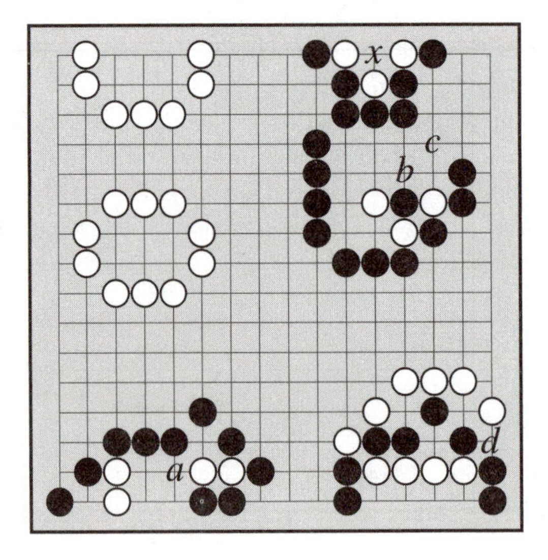

Figure 9.22. Orbits, half-orbits, and territory.

The game ends when either the board fills up or both players pass in succession, and is won by the player with the most territory. This is given by the number of empty points contained by players' orbits and half-orbits; captured pieces and empty points in shared territory are not counted.

Orbit was designed by Steven Meyers and was first described in *GAMES* magazine [Meyers 2001b]. Orbit was originally a family of over two dozen related games, but it was decided that Half-Prohibition Orbit, the game described above, should become the standard [Meyers 2002].

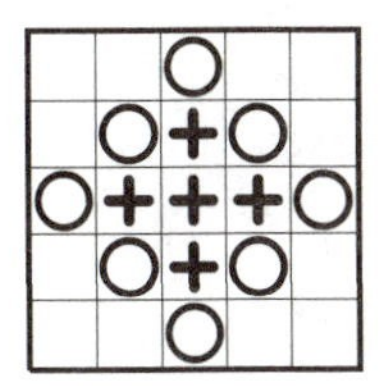

Waroway's Game

Waroway's Game is played using rectangular tiles two square units in size and marked with colored arc segments. Some examples are shown in Figure 9.23 (left).

Players take turns placing a tile to match existing tiles, and aim to complete cycles of their color such as those shown in Figure 9.23 (right). Players may score points based on cycle length or the amount of territory thus enclosed [Waroway 2001].

Figure 9.23. Waroway's Game tiles and a game in progress.

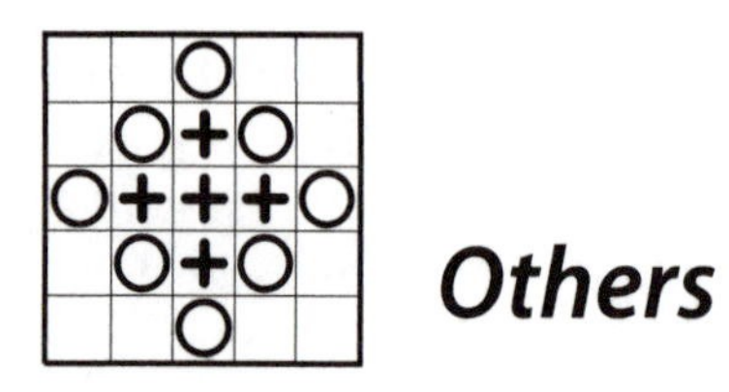

Others

Pathfinder

Pathfinder is played on a large square grid. Players share a common pool of square tiles marked with straight paths, right-angled paths, and path crossings. Players take turns placing a tile on the board, and score points by enclosing territory and completing paths between scoring areas. Pathfinder was published by Milton Bradley in 1954.

Steppe/Stak

Steppe/Stak is played on a 5 x 5 square grid that is initially empty. Each player owns a number of tiles of their color that cover one, two, or three squares in a row, plus three square neutral tiles. Players take turns placing a tile of their color until the lower level is full. Players then take turns placing tiles on level two until it is full, then level three. Upper level tiles can only be played to sit exactly over same-colored tiles on the level below. Players score a number of points per tile multiplied by the number of the level on which they sit.

Steppe/Stak was designed by David Rea and published by DMR Games in 1985. There is some confusion as to whether the game described above is actually called Steppe or Stak.

Rum's Game

Rum's Game [1996] is played on an 8 x 8 square grid. Grid lines are marked by grooves etched into the board. Each player owns a number of single unit rods called *pawns* and smaller square pieces called *connectors*. Pawns fit into the etched edge grooves, and connectors fit into the etched intersections where grooves meet. Connectors may contain straight, L-shaped, T-shaped, or X-shaped paths. The game starts with a centrally placed X-connector.

Players take turns placing a pawn and a connector on the board, such that the connector sits at an intersection and traces a path back to the central X-connector via the pawn. A connector need not be placed if the pawn is placed between two existing connectors. If a player encloses an area of the board, then all enemy pieces within that area are captured (under certain restrictions) and returned to their owner. Only the owner of an enclosure may play within it. Once all legal moves have been made, the player with the least remaining pawns in hand wins the game.

Through the Desert

Through the Desert is played on a hexagonally tiled board that is initially empty. A number of oasis tiles are placed randomly on the board, and a number of water hole tiles are placed at predesignated points. Players then take turns placing tokens representing camels of various colors bearing their riders, and the game proper begins.

Players take turns placing two camels of various colors on the board, subject to certain conditions. The camels must connect with the same-colored group bearing the player's rider, such that same-colored caravans grow as the game progresses. Once all camels are placed each player's score is calculated. Players score points for having the longest camel train, for having camels adjacent to key cells (oases and waterholes), and for enclosing territory.

Through the Desert was designed by Reiner Knizia, and was originally published by Kosmos as Durch die Wüste in 1997.

Nasca

Nasca is played on a 10 x 11 square grid. Each player owns a number of planks of their color, ranging from one to five square units in length. Neutral planks of length one to three may be placed around the game to create variable starting positions.

Players take turns placing a plank of their color. All planks on the board must align with the board grid. Planks must always sit flat and may be stacked on top of existing planks, provided that no two planks overlap by more than one square, and stacked planks rest on top of at least one same-colored plank. This method of piece stacking is similar to that of Druid.

The game ends when no player can place any further planks. Players score points according to the amount of territory enclosed by orthogonally connected paths of their planks. Nasca was designed by Corné van Moorsel and published by Cwali in 1998.

Blokus

Blokus is played on a 20 x 20 square grid that is initially empty. Each player owns a corner of the board and has 21 tiles of his color, each composed of two to five squares joined in various configurations. Players start by placing one of their tiles in their corner. Thereafter players take turns placing one of their tiles such that it touches at least one of their existing tiles at a corner (no tiles of the same color may meet at an edge). The aim is to place as many tiles as possible.

Orthogonal connections between tiles are forbidden within color groups, but diagonal connections are allowed (and in fact compulsory). This produces an unusually porous connection pattern that allows the opponent to push his connection through if a player is not careful, leading to tactical battles. Blokus has a territorial aspect in that players tend to define areas with their spreading connection, even though these are likely to be invaded by an enemy connection at some point.

Blokus was designed by Bernard Tavitian and published by Sekkoia in 2000.

HexGo

HexGo is a Go variant played on the intersections of the hex hex board's dual; a hexagon tessellated by triangles. Two players, Black and White, take turns placing a piece of their color on an empty intersection. Players strive to connect specified pairs of sides to enclose territory, and capture enemy pieces by preventing them from making similar connections.

HexGo was devised by Gregory van Patten in February 2001. It shares many similarities with Anchor, which was invented independently at around the same time.

Block

Block is played on an 8 x 8 square grid. Two players, Black and White, take turns claiming an empty square. When all squares are filled in, the player with the single largest orthogonally connected group wins the game. Block is Joris game #23 [2002].

Occupier

Occupier is played on an 8 x 8 square grid with the middle four squares removed. The board is initially empty. Two players, Black and White, take turns claiming an empty square. A player who completes a straight path between any two board sides, using orthogonal or diagonal connections, may begin counting the territory enclosed. Occupier is Joris game #24; see Joris [2002] for further details.

Parcel

Parcel is played on a 10 x 10 square grid that is initially empty. Two players, Black and White, take turns claiming either one empty square or two orthogonally adjacent empty squares. As soon as an area of empty squares is completely surrounded by an orthogonally or diagonally connected border of a player's pieces, then that player scores one point for every empty square enclosed. Parcel is Joris game #92 [2002].

Great Walls

Great Walls is played on a 20 x 20 square grid that is initially empty. Two to four players each own a number of stones of their color, and share a common pool of 116 cards. Each card shows a 5 x 5 grid with between one and five squares marked in a pattern.

Players are dealt five cards each, which are kept hidden from opponents. Each turn the current player must either pass, or use one of his cards and draw a fresh one to replenish his hand. A player uses a card by placing a number of his stones on the board in the same formation as that shown on the card. All stones must occupy empty board squares, and may not be placed in enclosed areas. An area is *enclosed* if surrounded by an orthogonal chain of same-colored pieces (board edges do not count as part of an enclosure).

Any stones surrounded by an enemy enclosure are captured and removed from the board, even if they form an enclosure themselves. The game ends when all players pass in succession, and is won by the player who has enclosed the greatest number of empty squares.

Great Walls was designed by Benjamin Cedarberg and published by Microcosm Games in 2002.

Feurio!

Feurio! is a boardless tile-laying game played with 36 hexagonal tiles. Two to four players each have access to a common pool of tiles and own a number of firefighters of their color. Some tiles show water, but all show a portion of a forest fire and a number between one and six indicating the severity of the fire. Lower numbers indicate the presence of more water. Each tile also contains between one and three locations where firefighters may be placed.

The forest tiles are shuffled and placed face down, then four tiles are turned up to form a 2 x 2 rhombus to start the game. Players take turns randomly selecting a forest tile and placing it at the point along the perimeter with the highest score; thus the fire always spreads from its hottest point. After placing a forest tile, the player may then place up to three firefighters on any tile, provided that the tile's locations are not already fully occupied, and the total number of firefighters upon it does not exceed its number of freedoms (free edges).

Once all tiles are placed, then players continue to place their firefighters until there are none left or all players pass. Players' scores are then calculated as follows: connected sets of tiles containing same-colored firefighters constitute *regions* that score for that player. A region is worth the total number shown on all member tiles, divided by the smallest number shown on any member tile with a free edge. Completely enclosed regions with no free edges score zero points. Players therefore score highly if their firefighters have a path to safety, and score poorly if their firefighters are enclosed by fire.

Feurio! was designed by Heinrich Glumpler and published by Edition Erlkönig in 2003.

Connection-Related Games

A number of board games contain a significant but not exclusive aspect of connection. These are not connection games as such, but are worth mentioning to further define the point at which a game is deemed to belong to one of the strictly connective categories. The coverage of Connection-Related games is deliberately kept brief, so as not to distract from the main topic of strictly connective games.

Figure 10.1 shows four broad categories into which Connection-Related games tend to fall. Please bear in mind that the games described in this chapter are *not* connection games as defined by the proposed classification scheme, and that the importance of connection varies significantly within this group.

10.1 Connection Quality

The Connection Quality group involves games in which some connections are better than others; the winner is the player with the longest, strongest, or best connection, or connections worth the greatest score. It is not the fact of connection that counts, as in true connection games, but the quality of connection.

Connection-Related: Connection plays a significant but not exclusive role.

- **Connection Quality:** Biggest or best connection wins.
  Alhambra, Aquarius, Ataxx/Hexxagon, Collector, Dominoes, Dos Rios: Valley of Two Rivers, Elcanto, Fresh Fish, High Kings of Tara, Ish, Medina, Nomadi, Poison Pot, Power Grid, Scrabble, Spell Sport, Splitter, Take It Easy, Zenix, Cube Farm, Steam Tunnel
  - **Empire Building:** Best empire wins.
    Acquire, The Ark of the Covenant, Attika, The Bridges of Shangrila, Carcassonne, Conflict, Drunter & Druber, Euphrates & Tigris, Illuminati, Kahuna, Magna Grecia, Mexica, Morisi, Ozymandia, Risk, Settlers of Cataan, Vinci, Diplomacy, Domaine
  - **Rail Network:** Best railway network wins.
    1830, Crayon Rails, Stephenson's Rocket, Streetcar/Linie 1, Ticket to Ride, Union Pacific
  - **War Games:** Best army wins.
    Advanced Squad Leader, Afrika Korps, Air Assault on Crete, Battle Cry, Battle of the Bulge, Burma, D-Day, Fortress: Europa, Hammer of the Scots, Napoleon, PanzerBlitz, Quebec 1759, Russian Front, Stalingrad, Third Reich
- **Reach a Goal:** First player to reach a specific goal with his piece(s) wins.
  Breakthrough, Chinese Checkers, Dodgem, Dragon Delta, Expedition, Mouse Island, Mutternland, Octiles, Peaceful Resistance
  - **Maze Games:** First through the maze wins.
    The Amazeing Labyrinth, Barrier (Joris #38), Continua, Elfenland, GOOTMU, Labyrinth, Magalon, Master Labyrinth, Minotaur Maze, Pathfinder, Porto (Joris #8), Quoridor, Rally, Ruhmreiche Ritter, Rush Hour, Shuttles
- **Form a Pattern:** First to achieve a specific target pattern wins.
  Amoeba, The Antoni Gaudi Tile Game, Canoe, Dao, Deduction, Hexade, Imago/Mem, Kensington, Plotto, Polygon, Quadrature, Reversi/Othello, Rings, SanQi, Spangles, Susan, Territoria (Joris #13), TriHex
  - ***N*-in-a-Row:** First to achieve *n* pieces in a row wins.
    Black & White, Boku, Bolix, Connector (Joris #97), Fire & Ice, Gomoku, Gomullo, Interplay, Neutreeko, Pente, Quarto!, Qubix, Rapid 4, Space Marbles, Tic-Tac-Toe, Time-Vectors, Yinsh
- **Limit the Opponent:** The player to limit or kill his opponent's piece(s) wins.
  Amazons, Atta Ants, Axiom, Breakthru, Cathedral, Chase, Chex, Connected Chess, Dots & Boxes, Dvonn, Fibonacci, Gro: Battle for the Petri Dish, Guerilla, Hive, Ichishiqi, Isolation, Killer Beams (Joris #25), Labyrinth (Joris #35), Lotus, Medusa, Northcott's Game, Orion: Draco, Quivive, Rondo, Sprouts/Brussels Sprouts, Timeline Chess, Terrace, Topple, Uisge, Universe

Figure 10.1. General types of Connection-Related Games.

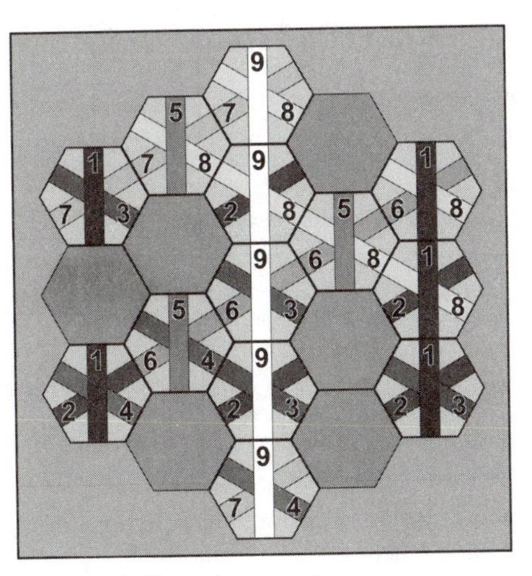

Figure 10.2. A Take It Easy tile and a game in progress.

For instance, Figure 10.2 shows the popular tile-placement game Take It Easy. Each player controls his own copy of the board, in addition to 27 tiles each showing three path segments of a particular color and a corresponding score. Each turn, one player randomly selects a tile, then all players must place their copy of that tile on any empty cell on their board.

Upon completing any straight line of a single color between any two sides of the board, a player scores the total number of points along that line. For instance, the game shown in Figure 10.2 is currently worth (5 x 9) + (4 x 8) + (3 x 7) + (3 x 1) = 101 points for that particular player. Take It Easy was designed by Peter Burley and released by FX Schmid in 1994. It is commonly described as a game of spatial bingo.

Another aspect of Connection Quality is found in Topple and the curious games described by Soriano [1981] and Brotz [1988] that involve domino tumbling. Players take turns placing a domino or tile on the board so that it balances on its edge. Either player may choose to push a domino over at a certain point in the game to initiate a chain reaction, creating a connected path of toppled dominoes. Players may score points based on criteria such as the number of dominoes toppled, whether the toppled path spans the board sides, and so on.

Three further subcategories of Connection Quality may be defined: Empire Building, Rail Network, and War Games. These tend to include strongly themed games that hide a connective element underneath.

10.1.1 Connection Quality > Empire Building

Many games of this type involve the development of a connected empire spreading across the board. However, connection is not the overriding concern of the game, and is secondary to aspects such as the size and strength of the empire, monuments built, key resources owned, and so on. Play tends to be adjacency-based rather than connection-based, as in Risk and its relatives, and often features trade and warfare.

10.1.2 Connection Quality > Rail Network

As the name suggests, this group involves the expansion of rail networks. Connection plays an important part as players compete to establish tracks between key points; however, track development and the deployment of trains is usually complicated by the logistics implied by the railway theme. Metro and TransAmerica are rare examples with strictly connective play.

Rail Network games usually involve a strong aspect of trade, barter, and economics, and are won by the player with the largest or strongest network, or the most money.

10.1.3 Connection Quality > War Games

War Games simulate real or imagined battles, campaigns, or wars. The rules do not usually specifically mention connection; however, strong connective elements can often be found if they are examined closely enough. Examples include

- invasions along a coast in which armies must establish beachheads then connect up with other units invading at other points;
- the development of a connected network of roads or rails, and the disconnection of enemy lines by sabotage;
- supply lines that must remain connected back to a headquarters or depot;
- the encirclement and capture of enemy armies;
- the congregation of allied units to form stronger units;
- the maintenance of a solid front against an enemy; and
- line-of-sight connection for communication networks, or providing cover fire for advancing troops.

Most war games are played on a hexagonal grid superimposed on a stylized map of the field of battle, and benefit from its trivalent nature. Richard Reilly's treatise "Connection Games and War" examines this substantial group of games in detail [2004].

10.2 Reach a Goal

Reach a Goal games are those in which players strive to reach a particular goal with their piece or pieces. Connection plays a part in defining players' paths to their goals. Reach a Goal games include the subcategory of Maze games.

10.2.1 Reach a Goal > Maze Games

Maze games are a well-known genre. Players must negotiate a maze to reach a goal from a specific starting position. Typically the path may change over the course of the game.

Figure 10.3 illustrates the game of Quoridor. The game starts with players' pieces facing each other across an open board (left). Each player owns ten fences, each two units long, which define a maze of walls as they are placed on the board. Each turn the current player must either

- move his piece to an adjacent square, possibly hopping over the opponent's piece in the process, or
- place a fence, provided it does not completely disconnect the opponent from his target edge.

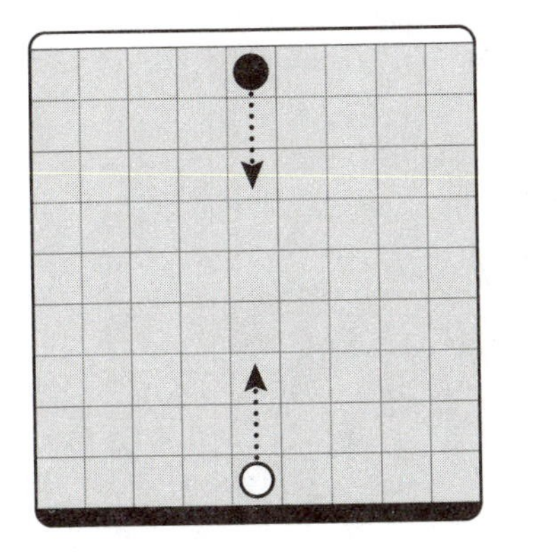

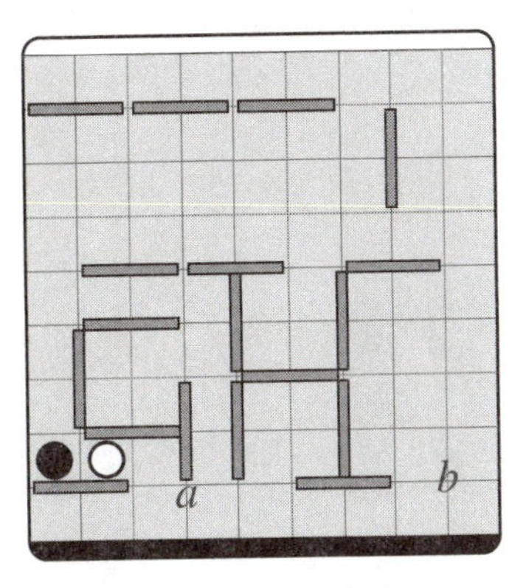

Figure 10.3. The Quoridor starting position and a puzzle: Black to play (each player has one fence remaining).

A player wins by reaching his target edge on the far side of the board. The number of ways for a player to reach his target edge decreases as more and more fences are placed on the board; Quoridor begins in a maximal state of connective potential that converges as the game progresses.

Quoridor is strongly connection-based while players have fences in hand, but becomes a simple maze-solving race once the last fence is played. Figure 10.3 (right) shows a game nearing the race stage, with Black to play and each player with one fence in hand. Black *must* play his last fence horizontally at ***b*** to win the game. This is the only move that stops White playing his last bridge horizontally at ***a*** and setting up a race that he will win.

Quoridor was designed by Mirko Marchesi and manufactured by Gigamic in 1997. It was previously released under a slightly different form as Pinko Pallino, which in turn appears to have been derived from the 1970s game Blockade.

Board games based on the Dungeons & Dragons role-playing theme also fit into this category, as a dungeon is a maze of connected areas.

10.3 Form a Pattern

Form a Pattern games are those in which players strive to form a particular pattern with their pieces. Such patterns are often connected sets of pieces; however, connection is little more than a side effect of achieving these patterns.

For instance, Figure 10.4 (left) shows a pattern called a *canoe* that players must create in the game of the same name. Players take turns either placing one of their 13 pieces on the board, or moving an existing piece to any empty adjacent point (orthogonally or diagonally).

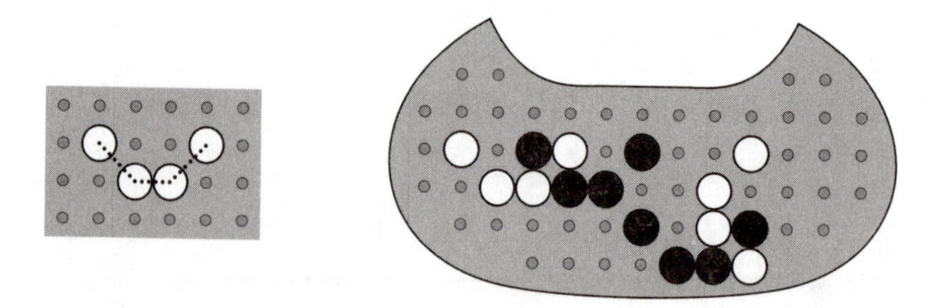

Figure 10.4. A canoe pattern and a game of Canoe won by White.

A player wins by creating two canoes of his color, which can be at any of the four possible rotations, but must be disconnected from all other same-colored pieces. Figure 10.4 (right) shows a game won by White. Black has also achieved two canoes, but these are connected so do not count as a win. Canoe was published by Pin International in 2004. It has a strong anticonnective element though it is not itself a connection game.

Form a Pattern games include the subcategory of N-in-a-Row games.

10.3.1 Form a Pattern > *N*-in-a-Row

Games involving the formation of straight line patterns of n pieces in a row should be familiar to most board game players and do not warrant further explanation. Popular games of this type include Tic-Tac-Toe and Gomoku.

10.4 Limit the Opponent

Limit the Opponent games are those in which players attempt to limit the opponent's options until he is captured or has no positive moves left. Connection typically manifests itself as defining boundaries in territorial disputes or imposing constraints on previously defined piece movement.

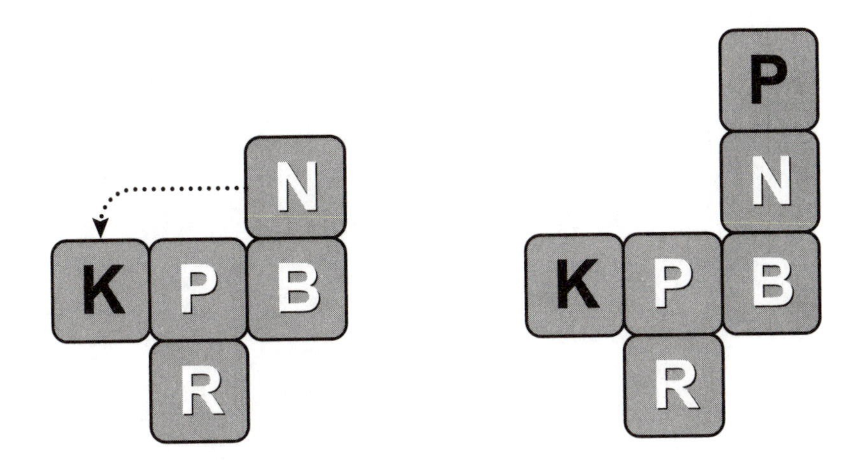

Figure 10.5. Black's king is saved by a pawn in Chex.

For instance, Figure 10.5 shows the game of Chex, in which each player has 16 square tiles, each showing a piece from a standard Chess set. Each turn the current player must either

- randomly draw one of his unused tiles and place it on the playing area adjacent to at least one existing tile (orthogonally or diagonally), or
- move one of his existing tiles as the equivalent Chess piece would move, including captures, provided that all played tiles form a single connected group at all times.

The concept of check is necessarily redefined in Chex, and the game is not won until the opponent's king is captured. For instance, Figure 10.5 (left) shows the black king in check from a white knight. However, if Black then places a pawn as shown on the right, then the white knight is pinned (any knight move would disconnect the pawn) and the king is no longer in check. Chex was designed by David Smith in 1994. The official set shows a rendered picture of the relevant Chess piece on each tile.

The Connection-Related categories describe general trends rather than hard distinctions, and many games will cross over into more than one group. For instance, Lotus is a territorial Limit the Opponent game with a Form a Pattern component.

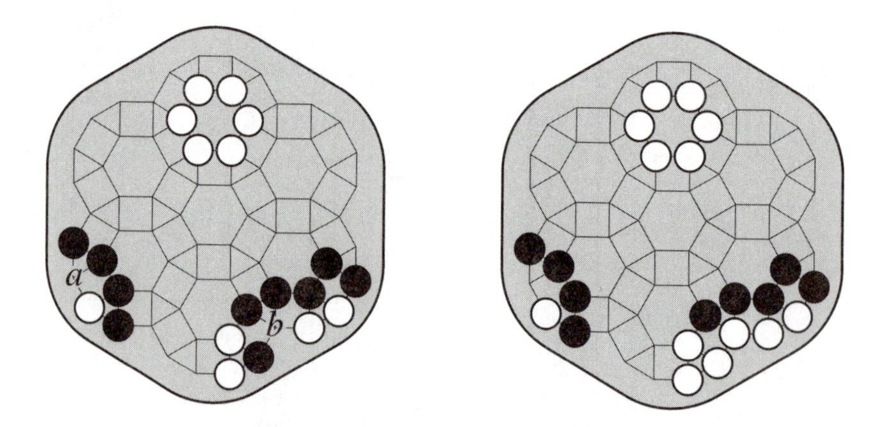

Figure 10.6. A surround-flip capture in Lotus.

In an interesting twist on the usual surround-capture rule, Lotus features a surround-flip rule; enemy pieces with no freedoms are flipped to the capturing player's color. For instance, if Black moves at point ***a*** in Figure 10.6 (left) then the white piece is flipped to black. White cannot move at ***a*** as suicide is not allowed. However, White can move at ***b***, as that move flips a black piece and creates liberties for the newly formed white group, as shown on the right.

A player forms a *lotus* by placing six friendly pieces at the corners of any of the seven hexagons on the board, as shown towards the top of Figure 10.6. Lotuses and all pieces connected to them are safe from being flipped. Lotus was designed by Christian Freeling and is described in *New Rules for Classic Games* [Schmittberger 1992]. The lotus formation is identical in form and function to the *rosette* formation of Mark Berger's 1975 game Rosette.

Part III

End Play

Part I examined the basic nature of connection, and Part II presented a catalog of games. Part III now touches on some of the more speculative aspects of connection games including game design and the psychology of connection.

Rolling Your Own

Creating new connection games can be an interesting exercise in itself. This chapter describes some general observations on the game design process (with an emphasis on connection games), methods of evaluating games, and some case studies of game designs that didn't quite work.

11.1 The Design Process

Recall from Section 4.1 that ludemes are the conceptual elements that make up a game. Game design can be seen as the process of creating new combinations of ludemes that work together harmoniously.

Such combinations may click into place immediately and a complete game may develop within minutes (see Quintus), or may take years of iterative refinement as a rule set is optimized (see Onyx). Ideally the rule set will converge to a simple, tight combination resulting in a game that is correct and of high quality. These concepts will be defined shortly.

Schmittberger's *New Rules for Classic Games* [1992] provides some tips for getting extra mileage out of familiar games. Adding connection metarules to existing games that are not already connection games is difficult; however, players may warm to a new game that is at least partially familiar to them.

Most new connection games are extensions of existing connection games, and in a broader sense, all connection games are extensions of the basic connection idea. Even Hex itself was inspired by work in the field of map coloring [Gardner 1957].

11.2 Evaluating a Game

Games can be evaluated according to two basic criteria: correctness and quality.

11.2.1 Correctness

A game is described as being *correct* if it provides a contest between players with a guaranteed conclusion and no serious flaws. Such flaws might include unachievable goals, frequent deadlocks (or *any* deadlock, depending upon how much of a purist the designer is), or awkward play mechanics that make the game unworkable.

For example, Figure 11.1 shows a generic connection game on a hex hex board in which players aim to connect three nonconsecutive sides with a chain of their color. It's not known how many times this fundamental idea has been tried and discarded over the years, but there is a good reason for its conspicuous absence: the game can be spoiled so that neither player can win (Figure 11.1, right). The connection of two opposite sides is at least as easy to achieve as the connection of three nonadjacent sides, hence it can be expected that this game will usually end in a draw.

One of the first things to test when evaluating a game is the ease with which an obstructionist opponent can spoil the game. This is generally not an issue for games with complementary goals.

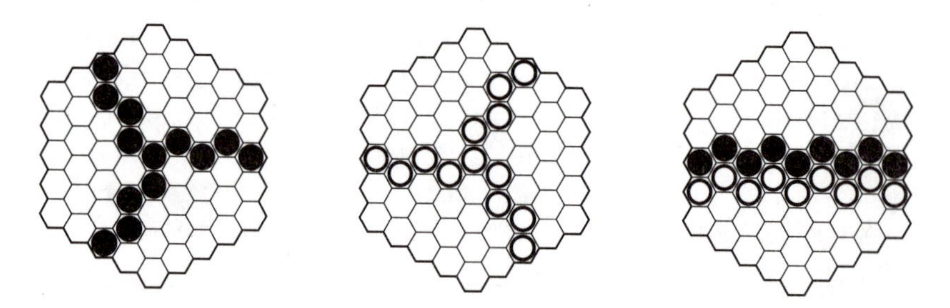

Figure 11.1. A game won by Black (left), White (center), and neither player (right).

Inconclusive end games can also lead to unachievable goals. For instance, if the aim of the game is to capture all enemy pieces but both players have been reduced to a single piece, then it may be possible for players to elude capture indefinitely. In this case a simple rule change specifying the winner to be the first player to reduce the opponent to two or fewer pieces might solve the problem.

11.2.2 Quality

Once it is established that a game is correct then its *quality* should be evaluated. Mark Thompson, in his excellent article "Defining the Abstract," points to four basic qualities that a game should possess to have lasting merit [2000]:

- depth,
- clarity,
- drama, and
- decisiveness.

Depth describes how amenable the game is to being played at different levels of skill. A game should be as enjoyable to the beginner at a superficial level as it is to an advanced player at a much deeper level. In a deep game, new strategies should open up as the game is examined in greater detail.

Clarity is the ease with which a player can understand the rules and goals of a game and plan ahead. Games with excellent clarity are described as *transparent*, while those at the other extreme of the clarity scale are described as *opaque*, due to excessively complex rule sets or winning conditions that muddy the strategic waters. Good connection games tend to have excellent clarity, as discussed in Section 4.3.1.

Drama is the ability of a player to recover from a bad position and win the game, or, as Schmittberger describes it, the tendency of a game to have reversals of fortune and close finishes [1992]. This means that a losing player should still have a vested interest in the game.

Drama will be lacking if the slightest advantage will eventually lead to victory unless an error is made. First-move advantage is a common problem that must be addressed; see Section 4.3.2 for ways to resolve this problem. Games subject to trivial winning strategies (Appendix E) also lack drama.

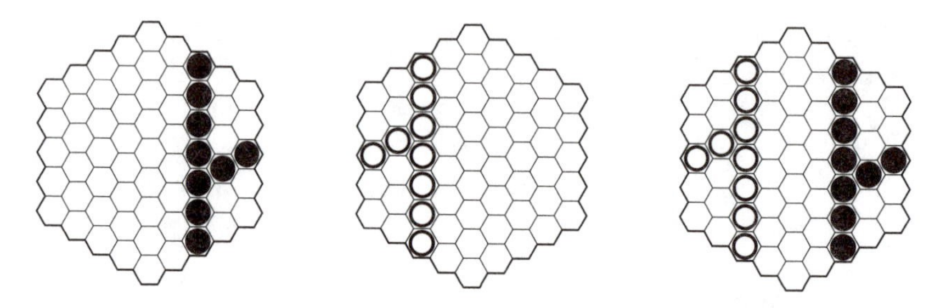

Figure 11.2. A game won by Black (left), White (center), and both players (right).

Returning to the generic hex hex game, the spoiling problem may be corrected by relaxing the winning condition so that any three sides may be connected. This game is now correct but lacks drama, as it has become a race easily won by the first player.

The connection of three sides of a hex hex board is a natural and appealing idea, and although it does not stand on its own as a game, it has been successfully incorporated as an element of games such as Havannah, Unlur, and Triskelion.

Decisiveness means that a player can achieve an advantage from which the opponent cannot recover. An attractive feature of connection games is that they generally converge to increasingly well-defined solutions as they progress. In Richard Reilly's terms, one player runs out of ways to stop the winner.

A further quality can be added: *interaction.* Meaningful interaction between players is the key to success in most nonsolitaire games. In nonabstract games this might involve players bidding, trading pieces, swapping property, and so on. In abstract board games this manifests itself as interactions between pieces on the board, and in connection games more specifically as the interference between players' connections. In terms of connective flow, this can be seen as the turbulence of flow within a game.

If players are discouraged from engaging enemy forces (perhaps due to clumsy move mechanics or overly powerful symmetrical capturing rules) then little interaction will take place and the game will lack interest. If a game lacks interaction then it may be a simple matter of changing the initial board set-up so the game starts with enemy pieces already in conflict.

11.2.3 Other Factors

Abstract board games and puzzles are intimately related, and a game may be seen as a series of puzzles that each player presents to the other. Thompson says, "the design of a good abstract game must therefore allow an inexhaustible supply of puzzles to be discovered in the possible positions of its board" [2000].

This observation leads to a simple rule of thumb for further evaluating a given game: *how easy is it to create interesting puzzles based on the game's rules?* The existence of such puzzles does not in itself indicate the quality of a game, as such puzzles may never occur naturally during normal play. However, a lack of such puzzles is usually a warning sign that there is something wrong with the game. If a tester cannot construct an interesting board position, then it is unlikely that players will find much of interest in the game either.

Other views on evaluating games can be found in *New Rules for Classic Games* [Schmittberger 1992] and "What Makes a Game Good?" [Kramer 2000]. These works emphasize different aspects of game design; however, depth appears to be a universal and overriding concern.

Abbott's observation that "offense must generally be more powerful than defense" [1988] does not apply to the same extent to complementary-goal connection games, where strong defense is as good as strong offense.

11.3 Convergence of Rule Sets

It is easy to create new games with complex combinations of ludemes. However, rule complexity does not necessarily mean that a game is better; it may simply mean that the game is harder to play.

Converging to the simplest possible rule set by applying Occam's Razor to strip away superfluous rules will generally improve the elegance and clarity of a game. Connection games are peculiar in that exceedingly deep games can be achieved with the simplest of rules. One reason for this may be the very presence of the concept of connection, which implies relationships between pieces that do not need to be explicitly stated in the rules. The implications of connection can be bewildering on a tactical level yet easily understood by most players. As Phil Bordelon says, it's like you get extra rules for free.

The history of connection games contains many examples of games being independently reinvented, often at practically the same time. Hex is a famous example, but it is likely that many such rediscoveries have not even been recorded. This may be coincidence as the territory of connection game variants becomes increasingly well trodden, or perhaps this points to optimal basic forms to which rule sets tend to converge within given frameworks.

If games are combinations of ludemes, and subject to the same laws of propagation that govern other types of memes, then this may be "*memetic*" *convergence* in action.

11.4 Case Studies

Examples of successful game design can be found aplenty in Part II of this book. However, examining game ideas that did not quite work can give a glimpse into the game design process.

The most readily available cautionary tales include some designs of my own that did not work out. I prefer to think that this is because of convenient access to their history (game designers will generally not share their failures with others) rather than any talent on my part for discovering bad ideas.

11.4.1 Hexma

The first example demonstrates how correctness can be achieved easily, but quality is more elusive.

The inspiration for Hexma was to create a connection game based on Chinese Checkers. It was hoped that the resulting game would benefit from the freedom of movement found in Chinese Checkers, as well as the fact that most board game players would already have ready access to that game's equipment and rules.

After some experimentation it was found that a basic game could be played by two or three players using 13 pieces each to connect opposite sides of the board. Figure 11.3 (left) shows the starting position for White and a winning chain for Black who has the connected the two opposite sides of the board (marked in gray). Piece movement is according to standard Chinese Checkers rules: a piece may move either to an adjacent empty cell, or as a series of single piece hops over adjacent pieces of either color to empty points beyond.

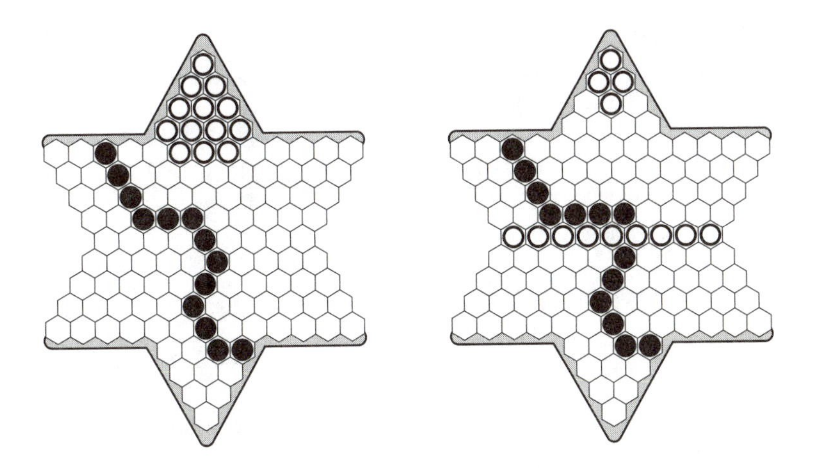

Figure 11.3. A game of Hexma won by Black (left) and a game spoiled by White (right).

However, it was soon found that the board geometry made it easy for either player to spoil the game, as White has done in Figure 11.3 (right). Hexma in this form was not correct. Artificial rules specifically forbidding barricades, and even different board shapes on which such barricades are impossible, easily corrected this problem.

The game was then correct according to the definition above, but further testing revealed serious problems with quality. Mobility became increasingly constrained as players were loathe to move pieces crucial to their connection. Even worse, interference between opposing pieces clogged up the board, and it proved more effective for each player to just ignore the opponent and pursue an independent parallel connection. There was little incentive to engage enemy pieces and minimal interaction between players took place.

The Chinese Checkers rule set rejected attempts to graft a connective goal onto it; consequently Hexma has been noted as a curiosity and shelved.

11.4.2 Chikadou

The second example demonstrates how rules tend to converge to optimal sets.

Chikadou is a game by Phil Bordelon that combines the connective growth mechanism of Tanbo with the connect-alternating-sides metarule of Hex.

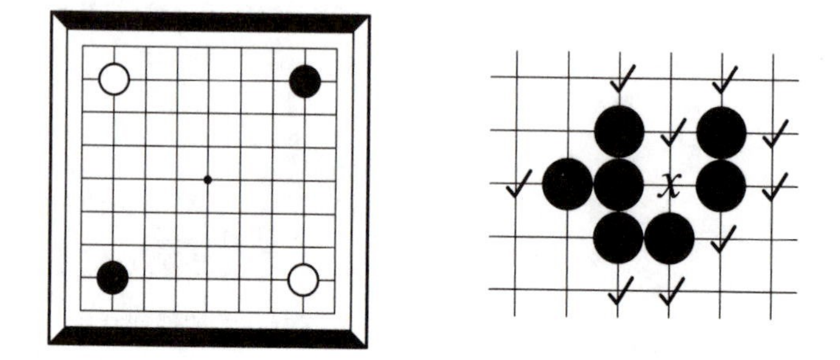

Figure 11.4. The Chikadou starting position (left) and examples of valid moves (right).

Figure 11.4 (left) shows the Chikadou starting position on a 9 x 9 board. Note that the game is played on the board intersections. Each turn the current player must place a piece of his color to be squarely adjacent to exactly one of his existing pieces, as per Tanbo. Enemy chains with no freedoms are captured and removed from the board.

To achieve its connective goal, Chikadou is played with an additional rule: a piece may be placed so as to join two or more friendly chains if it is adjacent to exactly one piece from each chain. For instance, all of the ticked positions in Figure 11.4 (right) are valid Chikadou moves. The position marked *X* is not a valid move as it is adjacent to two pieces in the chain to its left.

This rule set looked promising. The Tanbo capture rules cleared any potential deadlocks and the game was correct. However, games tended to devolve into a simple race won by the first player, who was always a step ahead and able to push his connection forward; the second player was unable to impede this progress due to the constrained placement rules. In addition, the fact that no player could ever create more groups than the initial number made the game very limiting for all but larger boards.

The obvious solution to these problems was to relax piece placement constraints, and allow pieces to be placed at any vacant square that was not adjacent to more than one piece from any friendly chain. This new combination of rules proved workable, but it was hard to ignore the fact that Chikadou Redux was now really just Gonnect with directional goals and limited piece placement.

Gonnect had been rediscovered via a significantly different route. This suggests that Gonnect may be the epitome of surround-capture connection games, to which others of this type will tend to converge.

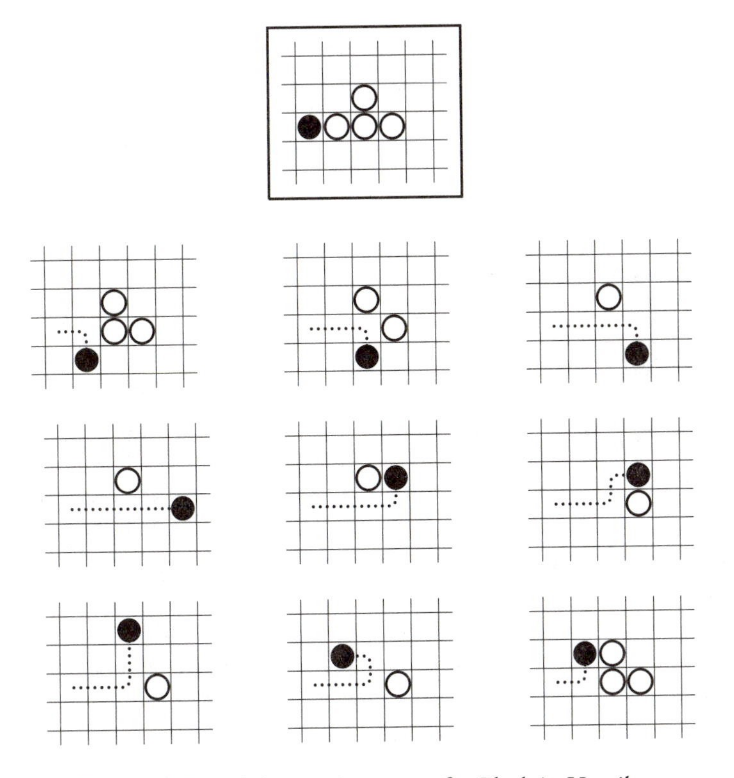

Figure 11.5. Valid capturing moves for Black in Hamilton.

11.4.3 Hamilton

The third example demonstrates that a good idea in isolation is not sufficient.

Hamilton is a path-based capture metarule, as illustrated in Figure 11.5. At the top is shown a board position with Black to play. The nine diagrams below this board position show legal capturing moves for Black, who can hop over any number of enemy stones in an orthogonally connected path.

The black piece must land on an empty cell adjacent to the white chain, capturing any white pieces that it passes over in the process. No point can be revisited that turn. Like Checkers, a capturing piece may continue hopping over multiple adjacent enemy groups per turn. No group can be revisited that turn.

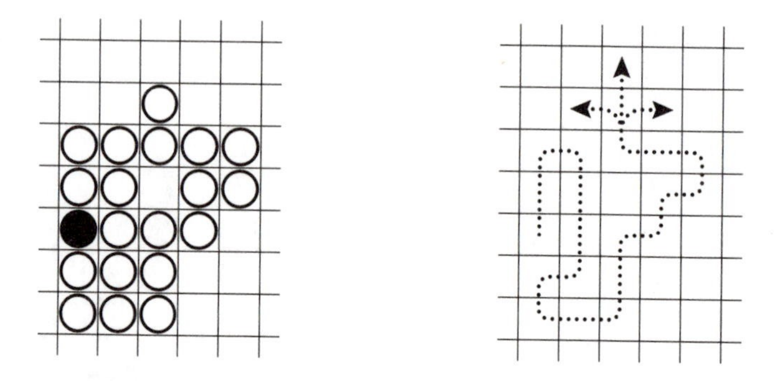

Figure 11.6. Capturing all pieces in a chain involves tracing a Hamiltonian path through it.

Hamilton is so named because any move that completely captures an adjacent chain must trace a *Hamiltonian path* through that chain (defined in Appendix A, Basic Graph Theory). Figure 11.6 shows the only path that completely captures the white chain shown. Black may choose to end the move at any of the three destination points indicated.

Hamilton was tested on an 8 x 8 Chess board with 16 pieces per player initially placed along each player's first two ranks. Each turn the current player could move one piece either to an empty adjacent point, or as a series of one or more captures. The winner was the first player to reduce the opponent to two pieces or less.

However, a serious flaw immediately emerged. Due to the symmetric nature of captures, no player would dare place any piece adjacent to an enemy piece. Failure of interaction ruined this rule set. Many artificial rules were tried, such as

- misère (the first player to lose all pieces wins),
- mandatory maximal capture, and
- asymmetric capture (for instance, only larger/smaller chains may be captured).

No combination of rules was found to encourage players to exploit the Hamiltonian style of capture to its full extent without introducing undue complications. Hamilton remains a metarule in search of a good home.

12 The Psychology of Connection

This chapter explores some of the cognitive aspects of connection games, such as why some players (usually male) seem naturally adept at them, and their possible role as educational tools.

12.1 Everyday Connections

One reason for the inherent appeal of connection games may be the familiarity of the concepts involved. We deal with different sorts of connections on a daily basis, be it joining the dots of a scribble, looking for constellations in the night sky, solving a maze, or even just planning the best route to drive home from work (Figure 12.1).

The concepts used in interpreting the board position of a given connection game have their basis in everyday skills. Even beginners will have some life knowledge that may assist them in understanding a new connection game, making it feel intuitively familiar.

Connection as an art form is not a new concept. The ornamental art style known as Celtic knotwork has been practiced for over 1,500 years, and is little more than a connection puzzle. The true skill in this art form lies in creating a complex design from a single continuous cord. There is little evidence to suggest that this desire to complete patterns with a single cord has any spiritual or cultural significance, despite popular

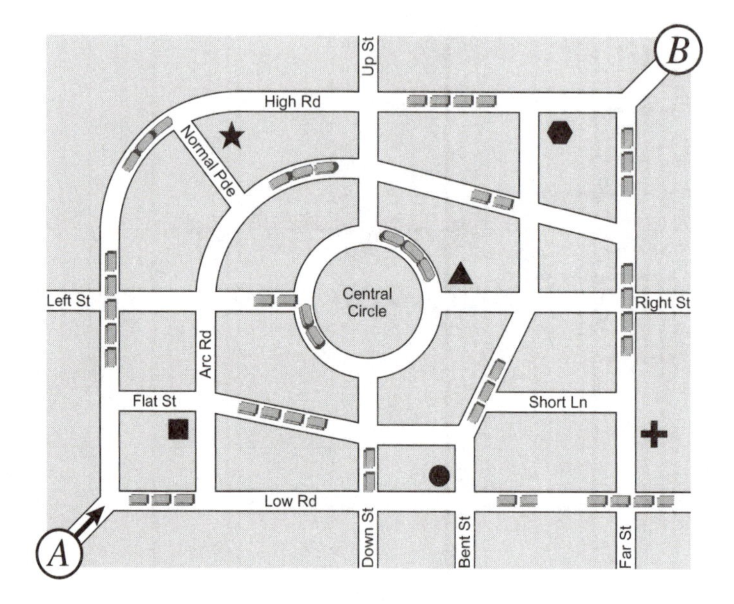

Figure 12.1. Planning a route from A to B in peak hour traffic.

opinion to the contrary [Ivan 1999]. More likely it is just a testament to the artist's skill.

It has recently been shown that traditional Celtic knotwork designs can be stripped down to their underlying graphs, much like the games discussed in this book [Mercat 1997].

Taking the simple graph of Figure 12.2 (left) and placing a crossing at the midpoint of each edge, nearby crossings can be linked to define a path (middle). Introducing an over/under weave pattern and filling this path gives the final plait (right).

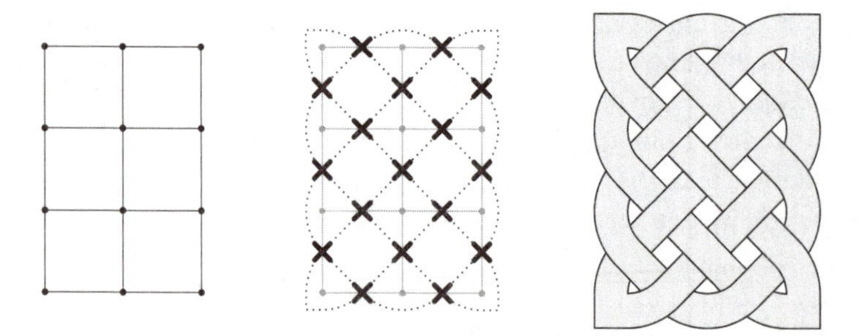

Figure 12.2. Celtic knotwork from a graph.

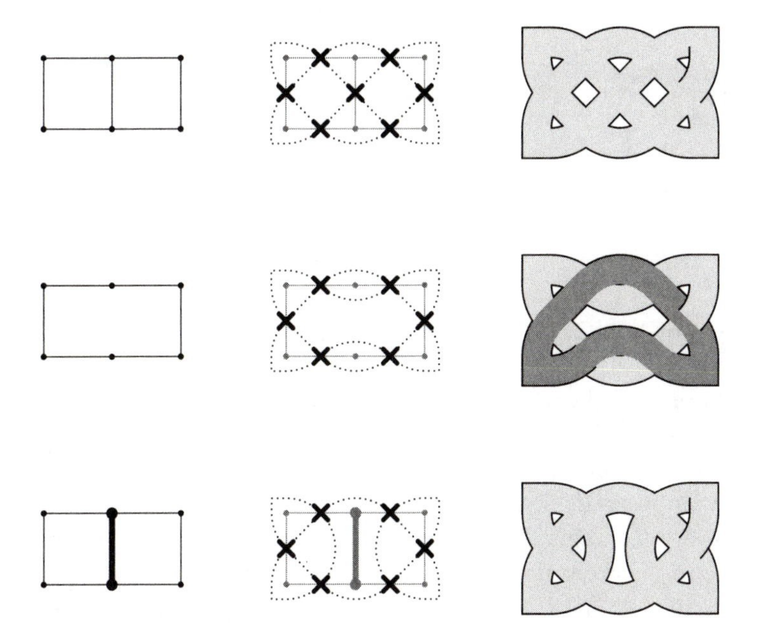

Figure 12.3. Edge operations introduce breaks into the pattern.

The design in Figure 12.2 fortuitously gives a single closed cord. This is not always the case, and breaks are often introduced into the weave pattern to split or combine cords to achieve some artistic effect. These breaks are defined by operations on graph edges similar to moves on the graphs of connection games, as shown in Figure 12.3. The top row shows a basic plait. The middle row shows the same design with one crossing removed by deleting an edge. Note that the original plait has been split into two interwoven cords by this break. Claiming the edge rather than deleting it (bottom row) introduces a perpendicular break into the weave.

Just as decorative knotwork forms an integral part of some connection-related games (for instance, High Kings of Tara and Knots), the edge break rules shown in Figure 12.3 allow any board position reducible to a planar graph to be turned into Celtic knotwork. However, a good game does not necessarily make good artwork; aesthetically pleasing designs tend to be strongly symmetrical and pattern-based, whereas connections on game boards do not necessarily follow these constraints.

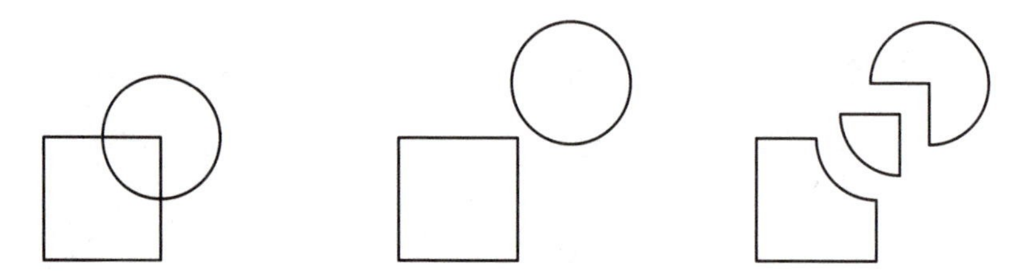

Figure 12.4. We tend to perceive the simplest interpretation of ambiguous figures.

12.2 Visualizing Connections

We have a natural tendency to mentally complete any unfinished pattern, be it a half-drawn picture, a mumbled sentence, a melody cut short before the final note, or the flow of connection across a game board. This is more than just a tendency to complete known patterns; it also applies to higher-level concepts such as the continuity and form of a system, and may be another reason why connection games can be inherently appealing to some people. The study of such phenomena is within the domain of the Gestalt school of psychology, which is based on the idea that perceived forms can transcend the stimuli used to create them.

One of the basic tenets of Gestalt theory is *pragnanz*, the law of good form, which suggests that we tend to perceive the simplest interpretation of an ambiguous figure. For instance, most people would interpret the left-most figure of Figure 12.4 to be composed of an overlapping circle and square (middle). The right-most interpretation of three nonoverlapping irregular shapes is equally valid but not often made, as a believable explanation can be found using simpler shapes. In terms of Occam's Razor, the perceived explanation of an ambiguous figure is the one that relies on the fewest assumptions about the objects in question.

We generally perceive shapes that share some common feature such as *proximity* or *similarity* as being grouped together. Consider the regular grid of dots shown in Figure 12.5 (first grid). If we push some of these dots closer together then we perceive them as forming horizontal lines (second grid) or vertical lines (third grid) rather than a uniform field. Similarly, if we replace every second dot with a square we see distinct alternating columns, not a mixed set of dots and squares (fourth grid).

More important to the context of connection games is the law of *good continuation*. This is the brain's tendency to choose the interpretation of

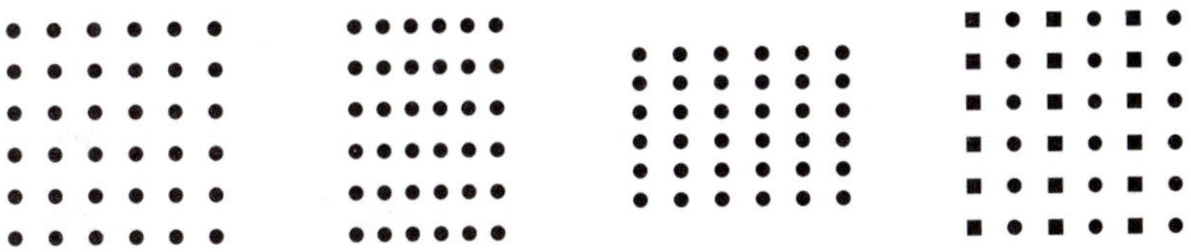

Figure 12.5. Grouping by proximity and similarity.

a path that maximizes continuity. Figure 12.6 (left) shows an ambiguous figure with four loose ends. This figure will almost universally be interpreted as two smoothly continuous paths crossing at right angles, rather than two paths with sharp bends that meet at the corner (right).

This phenomenon is important to our understanding of maps. For instance, it should be obvious to any reader of the map shown in Figure 12.1 that Arc Rd runs from the bottom left of the page and curves around to the upper right in a single path. It is assumed to be continuous across intersections. This is also the principle that defines continuity in tile-based games such as Black Path and Turnabout, in which paths that cross within a tile are assumed to continue across the intersection rather than turning at right angles.

Not only do we interpret ambiguous shapes to preserve continuity, but we are even adept at filling in missing information. For instance, we understand the object on the left of Figure 12.7 to be a single continuous curve rather than a random set of dots. Firstly we perceive each group of five to ten dots as a curve segment, then we interpret the four segments as a single continuous curve with some intermediate dots missing.

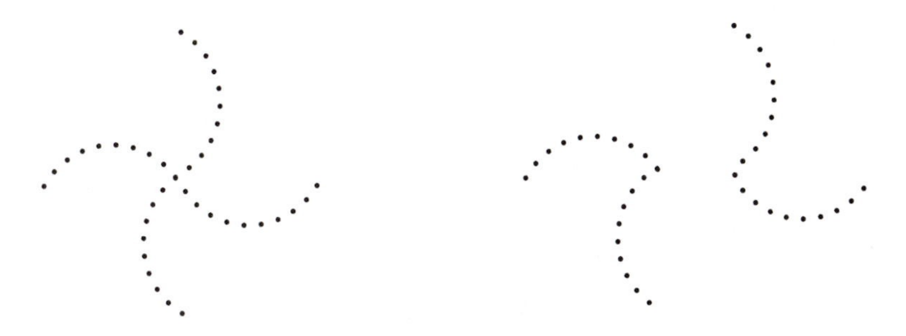

Figure 12.6. Good continuation: two smooth lines crossing, or two sharp corners meeting?

Figure 12.7. Interpolating continuity and the Gillam distance.

Barbara Gillam [1984] suggests that element similarity probably leads to *aggregation* (the perception of related units as belonging to a class) while good continuation probably leads to *unit formation* (the perception of related units as belonging to a single unit in which the parts are no longer distinguishable). Masin [2002] investigated similar concepts in relation to the grouping of disks, such as those that might be used as pieces in a board game. He defined the Gillam distance to be the ratio of the inner and outer distances of two disks (d/D in Figure 12.7, right) and confirmed that this distance plays a significant role in their perceptual grouping.

In the context of connection games, these factors suggest that nearby pieces of similar color are perceived as belonging to groups, depending on their proximity. Moreover, they suggest that we automatically perceive connective paths running through these groups, which may then be perceived as single units. This is demonstrated in the game of Hex shown in Figure 12.8.

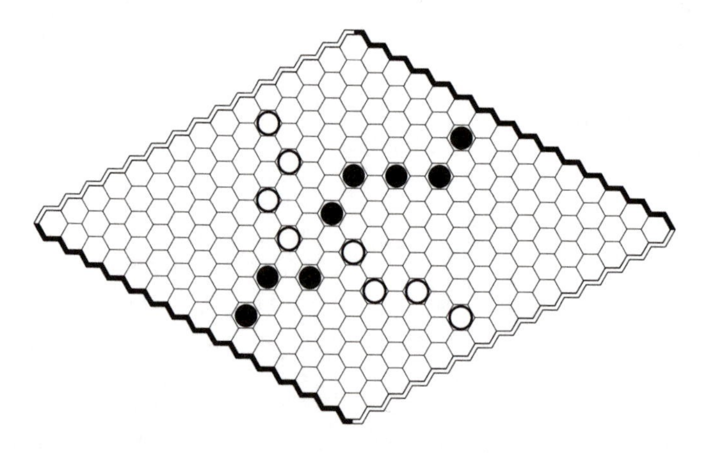

Figure 12.8. Continuity and flow in a game of Hex.

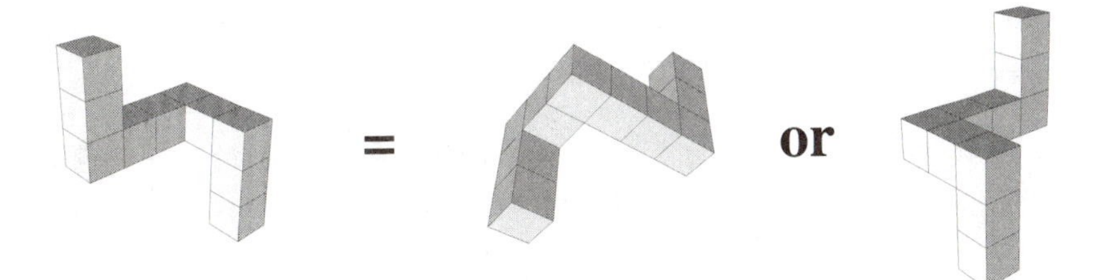

Figure 12.9. Which of the last two shapes is a rotation of the first one?

Firstly, note that the distinct colors immediately segregate this distribution of pieces into black and white camps (similarity). Within these camps, each player's pieces appear to form two main groups, each leading to an edge (proximity). Each pair of groups appears to form a single connected path across the board (good continuation) that intersects the opponent's path near the center. Connecting the dots is something that most of us do throughout our lives in one form or other, and it has immediate relevance to most connection games.

The brain is adept at visualizing connective flow across the board. The relative strength of each player's position stands out immediately, as do board areas that warrant closer examination. These should be evident even to someone with little knowledge of the game.

To summarize, we have a tendency to mentally complete unfinished connections, and these are inherently easy to visualize in most connection games. Given an unfinished connection game with clearly understood goals, there is a natural urge to resolve it.

12.3 Games for the Boys?

While performing research for this book (that is, playing games for years) it became obvious that very few women take an interest in connection games, far fewer than would be expected even given the male preference for abstract board games. There may be tangible reasons for this even more pronounced gender discrepancy in the connection game community.

It has been established that men consistently perform better at certain spatial tasks, such as finding their way through a maze or mentally rotating objects [Kimura 1999]. One such mental rotation test is shown in Figure 12.9. A possible explanation for this skill is that men may be more adept at creating cognitive or mental maps from descriptions of objects.

In addition, men are generally better at navigating a route on a map, as demonstrated by Galea and Kimura in their intriguingly titled paper, "Sex Differences in Route-Learning" [1992]. An important finding of this study was that women appear to navigate by landmarks while men appear to navigate using Euclidean properties such as direction and distance. Women had better recall of landmarks and tended to make more errors in regions with fewer landmarks.

For example, when navigating a route from ***A*** to ***B*** on the map shown in Figure 12.1, a woman is more likely to give instructions such as "turn left, then right around the square, then left for a while until the circle," and so on, while a man is more likely to give instructions such as "head North for a block, East for a block, South for a block, then East along Low Rd for four miles," and so on.

These results correspond with related findings that women generally outperform men at tasks involving recall of patterns or object locations, such as restoring a shuffled set of objects to their original positions, or finding matching pairs of cards in a deck spread face down on a table. Women tend to have a stronger innate ability to match patterns. Kimura [1999] suggests some possible cultural reasons for these different spatial skills in men and women.

The hypothesis that women tend to be landmark navigators is supported by recent virtual reality experiments conducted by Czerwinski et al. [2002]. Although men have traditionally outperformed women at virtual environment navigation, it was found that women can achieve similar levels of performance to men when using large displays with wider fields of view. Czerwinski et al. believe that this is because the cognitive task of building a mental map then becomes less important, and landmark navigation is optimized with a wider field of view.

These results suggest a possible reason for the male preference for connection games. Connections, by their very definition, are not fixed in space and follow no particular pattern or landmark (apart from the goal areas). The female skill of finding patterns and landmarks has little application in this context, except perhaps for identifying small connective subpatterns such as edge templates in Hex. The male skill of constructing cognitive maps to trace routes, on the other hand, is ideally suited to the task of identifying potential connections in the typical connection game. It is interesting to note that the Pure Connection games described in Chapter 7 were designed almost exclusively by men.

Of course, there are always exceptions to such generalizations. The staff at De Speelstijl observe that women often outperform men when playing Meander [2004]. They suggest that this is because men tend to play more aggressively and invest more energy in obstructing the opponent. However, as seen in Sections 5.3 and 5.6, overly aggressive play in connection games is usually not as strong as a style that balances both defense and attack.

On the broader topic of gender differences in board gaming in general, Tom Vasel reports that women make up a surprising 65 percent of the clientele of the popular board game cafés in Korea [2004]. Vasel attributes this trend reversal to the gregarious nature of Koreans, and the fact that café games are often just an excuse for players to meet for social reasons. This observation is supported by the fact that the most popular café games are overwhelmingly themed "social" games rather than abstract ones.

12.4 Child's Play

Browse the shelves of any educational equipment supplier; most of the available board games will have a mathematical bent, and it's likely that a significant number will involve connection in some way. It has long been understood that playing with spatially oriented toys promotes spatial skills in children [Serbin and Connor 1979]. Boys tend to benefit from this most, since such toys are generally deemed to be masculine and hence more likely to be given to boys.

This claim is supported by Tracy [1990] who found that boys had significantly higher spatial skills than girls, and furthermore that students with high spatial ability also had significantly higher science achievement scores than students with low spatial ability. Tracy suggests that this may be because spatial toys (three-dimensional manipulation toys such as Lego and Tinker Toys) provide concrete representations and manipulations of objects that children may then be asked to make mentally, usually at school. Brosnan [1998] also points out that skill with Lego is a good indicator of spatial skill. Children who completed a particular Lego model scored significantly higher in spatial ability than those who did not.

By contrast, students who played more with two-dimensional and proportional arrangement toys tended to have significantly lower science achievement scores. The majority of proportional arrangement toys

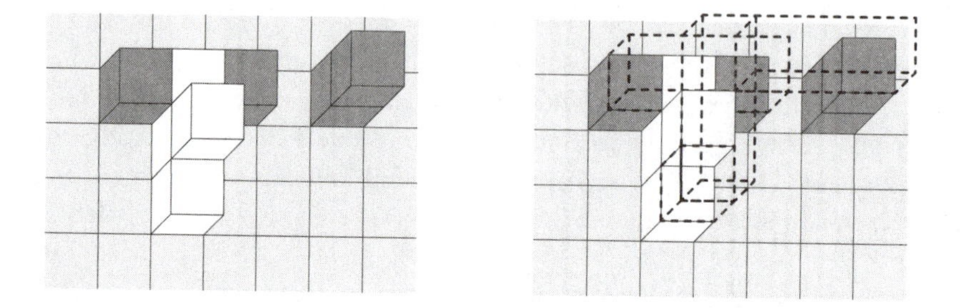

Figure 12.10. Using a three-dimensional mental map to plan Druid moves.

such as tea sets, dollhouses, and kitchen toys are marketed towards girls [Fleming-Davies 2004]. As an aside, Spertus points to societal factors that may contribute to systemic gender biases that discourage girls from technical study even at the highest level of education [1991].

Despite their two-dimensional nature, most connection games arguably belong in the spatial toys category. They do not rely on pattern recognition so much as the development of mental maps of potential connections, and may confer to children the benefits that devising such maps imply.

However, there can be no doubt regarding the description of three-dimensional connection games as spatial toys. Although the list of three-dimensional connection games is short, it is notable that one of them, Round the Bend, is also one of the few connection games specifically targeted at children.

The benefits of three-dimensional play are highlighted in Figure 12.10, which shows a Druid board position and the most likely development of play over the next few moves (dashed lines). This level of planning ahead obviously requires some spatial skill on the player's behalf, and encourages the development of mental maps for estimating potential choices and threats. The fact that Druid is well suited to being played with Lego and other types of children's building blocks also points to its spatial nature.

Apart from these issues, there are strong indicators that simply playing board games in general improves children's mathematical skills. Hallett [2003] describes how fifth-grade students encouraged to play board games performed significantly better at a statewide mathematics test than students of previous years—and had a lot of fun along the way. The first game introduced to the students in this program was the connection-oriented Take It Easy.

13 Afterword

Every connection game is a variation on the same basic theme. However, the rich variety of ideas encompassed by the games in this book cannot be denied. This is a testament to the ingenuity and creativity of the game designers involved.

There are many connection-related avenues yet to be explored. So far there have been few successful marriages of the connection theme with traditional board games, and we have only begun to scratch the surface of at least one promising area, three-dimensional connection games.

So what are the best connection games invented to date? Based on criteria such as elegance, clarity, depth, and importance, in addition to just being fun to play, likely contenders for the top five would be Hex, Y, Gonnect, Star, and Conhex.

Schmittberger describes connection games as one of the most important developments in strategy games over the last century [2000]. They have had a subtle but profound impact on different genres of board games, and have developed into a substantial genre themselves.

Part IV

Appendices

A Basic Graph Theory

This whirlwind introduction to the field of graph theory defines some terms touched on briefly in the main text. The following definitions are taken from *Introduction to Graph Theory* [West 1996].

A *graph G* with *n vertices* and *m edges* is described as having a vertex set $V(G) = \{v_1, \ldots, v_n\}$ and an edge set $E(G) = \{e_1, \ldots, e_m\}$ using set notation. A graph is *planar* if it can be drawn without any edges crossing.

Each edge consists of two vertices called its *endpoints*. The edge and its endpoints are described as *incident*. If the endpoints are equal, then the edge is a *loop*. The edge $e = \{u, v\}$ is written uv.

Two vertices are *adjacent* if they are both coincident with the same edge. We write $u \leftrightarrow v$ to describe adjacency between vertices u and v.

The *degree* of a vertex v is the number of nonloop edges containing v plus twice the number of loop edges containing v. A vertex of degree k is described as *k-valent*. The neighborhood of v, $N(v)$, is $\{x \in V(G) : x \leftrightarrow v\}$.

A *face* is a closed region bounded by edges. For instance, in Figure A.1 the left graph contains five faces, the middle figure contains 16 faces, and the right graph contains no faces.

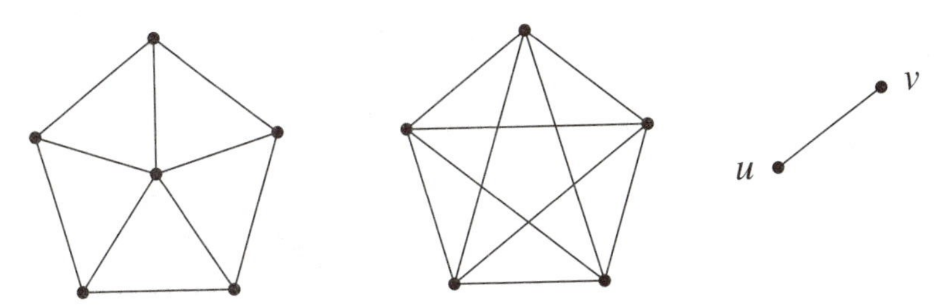

Figure A.1. A planar graph, a nonplanar graph, and an edge uv.

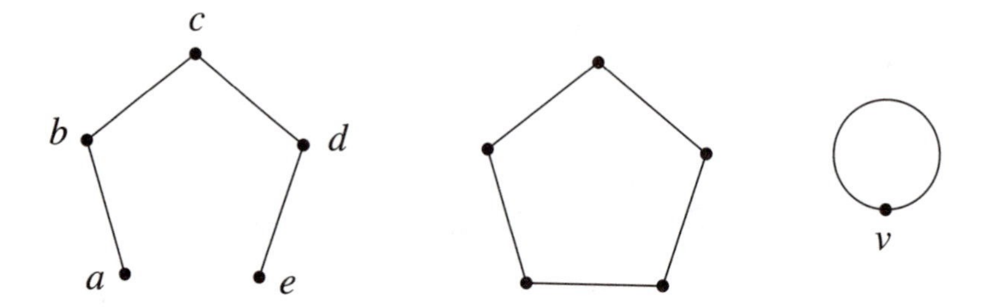

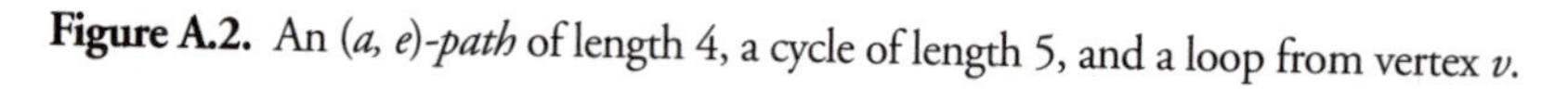
Figure A.2. An *(a, e)-path* of length 4, a cycle of length 5, and a loop from vertex v.

A *directed graph* or *digraph* G consists of a vertex set $V(G)$ and an edge set $E(G)$, where each edge is an ordered pair of vertices. We write $u \to v$ to describe an edge from vertex u to vertex v.

A *subgraph* of a graph G is a graph H such that $V(H) \subseteq V(G)$ and $E(H) \subseteq E(G)$. This is written $H \subseteq G$ and described as "H contains G."

A *walk* of *length k* is a sequence $v_0, e_1, v_1, e_2, \ldots, e_k, v_k$ of vertices and edges such that $e_i = v_{i-1} v_i$ for all i. A *trail* is a walk with no repeated edge. A *(u, v)-walk* has first vertex u and last vertex v, called its *endpoints*. A walk (or trail) is *closed* if it has a length of at least 1 and its endpoints are equal.

A *path* is a walk with no repeated vertex. A *cycle* is a closed trail in which $v_o = v_n$ is the only vertex repetition. A loop is a cycle of length 1.

If G has a (u, v)-path, then the *distance* from u to v, written $d(u, v)$ is the length of the shortest (u, v)-path. If G has no such path, then $d(u, v) = \infty$.

A graph G is *connected* if it has a (u, v)-path for each pair $u, v \in V(G)$, otherwise it is *disconnected*. If G has a (u, v)-path then u is connected to v in G. The *connection relation* in a graph consists of the vertex pairs (u, v) such that u is connected to v.

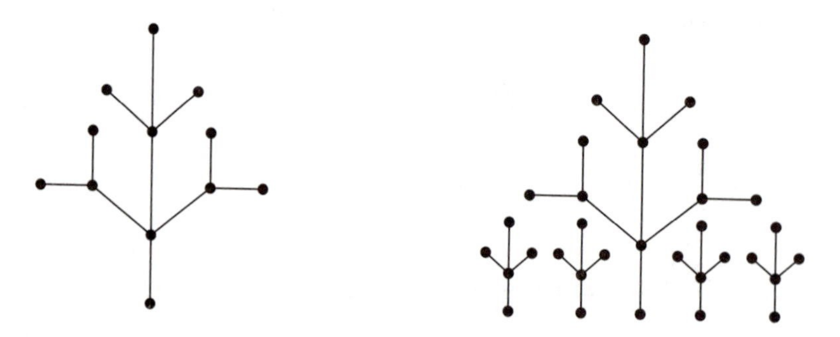

Figure A.3. A tree (left) and a forest (right). A tree is a connected acyclic graph.

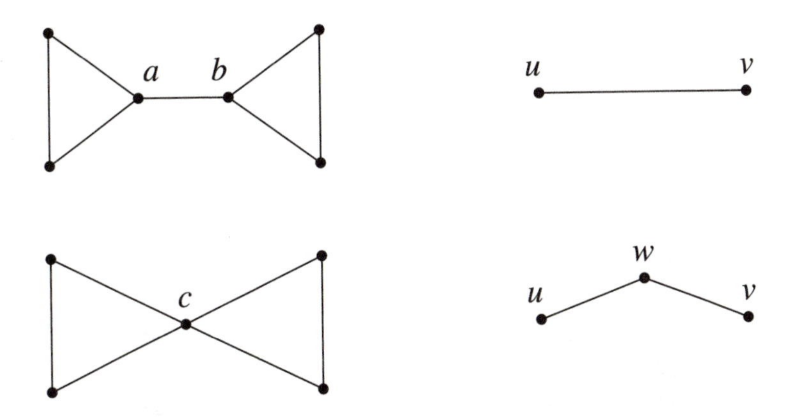

Figure A.4. Vertices a and b contracted to vertex c (left), and edge uv subdivided by inserting vertex w (right). Vertices a and b are of degree 3, while vertex c is of degree 4.

A graph having no cycle is *acyclic*. A *forest* is an acyclic graph, and a *tree* is a connected acyclic graph. A *leaf* is a vertex with degree of at most 1.

If $e = uv$ is an edge of G, then the *contraction* of e is the operation of replacing u and v with a single vertex whose incident edges are those edges other than e that were incident with u or v. Conversely, *subdividing* an edge uv is the operation of deleting uv and adding a path u, w, v through a new vertex w.

The *components* of a graph G are its maximal connected subgraphs. A component or graph is *nontrivial* if it contains an edge. A *cut-edge* or *cut-vertex* of a graph is an edge or vertex whose deletion increases the number of connected components. The resulting graphs are denoted $G - e$ and $G - v$, respectively.

A *disconnecting set of edges* is a set $F \subseteq E(G)$ such that $G - F$ has more than one component. A graph is *k-edge-connected* if every disconnecting set has at least k edges. The *edge-connectivity* of G, written $\kappa'(G)$, is the minimum size of a disconnecting set.

A *spanning subgraph* of G is a subgraph with vertex set $V(G)$. A *spanning tree* is a spanning subgraph that is a tree. *Edge-disjoint spanning trees* are spanning trees with no edges in common.

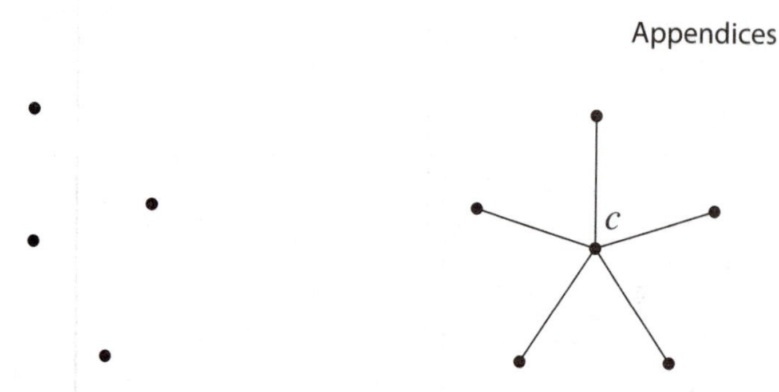

Figure A.5. A vertex set and one possible spanning path with its single cut-vertex c marked.

A *network* is a digraph with a nonnegative *capacity* $c(e)$ on each edge and a distinguished *source vertex* s and *sink vertex* t. Vertices are also called *nodes*. A *flow* f assigns a value $f(e)$ to each edge e. A *source/sink cut* partitions the nodes of a network into a source set S containing s and a sink set T containing t.

A *Hamiltonian cycle* is a spanning cycle (a cycle through every vertex) in a graph. Similarly, a *Hamiltonian path* is a spanning path. A graph is *Hamiltonian-connected* if a Hamiltonian path exists from each vertex to every other.

B Solving the Shannon Game

Contributed by Frederic Maire

Recall that in the Shannon game, the player Join tries to paint a path between the endpoints of an (unplayable) edge $e = xy$, securing edges by painting them, whereas the player Cut tries to disconnect x and y by deleting edges that have not been painted. The term *Shannon game* is understood to mean the Shannon game on the edges, as described in Section 3.2.1.

The Shannon game is a solved game. The elegant solution that Lehman proposed 40 years ago [1964] revolves around the notion of spanning trees that shall be explained shortly. Although Lehman presented his solution in the more abstract framework of matroid theory (the Shannon game can be extended to matroids), the essential ideas of the solution will be illustrated using graphs to avoid the heavy formalism of matroid theory. Refer to Novak and Gibbons [1999] and Bruno and Weinberg [1970] for more complete mathematical presentations.

B.1 Background

In the graph of Figure B.1, Join's objective is to complete a cycle containing edge $\boldsymbol{e} = (1, 7)$ by claiming unpainted edges, while Cut tries to prevent this by deleting unpainted edges.

Although the graph of Figure B.1 is small, a principled strategy is not obvious. On the other hand, the graph of Figure B.2 is a trivial win for Join, even if Join plays second. If Cut deletes the dashed edge, Join can play the dotted edge to win. Reciprocally, if Cut deletes the dotted edge, Join can play the dashed edge to win.

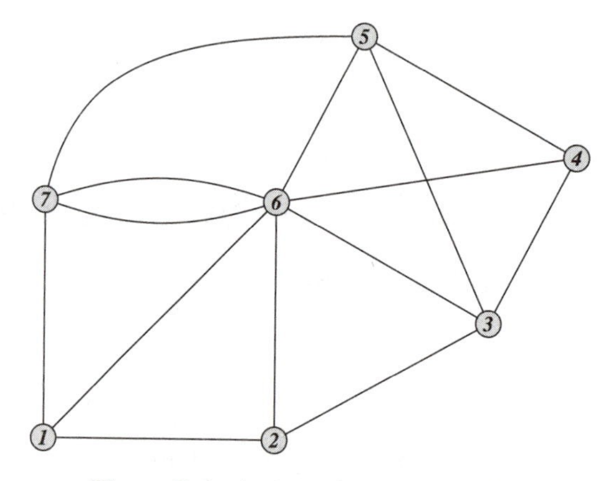

Figure B.1. An introductory example.

Join has two ways to complete the cycle containing e = (1, 2) in the graph of Figure B.2. Winning games for Join are characterized by the existence of two structured sets of edges that form an intertwined network connecting the endpoints of the distinguished edge e.

In Figure B.3, the dotted edges form a tree and the dashed edges form a second tree. Both trees are said to *span the edge* e = (1, 7) *with respect to the circuit* (a *circuit* is a minimal cycle). That is, the addition of the edge e to any of the trees creates a circuit.

Moreover, each dotted edge is spanned by the dashed tree and each dashed edge is spanned by the dotted tree, hence they are called *cospanning trees*. An elegant theorem of Lehman [1964] explains that a game can be won by Join playing second if and only if there exist two cospanning trees spanning the edge e. This is indeed the case for the graph of Figure B.3.

The winning strategy for Join consists of restoring this structure of cospanning trees after each move by Cut. In fact, the edges outside the cospanning trees, such as edge (4, 6) of Figure B.3, can be ignored by Join.

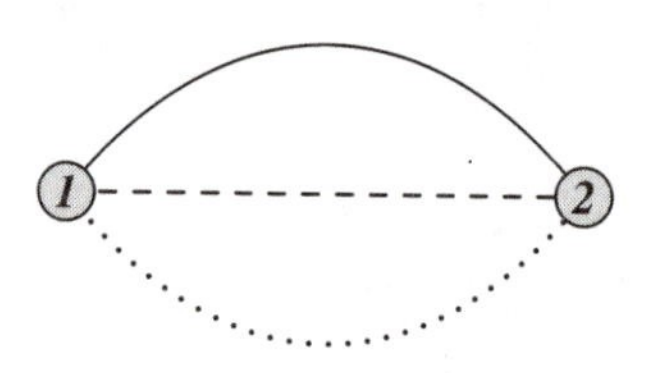

Figure B.2. An easy win for Join.

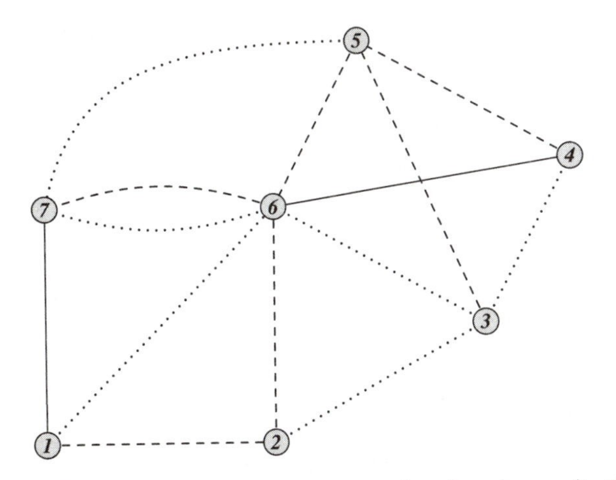

Figure B.3. Two cospanning trees spanning the edge $e = (1, 7)$.

Suppose that Cut plays edge (5, 6). The resulting dashed subgraph is displayed in Figure B.4.

As edge (5, 6) is spanned by the dotted tree, there must be a dotted path from vertex 5 to vertex 6. There must be a dotted edge along this path that bridges the two dashed components of Figure B.4. This bridging edge is the edge that Join must play, in this case edge (5, 7).

Instead of painting the edge, Join can contract the edge. The resulting graph, shown in Figure B.5, again contains two cospanning trees spanning the edge $e = (1, 7)$.

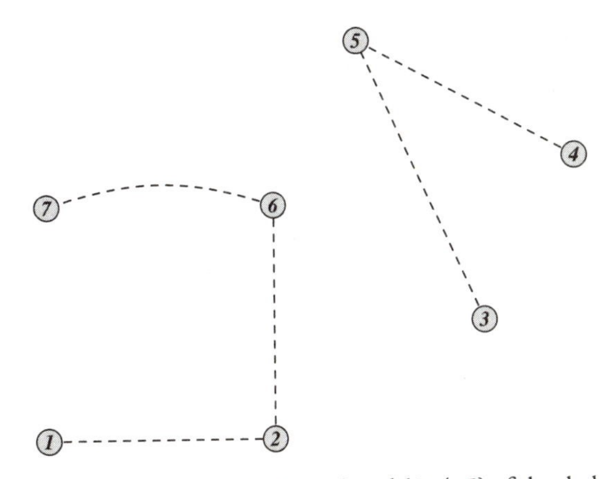

Figure B.4. The two components {1, 2, 6, 7} and {3, 4, 5} of the dashed subgraph.

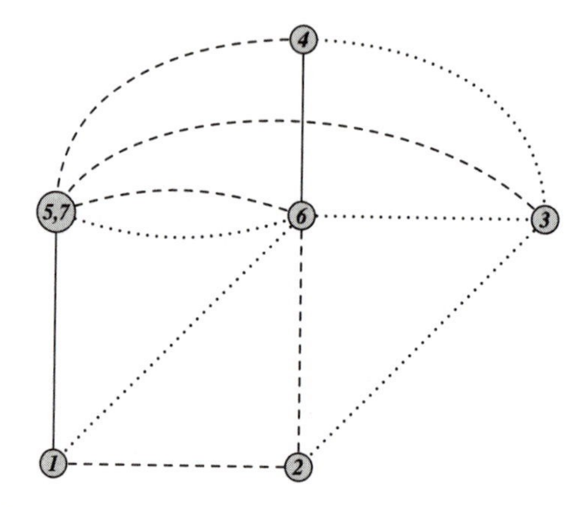

Figure B.5. The structure of two cospanning trees is restored after Join plays edge (5, 7).

A number of questions need to be answered to build a complete artificial player: What happens when there are no cospanning trees? When two cospanning trees exist, how can they be determined? Is there a similar strategy for Cut? These questions are answered in a more detailed discussion by Frederic Maire [2004].

C Hex, Ties, and Trivalency

A game of Hex can never end in a tie. For any Hex board completely filled with pieces, there will be at least one winning path for exactly one player.

The following informal proof was described by Frederic Maire, who first heard it from Claude Berge. The original author is not known, although a version of this proof was published by Beck [1969].

Firstly, the *outer neighbors* of a chain must be defined. These are the pieces adjacent to any member of the chain that are not themselves members of the chain or surrounded by it. Figure C.1 shows a black chain of pieces on a grid full of pieces (left) and its outer neighbors (right).

Note that the outer neighbors of a chain form a connected path, which is the necessary condition for the no-tie property to exist. This condition holds for the hexagonal grid and all *trivalent* grids (grids on which no more than three cells meet at each corner). This condition does not hold for the square grid where diagonal pieces may interrupt the outer neighbor connection, as shown in Figure C.2.

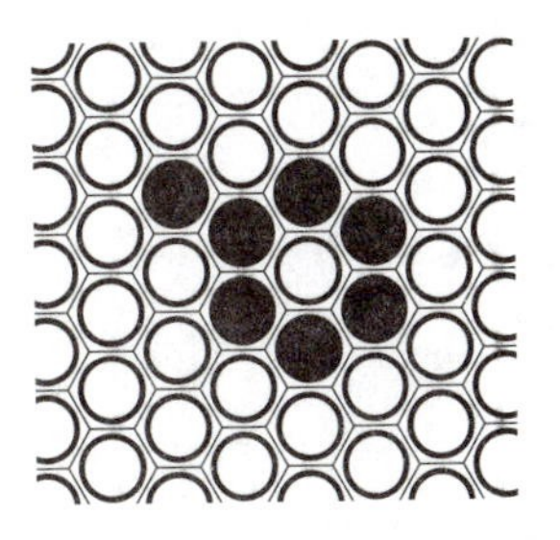

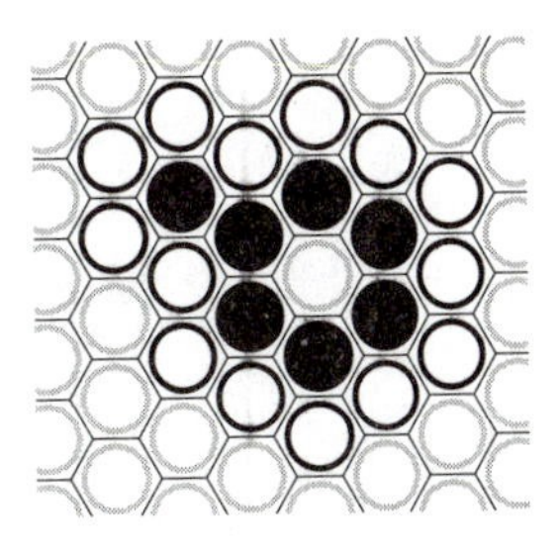

Figure C.1. A black chain on a full grid and its outer neighbors.

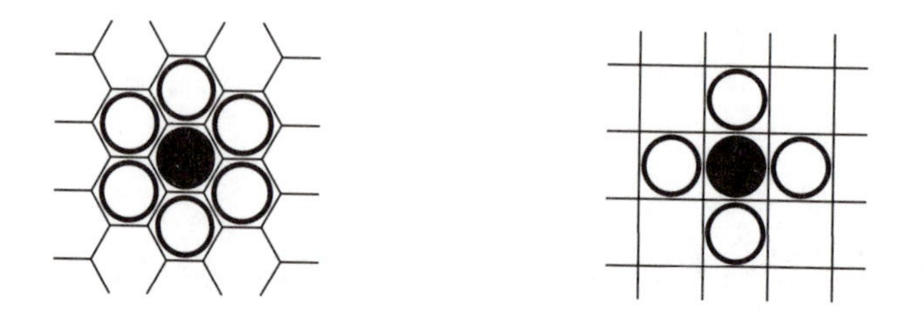

Figure C.2. Outer neighbors form a connected path on the hexagonal grid, but not necessarily on the square grid.

The no-tie property is demonstrated using the board full of pieces shown in Figure C.3.

Suppose that Black has no winning connection. This implies that none of the black chains connected to the lower left edge (shown in Figure C.4) are also connected to the upper right edge.

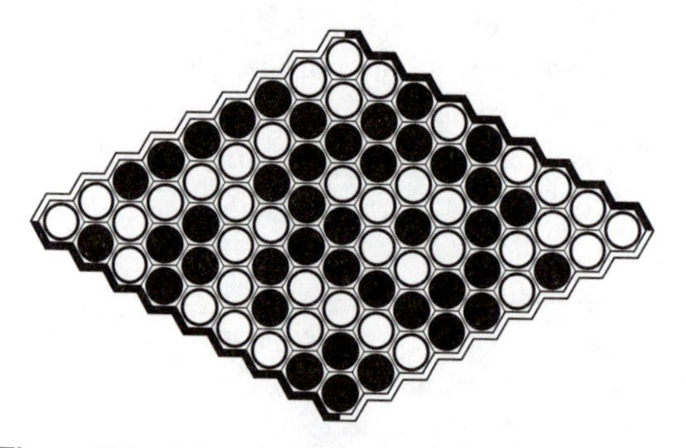

Figure C.3. A Hex board completely filled with pieces.

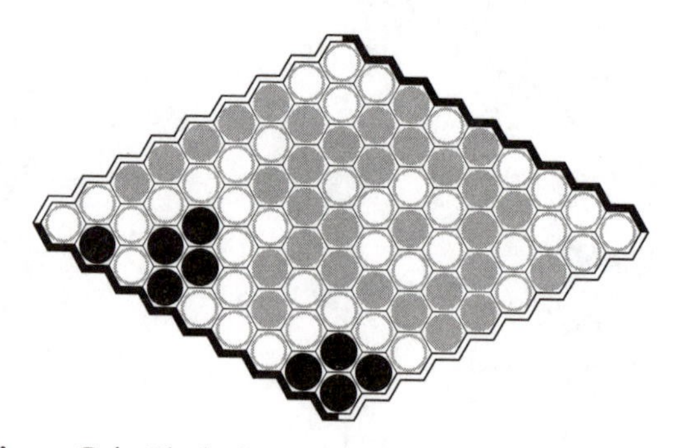

Figure C.4. Black chains connected to the lower left edge.

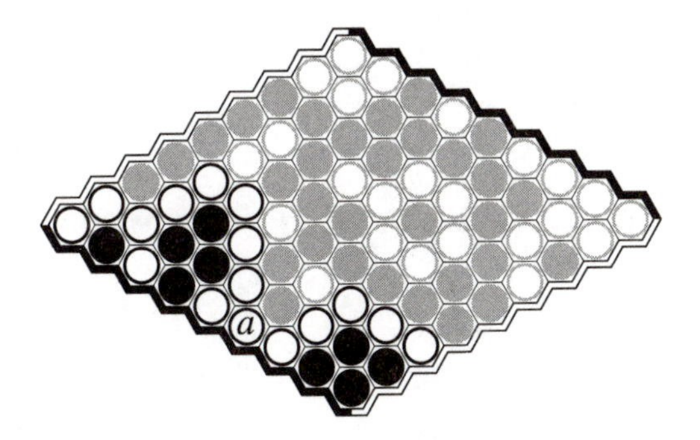

Figure C.5. The union of outer neighbors forms a winning connection for White.

Now consider the outer neighbors of each of these chains. Each of the paths formed by the outer neighbors must start and end at either White's upper left edge, White's lower right edge, or Black's lower left edge (Figure C.5). Observe that any cells along Black's edge not accounted for so far, such as piece ***a***, must be white.

The union of these outer neighbor paths will connect to form a single path touching both of White's edges; this point is further demonstrated by Beck [1969]. Hence White must win this game if Black does not.

Maire also points out that if the white cells are considered to indicate empty space and the black cells solid walls, then the winning path can be obtained by walking through the board with the right hand on the wall, starting from the bottom-most white cell on White's top left edge.

This informal proof is similar in spirit to the well-known Gale proof (see Pierce [1961], Binmore [1992], and Van Rijswijck [2004]), but approaches the problem from a different perspective.

The fact that a game of Hex can never end in a draw has been related to the Brouwer fixed point theorem by David Gale [1979]. This theorem observes that any continuous mapping from the unit square onto itself must contain a point that is mapped to itself. A clear discussion of this point can be found in Binmore [1992].

D Strategy-Stealing

Almost since the invention of Hex, it has been known that the first player has a theoretical win. The argument for this can be summarized as follows: *if there exists a winning strategy for the second player, then the first player can steal it to win the game.*

The strategy-stealing argument was first proposed in 1949 by John Nash [Gardner 1957], and was formally proven by Jales and Hewett [1963]. Curiously, although the strategy-stealing argument demonstrates that the first player *must* win with perfect play, it gives no clue as to what that perfect play is. As Beasely puts it, this is a case of "when you know who, but not how" [1989].

The following outline of the strategy-stealing proof relies on a result due to Zermelo [1912] that was used to show that a finite game of perfect information has a pure-strategy Nash equilibrium [Fudenberg and Tirole 1991]. Basically this means that if one player must win, then one player has a winning strategy.

Theorem: Hex is a theoretical win for the first player.

Proof: The proof is by contradiction.

I. Either the first player or the second player must win (shown in Appendix C).

II. The winning player must have a winning strategy (due to Zermelo).

III. Assume that the second player has a winning strategy.

IV. The first player makes some opening move (clarified shortly), and thereafter adopts the second player's winning strategy.

V. The first player therefore has a winning strategy. This contradicts III, hence the second player cannot have a winning strategy.

VI. Therefore the first player must have the winning strategy.

The first player's opening move is traditionally described in the literature as a random move. If the stolen strategy dictates that a subsequent piece be placed there, then another point is chosen at random, and so on. This relies on the fact that having an extra piece on the board can never harm a player's position.

However, Frederic Maire points out that the second player's hypothetical strategy may rely upon knowing the opponent's last move. If this is the case, then the strategy thief, faced with an empty board, can formulate a first move by imagining an enemy piece on a random board point. If the opponent later moves at this point during the game, then the strategy thief just imagines an enemy piece on another random board point, and so on.

The advantage of Maire's approach is that an extra piece on the board need not be beneficial. This is important for games such as Jade and Chameleon, where an extra piece of either color can be used in an opponent's winning chain. Similarly, an extra piece in surround-capture games such as Gonnect can reduce the player's liberties, and lead to the unwanted capture of pieces.

E Point-Pairing Strategies

Connection games are particularly susceptible to *point-pairing strategies*, in which moves based on the opponent's last move guarantee a win. The existence of point-pairing strategies is one of the first flaws to test for when testing any new connection game.

Martin Gardner [1959] points out a simple point-pairing strategy for Hex on any n x (n + 1) board. Consider the 6 x 7 board shown in Figure E.1. Black, playing second, can always win by simply playing in the cell with the same label as White's last move. It can be observed that any winning path for White must include two cells with the same label, which is not possible if Black adopts this strategy.

Fortunately, this strategy does not apply to Hex on the standard n x n board. A recent Hex-playing computer program uses an approach combining this point-pairing scheme with special handling for the blind row not covered by paired points. However, this program is little more than a parlor trick that may defeat an unwary player, but is unlikely to beat an experienced opponent [Browne 2001].

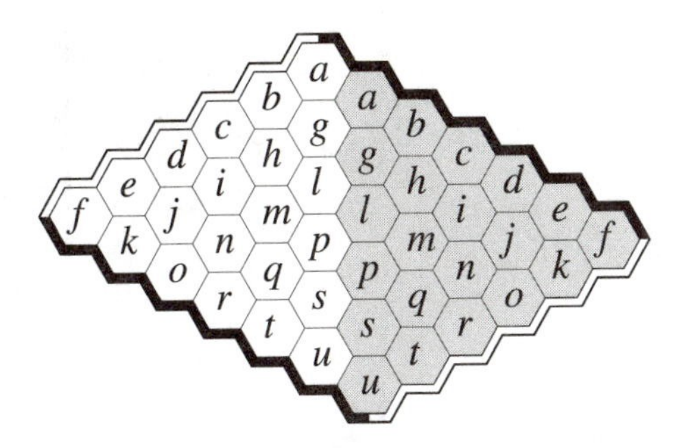

Figure E.1. The n x (n + 1) Hex strategy described by Gardner.

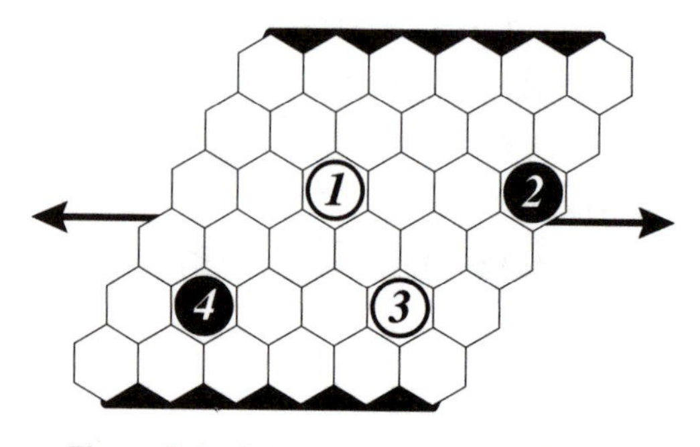

Figure E.2. Point-pairing in Cylindrical Hex.

Cylindrical Hex was introduced by Alpern and Beck in a paper [1991] that also includes its solution, shown in Figure E.2. Black, aiming to connect top and bottom, simply plays in the cell on the same row as White's last move and shifted right by half a board width (the board wraps around between the left and right edges). This result has been proven for any n x n board where n is even, but Alpern and Beck expect it to hold for odd n as well.

Escape Hex, played on a hemispherical board equivalent to the Cylindrical Hex board with one edge pinched to a point, also suffers from this point-pairing strategy. Antipod (Escape Hex's double hemisphere counterpart) is saved through the presence of equatorial handicap points that defeat the point-pairing mechanism. See the Antipod section in Part II for details of both games.

Figure E.3 shows two known symmetry strategies for Jade, as described by Paul van Wamelen.

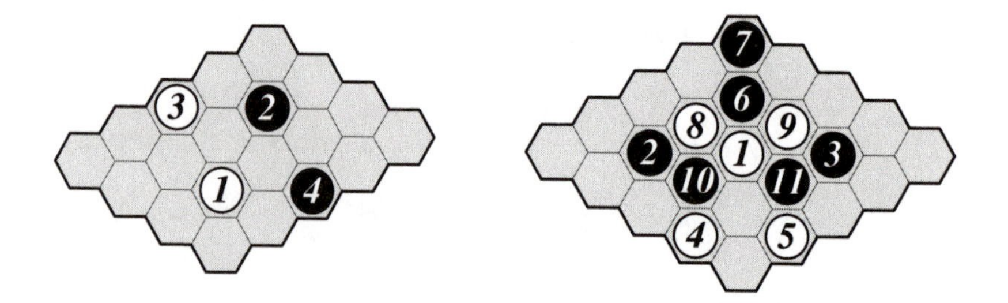

Figure E.3. Symmetry strategies in Jade.

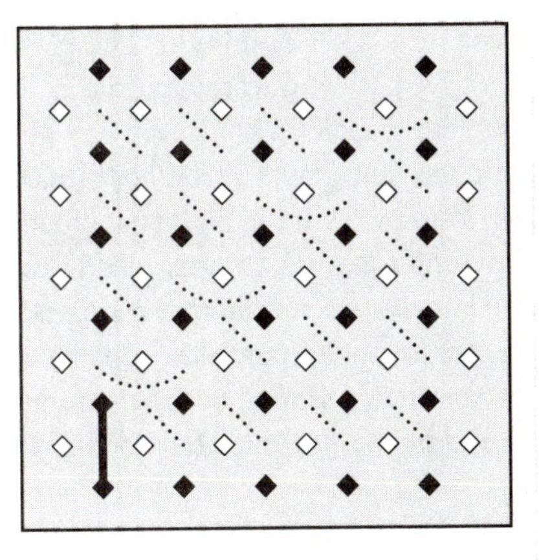

Figure E.4. The famous point-pairing solution to Bridg-It.

Parallel, moving second, can win on any m x n board, where either or both of m or n are even, by playing the opposite color to the opponent's last move at the rotationally symmetrical cell (left). Cross, moving first, can win on any n x n board, where n is odd, by playing the first move anywhere along the short diagonal, and thereafter playing the same color as the opponent's last move at the reflectively symmetrical cell (right).

Point-pairing strategies do not only apply to piece-based connection games. In fact, the most famous point-pairing strategy is probably Oliver Gross's solution to Bridg-It [Gardner 1966]. Black opens in the bottom left corner, and thereafter makes the move that touches the line touched by White's last move (Figure E.4). This strategy was proposed shortly after the game's invention (see Appendix B, Solving the Shannon Game).

Point-pairing strategies also apply to tile-based connection games. Figure E.5 shows a winning strategy for the Black Path game described by Berlekamp et al. [1982] almost 20 years after the game's invention.

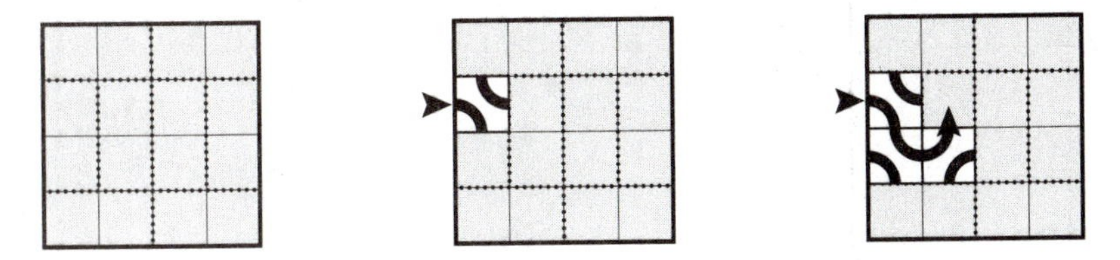

Figure E.5. The domino division strategy for Black Path.

The board is divided into 2 x 1 dominoes. The player simply makes the move that leaves the active path end in the middle of the next domino, safe in the knowledge that the middle of a domino can never be a board edge. This strategy wins for the first player if the board has an even number of squares, and for the second player if the board has an odd number of squares (in which case the opening square is not included in the domino division).

Harary et al. [2002] describe symmetry strategies for *graph-avoidance games*, in which the aim is not to complete a particular subgraph but to avoid doing so. The symmetry strategy of the second player ensures that both players' subgraphs are isomorphic after each round, regardless of the first player's strategy.

F Y Reduction

This appendix discusses an interesting property of Y that may also have some implications for Hex. The diagrams and text are based on material from Jack van Rijswijck's excellent paper "Search and Evaluation in Hex" [2002].

F.1 Hex and Y

The difference between Hex and Y can be summarized as the difference between each player trying to connect his own goals as opposed to each player trying to connect all goals. Although Hex is probably the most conceptually pure connection game (especially for beginners) it can be argued that Y is the most fundamental.

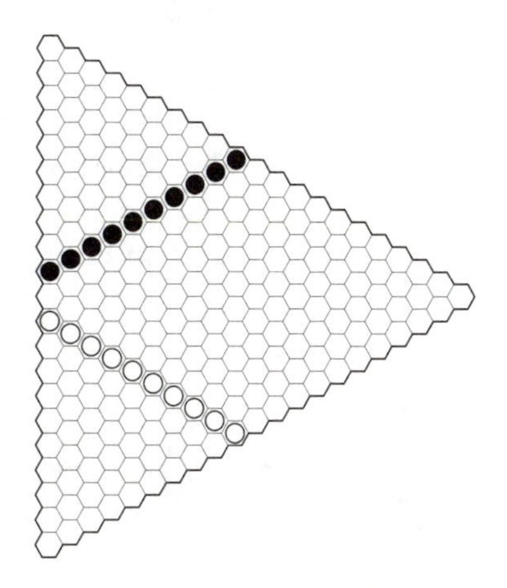

Figure F.1. Hex is a subset of Y.

This argument is based on the observation that a game of Y played on a board with pieces set as shown in Figure F.1 is then equivalent to a game of Hex. Black wins by connecting the line of black pieces with the parallel edge opposite, and White wins by connecting the line of white pieces with the parallel edge opposite it. A winning Y connection is not possible without also achieving a winning Hex connection.

In other words, the rules of Hex are a subset of the rules of Y, and any breakthrough regarding the game of Y should have some impact on our understanding of Hex. It is for this reason that the idea of Y Reduction is especially interesting.

Y Reduction comes in two dual forms: Macro Reduction and Micro Reduction.

F.2 Micro Reduction

Micro Reduction was discovered more than 30 years ago by Craige Schensted, one of the independent coinventors of Y. It is based on the following idea: *three pieces on coincident hexagons can be replaced by a single piece of the majority color.*

Figure F.2 shows all unique permutations of three pieces on coincident hexagons (ignoring rotations and reflections) and their Micro Reduced values.

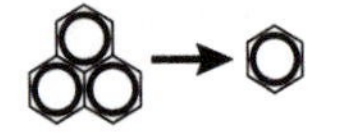

Figure F.2. The basic Micro Reduction steps.

A size n Y board completely filled with pieces can therefore be summarized by its size $(n-1)$ reduction, which can then be summarized by its $((n-1)-1)$ reduction, etc., until the size 1 reduction (a single piece) gives the winner of the game.

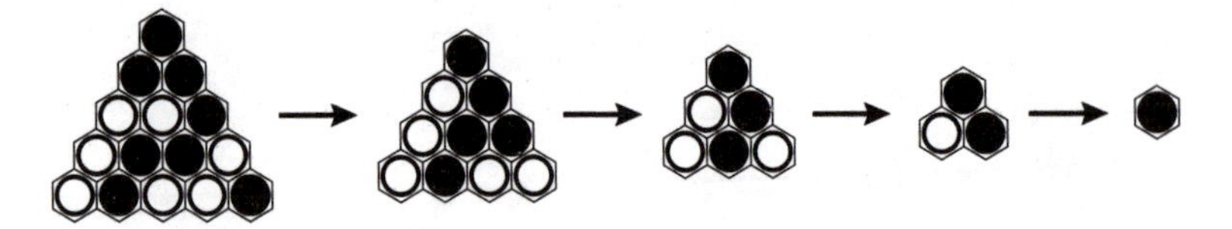

Figure F.3. A size 5 win for Black proven by Micro Reduction.

This process is shown in Figure F.3 where a size 5 board completely filled with pieces is successively reduced to a single black piece, hence Black has won the game. Micro Reduction will reveal the winner on any Y board completely filled with pieces.

F.3 Macro Reduction

Macro Reduction is the dual form of Micro Reduction. It was independently discovered by Steven Meyers in January 2002.

Suppose a game of Y is being played on a size n board (Figure F.4, top). This game may be described by three $n-1$ subgames, where each subgame is identical to the game with one row of edge cells removed (Figure F.4, bottom row). *If either player has won at least two of these three subgames, he has won the overall game.*

Figure F.4 shows this process applied to a completed size 5 game of Y. Of the three size 4 subgames derived from this board, the first two are won by Black. This corroborates the fact that Black has won the overall game.

This process can be applied recursively to each of the subgames to determine whether they have been won, until the size 1 subgame (a single piece) is reached. Obviously the owner of the piece has won the size 1 game. Hence any Y board can be decomposed to prove a win (or not) for either player. The board does not have to be full of pieces.

Note that the winners of the three Macro Reduced subgames correspond to the three cells of the penultimate stage of a Micro Reduction [van Rijswijck 2002].

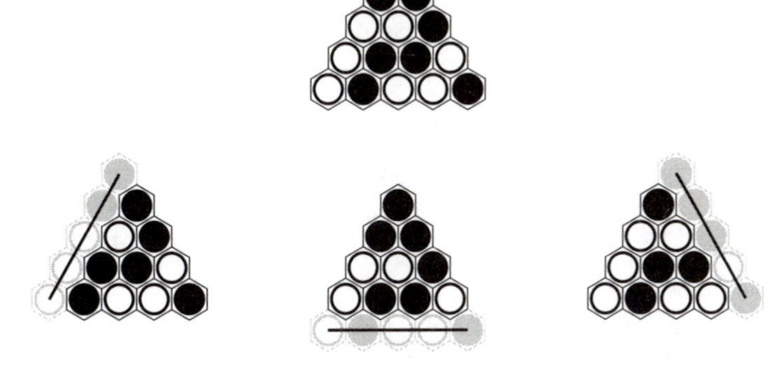

Figure F.4. Macro Reduction on a size 5 board.

F.4 Partial Reduction

Macro Reduction and Micro Reduction are interesting properties, but in themselves provide little information about a game beyond proving whether or not a player has won.

Jack van Rijswijck suggests a practical application [2002]. Given a partially board, a value is assigned to each board point as follows:

- Empty cell = 0,
- Black piece = −1, and
- White piece = +1.

These values are based on probabilities of cell ownership, but transposed to the range [−1, +1] to simplify calculations. The values of three coincident cells q_1, q_2, and q_3 can then be reduced to a single value q by

$$q = (q_1 + q_2 + q_3 - q_1 \times q_2 \times q_3) / 2$$

This reduction method generates a pyramid of values that can be evaluated quickly to yield a single value for a given game, indicating the favored player and the strength of the position. Van Rijswijck goes on to describe how each empty point can be evaluated as a possible move by calculating the partial derivative of the final evaluation based on each of the values in the reduction pyramid.

This method is novel and extremely fast, however, some experimentation has revealed that the final values only provide rough estimations. These estimations improve as the game converges to a solution, but further work is required before a serious Y-playing program can be implemented using this approach.

Sperner's Lemma

Contributed by Chris Hartman

A lemma first proven by Sperner [1928] is used to show that the game of Y always has a winner.

Consider a graph formed by a large triangle subdivided into smaller triangles. Such a graph is called *properly labeled* if the vertices are labeled with the numbers 1 to 3 such that each corner has a different number and one number is absent from the vertices along each side (in fact, each corner must be assigned the number missing from its opposed side). The internal vertices can be assigned any of the three numbers.

Sperner showed that a properly labeled triangle always contains a small triangle that has all three labels, called a *completely labeled triangle* (shaded in Figure G.1). A triangle's *dual graph* is used to prove that every properly labeled triangle contains a completely labeled triangle (Figure G.1, right). Dual graphs are described in Section 3.1.

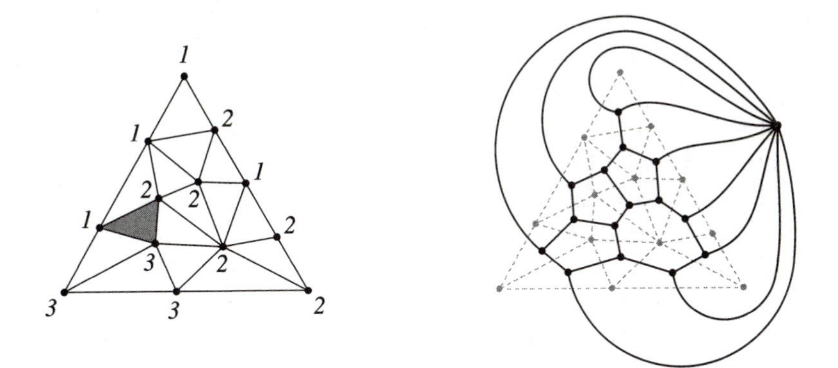

Figure G.1. A properly labeled triangle and its dual.

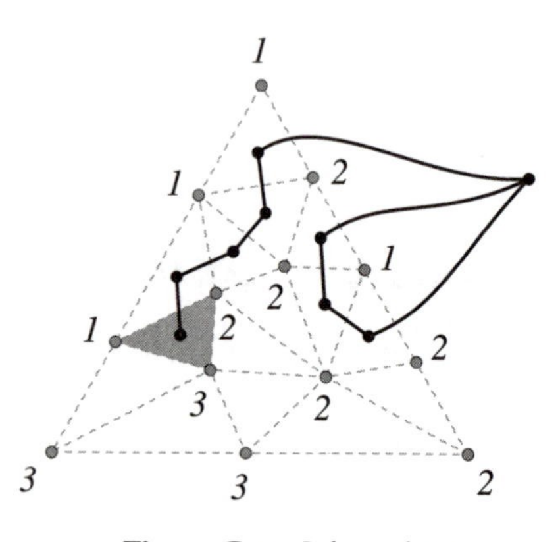

Figure G.2. Subgraph H.

Given a graph G that is properly labeled, we extract from its dual a subgraph H formed by edges (in the dual) that cross edges (in G) whose endpoints are labeled 1 and 2 (Figure G.2).

Consider the vertex (in the dual) corresponding to the outer face of G. Its degree in H must be odd, since edges from it must enter the large triangle on the side between the corners labeled 1 and 2, and there must be an odd number of changes from 1 to 2 (or back) along that side. Furthermore, since each edge in H contributes two to the total sum of degrees of vertices in H, this sum must be even. The vertex corresponding to the outer face has odd degree, so for the sum to be even, there must be at least one other vertex with odd degree. A vertex in H with odd degree corresponds to a triangle in G where an odd number of the edges have the labels 1 and 2 on the endpoints. The only such triangle must be completely labeled, as any combination of endpoints except {1, 2, 3} would not give an odd degree.

Sperner's Lemma can be used to show that the game of Y always has a winner, however, this must be done by contradiction. It will be assumed that there exists a Y board filled with pieces but without a winner; then it will be shown that such a situation leads to a mathematical contradiction. Thus, this situation cannot happen.

Label the three sides of the board with the numbers 1, 2, and 3. Then label each piece with the smallest label representing a side the piece *cannot* reach via a path of its own color. Since nobody has won the game, each piece has at least one side that it cannot reach, so a label exists for each piece. Furthermore, this labeling of the pieces gives a properly labeled triangle, since, for instance, each piece along side 1 is *not* labeled 1. According to Sperner's Lemma, there is a completely labeled triangle somewhere on the board. The vertices of this triangle must contain two pieces of the same color with different labels, but this is impossible, since two such pieces are adjacent and can therefore reach (via paths of their own color) exactly the same sides of the triangle.

This method of proving that the game of Y has a winner is given by Hochberg, McDiarmid, and Saks [1995]. A good introduction to Sperner's Lemma can be found in West [1996].

H Tessellations

Connection games are closely tied to the design of their underlying grid, more so than most other types of board games. The choice of grid has a dramatic impact upon a given game.

The standard Archimedean notation is used in the following examples; 3.4.6.4 means that four cells (a triangle, a square, a hexagon, and another square) meet at every vertex in that order.

Most connection games are played on a *regular* tessellation (a tessellation composed of a single type of regular polygon that fills the plane) as shown in Figure H.1. The hexagonal tessellation 6.6.6 is a common choice as it is trivalent, and games played on it are therefore safe from deadlock (see Appendix C).

Figure H.2 shows the eight *semiregular* tessellations, which are composed of two or more different polygons such that the same arrangement of polygons occurs around each corner.

The 4.8.8 grid is trivalent and hence is also a good choice for connection games. Quax and Stymie are played on this grid. The 3.12.12 and 4.6.12 tessellations are also trivalent, however, these are not used in any known

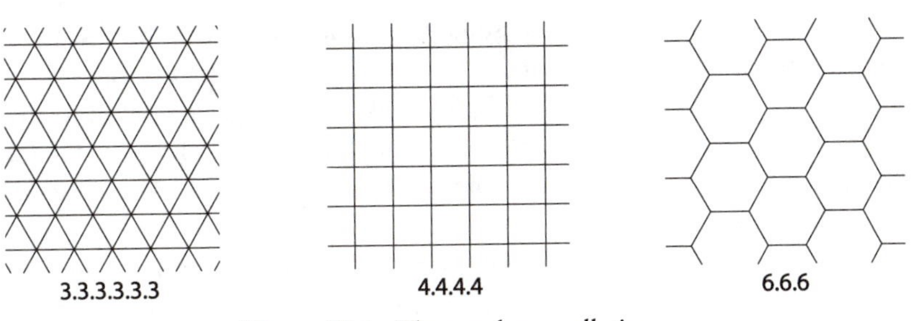

Figure H.1. The regular tessellations.

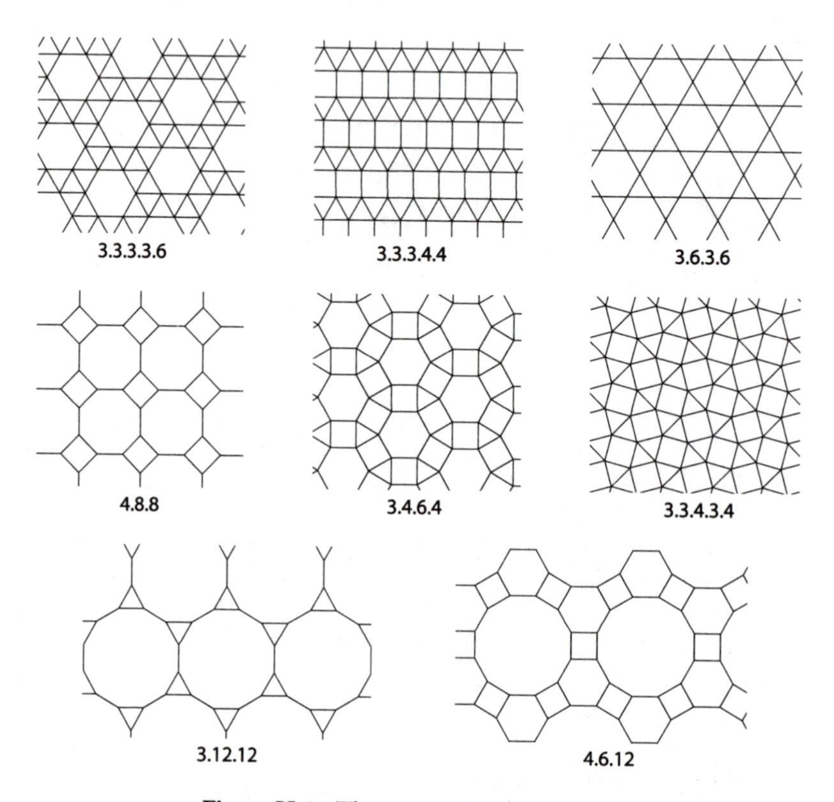

Figure H.2. The semiregular tessellations.

connection game. This is most likely due to the imbalance in cell sizes: playing in a dodecagon would be a much stronger move than playing in one of the lesser cells. Warp and Weft is played on a clever variant of 4.8.8 grid, reinterpreting it as a woven design.

Onyx is played on a variant of the 3.3.4.3.4 grid with connections across opposite corners of each square. Deadlocks are avoided through piece capture. Octagons is played on the dual of this variant tiling.

Grunbaum and Shephard [1987] also describe tilings according to the number of different polygons they include. The regular tilings in Figure H.1 are *monohedral,* while the tilings shown in Figure H.2 are *dihedral* and *tridhedral.*

So far only tessellations with exact edge meetings have been considered. However, if grid cells are offset relative to each other then the basic nature of the grid may change substantially.

Figure H.3. Offsetting a grid can change its basic nature.

For instance, Figure H.3 shows a regular 4.4.4 tessellation (left), a grid formed by a half-cell offset at each row making it topologically equivalent to the 6.6.6 grid (middle), and another grid formed by offsetting two out of every three rows (right). The right-most grid consists of three different arrangements of polygons around different corners (5.5.5/5.5.6/5.6.6) and has no regular equivalent.

Figure H.4 shows an interesting tessellation seen on a wall hanging in India. If each triangle is considered to consist of six half-unit edges, then three six-sided polygons meet at each corner; this grid is equivalent to the 6.6.6 hexagonal grid. It is possible to smoothly deform one into the other.

Moving now to *nonregular* polygons (that is, polygons with sides of nonuniform length), Figure H.5 shows an irregular tiling proposed by David Book [1998].

Figure H.4. An Indian tiling equivalent to 6.6.6.

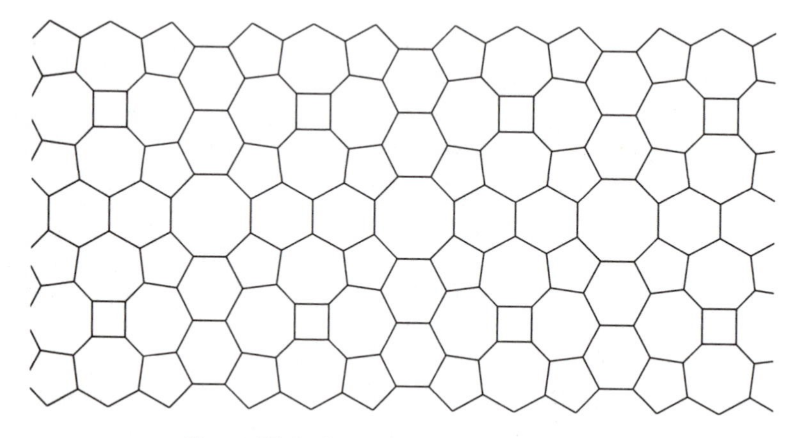

Figure H.5. An interesting irregular tiling.

Not only is this grid trivalent, but it consists of polygons with four, five, six, seven, and eight sides, allowing a wide variety of cell strengths without making one type too powerful compared to its neighbors. This tiling might make an interesting playing surface for a connection game.

The *phase space* of a tiling defines sets of cells relative to which pieces or connections are constrained. For instance, Figure H.6 (left) shows a square grid with a checkerboard pattern, defining two phase spaces. In the game of Chess, black bishops are constrained to the black phase space and white bishops are constrained to the white phase space. Knights may move to the closest nonadjacent cells of the opposite phase; bridge connections in Twixt are the connection game equivalent to a knight's move.

Figure H.6 (middle) shows the simplest phase space for the hexagonal grid, in which cells a bridge move away (that is, separated by a perpendicular edge) are in phase. Figure H.6 (right) shows a more complex four-phase pattern in which cells in phase are one cell removed from each other.

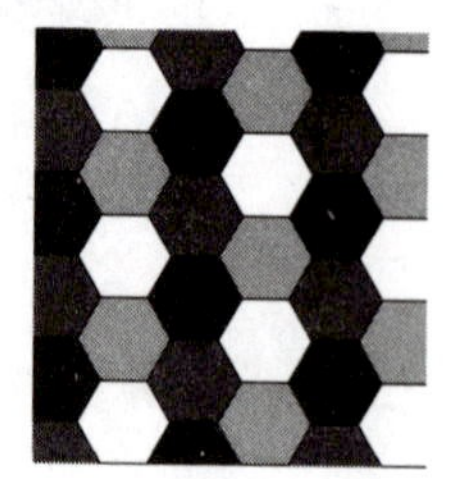

Figure H.6. Phase spaces.

References

Abbott, R. (1975) "Under the Strategy Tree," *Games & Puzzles*, 36, May, available at http://www.logicmazes.com/games/tree.html.

Abbott, R. (1988) "What's Wrong with Ultima," *World Game Review*, 8, 12–13.

Alpern, S. and Beck, A. (1991) "Hex Games and Twist Maps on the Annulus," *American Mathematical Monthly*, 98, 803–811.

Anshelevich, V. (2002) "A Hierarchical Approach to Computer Hex," *Artificial Intelligence*, 134, 101–120.

Back, L. (2000) "Onyx: An Original Connection Game," *Abstract Games*, 4, 9–12.

Back, L. (2001a) "Three Hex Variants," *Abstract Games*, 5, 24–25.

Back, L. (2001b) "Onyx: Strategy and Tactics," *Abstract Games*, 6, 22–27.

Bailey, D. (1997) *Trax Strategy for Beginners*, D. G. Bailey, Palmerston North.

Barton, W. I. (1939) "Board Game Apparatus," US Patent 2,162,876.

Beasley, J. D. (1989) *The Mathematics of Games*, Oxford University Press, Oxford, 141–143.

Beck, A. (1969) "Games", *Excursions Into Mathematics*, Eds. Beck, A., Bleicher, M., and Crowe, D., Worth, New York, 317–387.

Berlekamp, E. R., Conway, J. H., and Guy, R. K. (1982) *Winning Ways for your Mathematical Plays*, Volumes 1: *Games in General* and 2: *Games in Particular*, Academic Press, London.

Berlekamp, E. R. and Wolfe, D. (1994) *Mathematical Go: Chilling Gets the Last Point*, A K Peters, Natick, Massachusetts.

Bez, J. R. (1991) "Board Game," US Patent 5,058,896.

Binmore, K. (1992) *Fun and Games: A Text on Game Theory*, D. C. Heath, Lexington.

Book, D. L. (1998) "What the Hex," *The Washington Post*, September 9, H02.

Bovasso, B. X. (1972) "Board Game Apparatus," US Patent 3,643,956.

Bozulich, R. (1992) *The Go Player's Almanac*, Ishi Press, Tokyo.

Brosnan, M. (1998) "Spatial Ability in Children's Play with Lego Blocks," *Perceptual and Motor Skills*, 87, 19–28.

Brotz, G. R. (1988) "Board Game," US Patent 4,761,010.

Browne, C. (2000) *Hex Strategy: Making the Right Connections*, A K Peters, Natick, Massachusetts.

Browne, C. and Neto, J. (2001) "Gonnect: The Best of Go and Hex," *Abstract Games*, 6, 17–21.

Browne, C. (2001) "Hex Strategy: Computer Hex," *Abstract Games*, 8, 17–21.

Browne, C. (2003) "Akron: Connections in a Higher Dimension," *Abstract Games*, 14, 11–14.

Bruno, J. and Weinberg, L. (1970) "A Constructive Graph-Theoretic Solution of the Shannon Switching Game," *IEEE Transactions on Circuit Theory*, CT-17, 1, 74–81.

Bush, D. (2000) "Twixt Tactics: Part 1," *Abstract Games*, 4, 6–8.

Bush, D. (2001) "Twixt Tactics: Part 2," *Abstract Games*, 8, 14–16.

Christy, R. (1968) "Game Apparatus," US Patent 3,404,890.

Conway, J. H. (1976) *On Numbers and Games*, Academic Press, London.

Cornelius, M. and Parr, A. (1991) *What's Your Game? A Resource Book for Mathematical Activities*, Cambridge University Press, Cambridge.

Czerwinski, M., Tan, D. S., and Robertson, G. G. (2002) "Women Take a Wider View," *Proceedings of CHI 2002 Conference on Human Factors in Computing Systems*, ACM Press, 195–202.

Davies, H. E. L. (2002) "Board Game and Method of Playing," US Patent 6,460,856.

Dawkins, R. (1976) *The Selfish Gene*, Oxford University Press, Oxford.

De Speelstijl (2004) "Meander," http://www.meanderonline.nl/.

Deaton, C. U. (1978) "Method of Playing a Game," US Patent 4,078,805.

Dodge, W. L. (1963) "Board Game Apparatus," US Patent 3,075,771.

Doty, H. A. (1892) "Game," US Patent 471,666.

Estrin, G. (1970) "Board Game Apparatus with Path Forming Pieces," US Patent 3,516,671.

Evans, R. (1974) "A Winning Opening in Reverse Hex," *Journal of Recreational Mathematics*, 7:3, 189–192.

Evans, R. (1975) "Some Variants of Hex," *Journal of Recreational Mathematics*, 8:2, 120–122.

Even, S. and Tarjan, R. (1976) "A Combinatorial Problem Which Is Complete in Polynomial Space," *Journal of the ACM*, 23, 710–719.

Fleming-Davies, A. (2004) "Gender-Specific Toys: Increasing the Gender Gap?" http://www.stanford.edu/~zozo/gap/childhood/toys.html.

Frankel, R. (2000) "Rule Based Two/Three Dimensional Game," US Patent 6,120,027.

Freeling, C. (2003) "Havannah: Basic Tactics Part 1," *Abstract Games*, 15, 14–15.

Fudenberg, D. and Tirole, J. (1991) *Game Theory*, MIT Press, Cambridge.

Galdal, J. S. (1977) "Game Including Novel Board and Play Pieces," US Patent 4,047,720.

Gale, D. (1979) "The Games of Hex and the Brouwer Fixed-Point Theorem," *The American Mathematical Monthly*, 86:10, 818–827.

Galea, L. A. M. and Mimura, D. (1993) "Sex Differences in Route-Learning," *Personality and Individual Differences*, 14:1, 53–65.

Gardner, M. (1957) "Concerning the Game of Hex, which May be Played on the Tiles of the Bathroom Floor," *Scientific American*, 197:1, 145–150.

Gardner, M. (1959) "The Game of Hex," *Mathematical Puzzles and Diversions*, Penguin, Hammondsworth, 70–77. Reprint of his 1957 article with Addendum.

Gardner, M. (1963) "Four Unusual Board Games," *Martin Gardner's Sixth Book of Mathematical Games from Scientific American,* W. H. Freeman, San Francisco, 39–47.

Gardner, M. (1966) "Bridg-It and Other Games," *New Mathematical Diversions*, Simon & Schuster, New York, 210–218.

Gillam, B. (1984) "Aggregation and Unit Formation in the Perception of Moving Collinear Lines," *Perception*, 13, 659–664.

Goetz, R. H. (1962) "Game Board," US Patent 3,024,026.

Goldfarb, A. E. (1981) "Competitive Game Apparatus," US Patent 4,266,774.

Gómez, J. A., Ragnarsson, C. J., and Seierstad, T. G. (2002) "Unlur: Winner of the Unequal Forces Game Design Competition," *Abstract Games*, 12, 17–21.

Grunbaum, B. and Shephard, G. (1987) *Tilings and Patterns*, W. H. Freeman, San Francisco.

Hale-Evans, R. (2001) "Game Systems – Part I," *The Games Journal*, http://www.thegamesjournal.com/articles/GameSystems1.shtml.

Hallett, V. (2003) "The Game's Afoot at Henry, Teaching Toys Make Learning Fun at Alexandria School," *Funagain Games,* Kids' Games Section, http://kumquat.com/cgi-kumquat/funagain/marcia.

Handscomb, K. (2000) Afterword to "An Introduction to Twixt," *Abstract Games*, 2, 12.

Handscomb, K. (2001a) "Anchor: Redefining Life and Death," *Abstract Games*, 5, 6–7.

Handscomb, K. (2001b) "Octagons: Another Perspective on this Unusual Connection Game," *Abstract Games*, 7, 12–13.

Handscomb, K. (2001c) "8 x 8 Game Design Competition: Four More Games," *Abstract Games*, 8, 25–27.

Harary, F., Slany, W., and Verbitsky, O. (2002) "A Symmetric Strategy in Graph Avoidance Games," In *More Games of No Chance*, Ed. Nowakowski, R. J., Cambridge University Press, Cambridge, 369–381.

Hochberg, R., McDiarmid, C., and Saks, M. (1995) "On the Bandwidth of Triangulated Triangles," *Discrete Mathematics*, 138, 261–265.

How, M. (1984) "Beyond Hex," *GAMES*, July, 57–58.

Ivan, D. (1999) "The Origin and Meaning of Celtic Knotwork," http://www.thinkythings.org/knotwork/knotwork-meaning.html.

Joris, W. (2002) *100 Strategic Games for Pen and Paper*, Carlton, Dubai.

Keller, M. (1983) "A Taxonomy of Games," *World Game Review*, 1, 21.

Keller, M. (1984) "Lost and Little Known Games – Orion," *World Game Review*, 3, 21–24.

Kimura, D. (1992) "Sex Differences in the Brain," *Scientific American*, 267:3, 118–125.

Kimura, D. (1999) *Sex and Cognition*, MIT Press, Cambridge.

Koch, K. (1991) *Pencil & Paper Games*, Sterling, New York.

Kramer, W. (2000) "What Makes a Game Good?" *The Games Journal*, http://www.thegamesjournal.com/articles/WhatMakesaGame.shtml.

Lagarias, J. and Sleator, D. (1999) "Who Wins Misère Hex?" In *The Mathemagician and Pied Puzzler: A Collection in Tribute to Martin Gardner*, Eds. Berlekamp, E. and Rodgers, T., A K Peters, Natick, 237–240.

Lehman, D. (1964) "A Solution of the Shannon Switching Game," *Journal of the Society for Industrial and Applied Mathematics*, 12, 687–725.

Levy, L. (2002) "Strategy & Tactics," *The Games Journal*, http://www.thegamesjournal.com/articles/StrategyTactics.shtml.

Maire, F. (2004) "The Solution of Shannon Game," http://members.optusnet.com.au/cameronb/connection.games/maire-shannon-game.doc.

Mala, M. (1995) *Das Große Buch der Block- und Bleistift-Spiele*, Hugendubel, Munich, 170.

Masin, S. C. (2002) "Absolute and Relative Effects of Similarity and Distance on Grouping," *Perception*, 31, 799–811.

McGaughey, W. H. T. (1970) "Game Apparatus with Cards Played in Alignment Across a Board," US Patent 3,512,779.

McMurchie, T. (1979) "Squiggle Game," US Patent 4,180,271.

McNamara, T. (1993) "Board Game," US Patent 5,269,531.

McWorter, W. A. (1981) "Kriegspiel Hex," *Mathematics Magazine*, 54:2, 85–86.

Mercat, C. (1997) "Les entrelacs des enluminures celtes," *Dossier Pour La Science*, 15, available at http://www.entrelacs.net/.

Meyers, S. (2001a) "New Connection Games: Part One," *GAMES*, October, 67.

Meyers, S. (2001b) "New Connection Games: Part Two," *GAMES*, November, 68, 77.

Meyers, S. (2002) "Orbit: A New Game of Territory," *Abstract Games*, 12, 22–23.

Milnor, J. (2002) "The Game of Hex." In *The Essential John Nash*, Eds. Kuhn, H. W. and Nasar, S., Princeton University Press, Princeton, 29–33.

Minty, G. J. (1978) "Apparatus for Games," US Patent 4,067,577.

Morse, H. F. (1976) "Game Board Apparatus," US Patent 4,032,151.

Neto, J. (2000) "Mutators," *The Chess Variant Pages*, http://www.chessvariants.com/newideas.dir/mutators.html.

Neto, J. (2004) "The World of Abstract Games," http://www.di.fc.ul.pt/~jpn/gv/index.htm.

Novak, L. and Gibbons, A. (1999) *Hybrid Graph Theory and Network Analysis*, Cambridge University Press, Cambridge.

Panda, R. D. (1994) "Method of Playing a Board Game by Forming a Sequence of Words from Start to Finish," US Patent 5,324,040.

Parlett, D. (1999) *The Oxford History of Board Games*, Oxford University Press, Oxford.

Pierce, J. R. (1961) *Symbols, Signals, and Noise: The Nature and Process of Communication*, Harper Collins, New York.

Pierson, D. H. (1972) "Board Game Apparatus," US Patent 3,655,194.

Polczynski, J. (2001) "Lightning: A Connection Game from the 1890s," *Abstract Games*, 5, 8–9.

Reilly, R. (2004) "Connection Games and War," http://members.optusnet.com.au/cameronb/connection.games/reilly-war-games.doc.

Reisch, S. (1981) "Hex ist PSPACE-vollstandig," *Acta Informatica*, 15, 147–151.

Richards, M. D. (1978) "Suspension Game," US Patent 4,071,244.

Rudden, T. J. (1980) "Path Forming Game," US Patent 4,190,256.

Rum, W. M. (1996) "Game Board Game and Method Playing the Game," US Patent 5,553,854.

Ryan, S. A. (1974) "Game Board and Game Pieces Positionable upon the Board in a Limited Number of Positions," US Patent 3,804,415.

Sackson, S. (1969) *A Gamut of Games*, Random House, New York.

Sarrett, P. (1983) "Pipeline," *The Game Report Online*, http://www.gamereport.com/tgr3/pipeline.html.

Schaper, W. H. (1965) "Game Successively Utilizing Selectively Positionable Gear Playing Pieces Varying in Pitch Radii," US Patent 3,193,293.

Schensted, C. and Titus, C. (1975) *Mudcrack Y and Poly-Y*, NEO Press, Peaks Island.

Schmittberger, R. W. (1983) "Star: A Game is Born," *GAMES*, September, 51–54.

Schmittberger, R. W. (1992) *New Rules for Classic Games*, John Wiley & Sons, New York.

Schmittberger, R. W. (2000) "Making Connections," *GAMES*, June, 10–13, 44, 58–61.

Scott, W. L. (1938) "Educational Game," US Patent 2,177,790.

Serbin, L. A. and Connor, J. M. (1979) "Set-Typing of Children's Play Preferences and Patterns of Cognitive Performance," *The Journal of Genetic Psychology*, 134, 315–316.

Shannon, C. E. (1955) "Game playing machines," *Journal of the Franklin Institute*, 206:December, 447–453.

Shimizu, T. (1980) "Bridge-Linking Table Game," US Patent 4,226,421.

Shoptaugh, P. L. (1972) "Board Game Apparatus," US Patent 3,695,615.

Soriano, R. (1981) "Chain Reaction Falling Playing Pieces Board Game," US Patent 4,248,433.

Sperner, E. (1928) "Neuer Beweis fur die Invarianz der Dimensionszahl und des Gebietes," Abhandlungen aus dem Mathematischen Seminar der Hamburgischen Universitat, 6:3/4, 265–272.

Spertus, E. (1991) "Why Are There So Few Female Computer Scientists?" MIT Artificial Intelligence Laboratory Technical Report 1315, http://www.ai.mit.edu/people/ellens/Gender/pap/pap.html.

Straffin, P. D. (1985) "Three Person Winner-Take-All Games with McCarthy's Revenge Rule," *The College Mathematics Journal*, 16:5, 386–394.

Taylor, W. (2002) "Strategy and Tactics," posting to rec.games.abstract newsgroup, 30 May.

Thompson, J. M. (2000) "Defining the Abstract," *The Games Journal*, http://www.thegamesjournal.com/articles/DefiningtheAbstract.shtml.

Tracy, D. M. (1990) "Toy-Playing Behavior, Sex-Role Orientation, Spatial Ability, and Science Achievement," *Journal of Research in Science Teaching*, 27:7, 637–649.

Van Rijswijck, J. (2002) "Search and Evaluation in Hex," University of Alberta Technical Report, http://www.cs.ualberta.ca/~javhar/research/y-hex.pdf.

Van Rijswijck, J. (2004) "Hex," http://www.cs.ualberta.ca/~javhar/hex/.

Vasel, T. (2004) "Games in Korea," *The Games Journal*, http://www.thegamesjournal.com/articles/GamesInKorea.shtml.

Waroway, R. M. (2001) "Geometrically Patterned Tiles and Game," US Patent 6,305,688.

Weber, J. H. (1972) "Game Structure Employing Markers and Links," US Patent 3,695,616.

Weisstein, E. W. (1999) *CRC Concise Encyclopedia of Mathematics*, CRC Press, Boca Raton, Florida.

West, D. B. (1996) *Introduction to Graph Theory*, Prentice-Hall, Englewood Cliffs, New Jersey.

Yaeger, M. J. (1984) "Apparatus for Playing Game," US Patent 4,466,615.

Zermelo, E. (1912) "Uber eine Anwendung der Mengenlehre und der Theorie des Schachspiels," *Proceedings of the Fifth International Congress of Mathematicians*, Cambridge University Press, Cambridge, 501–504.

Note: All page numbers in the *List of Games* and in the *Index* need to be adjusted by minus 12.

List of Games

Akron 131
  Dipole 137
  Lakron 138
Alta 275
Anchor 321
Andantino 44, 301
Antipalos 179
Antipod 43,138
  Escape Hex 43, 143
Apex 274
Beeline 97
Bez's Game 283
Black Path 383, 211
  A Winding Road 214
  Squiggle Game 214
Block 328
Blokus 328
Bridg-It 16, 27, 170, 383
  September 173
  Web 170, 173
Caeth 39, 156
  Anu 160
  Cha 160
  Det 160
  Hex 157
  Noc 131, 160
  Y 157
Canoe 336
Chameleon 52, 54, 122, 241
Chex 337
Chikadou 349
Chinese Checkers 348
Conhex 125, 363
  Pula 130
Creeper 285
Crosstrack 265
Deaton's Game 311
Druid 60, 66, 149, 362
  Ryan's Game 154
  Span 155
Durch die Wüste
    See Through the Desert
Écoute-moi! 276
Eynsteyn 271
Fan 282
Feurio! 329
Gaia 122
Galdal's Game 310
Gale See Bridg-It,.
Go 17, 22, 27, 315
  Alak 282
  Gonnect 38, 49, 63, 204, 350, 363
    Three-Player 208
  One-Capture Go 317
Groups 291
Hamilton 351
Havannah 16, 46, 51, 193

Hex 16, 20, 30, 48, 60, 68, 80, 344, 348, 358, 363, 379, 381, 385
  Beck's 84
  Bipod 88
  Caeth 157
  Chameleon. *See* Chameleon.
  Cylindrical 43, 99, 382
  Diskelion. *See* Triskelion
  Eight-Sided Hex 249
  Escape 43, 143
  Head Start 86
  Kriegspiel 86
  Map 33
  Misère 85
  Nex. See Nex
  Pass 87
  Quadrant 247
  Scorpio 87
  Square 86
  Tex 85
  Three-Player 53, 85
  Vertical Vex 85
  Vex (1975) 85
  Vex (2004) 88
Hexa 224
HexGo 328
Hexma 348
Hextension 311
Iron Horse. *See* Metro
Jade 50, 118, 382
Kage 284
Kaliko 298
  Psyche-Paths 299
Knots 46, 233
Lazo, 257
Le Camino 310
Lightning 16, 185
  Hex 188
Lines of Action 27, 288
  Grupo 291
  Neighbors 290
  Slides of Action 290
  Twirls of Action 290
  Zen L'initié 290
Link 282
Lotus 338
Lynx 165
McGaughey's Game 280
Meander (1982) 268
Metro 312
Moloko 292
Mongoose Den 307
Morse's Game 281
Network 215
  Constellations 218
  Estrin's Game 218
  Killer Beams 218
Nex 162
Nexus 313
Nile 308
Notwos 285
Occupier 328
Octagons 394
Octagons 101
Octiles 286
  Team Up 286
Onyx 269, 343, 394
Orbit 323
Orion 263
  Banana Boat 287
  Hydra 265
Panda's Game 283
Parcel 329
Pathfinder 326
Pipeline 230
Poly-Y 16, 244
Projex 43, 251
Proton 273
Puddles 312
Quax 108, 393
  See also ZeN,.
Quintus, 278, 343

Quoridor 335
Ringelspiel 255
  Orthospiel 257
Rivers, Roads & Rails 308
Round the Bend 200
Rum's Game 326
Rumis 314
Schlangennest 312
Sisimizi 284
Snap 314
Speleo 306
Split (1966) 95
Split (1999) 285
Square 31
Square Board Connect 285
Square Off 281
Stak *See* Steppe/Stak.
Star 16, 295, 363
  *Star 298
  Maxi-Star 298
  SuperStar 298
Steppe/Stak 326
Stymie 146, 393
Suspension Game 281
Symbio 322
Tack 282
Take It Easy 332, 333
Tanbo 319, 349
  Three-Dimensional 321
Tantrix 300
Tara 282
Ta Yü 303
The Great Downhill Ski Game 308
The Legend of Landlock 314
The Very Clever Pipe Game 304
Thoughtwave 65, 219
  Barton's Game 224
  Davies's Game 224
  Pista 224, 237
Through the Desert 327
Tic-Tac-Toe 21, 337
Trails to Tremble By 309
TransAmerica 313
Trax 16, 22, 44, 59, 195
  8 x 8 200
  Chameleon 200
  Connexion 199
  LoopTrax 199
  Tantrix. *See* Tantrix
Trellis 105
TriHex2003 143
Trinidad 38, 272
Triskelion 238
  Diskelion 240
Troll 283
Turn 266
  Schaper's Game 267
Turnabout 227
Twixt 26, 174, 396
  Diagonal 178
Unlur 114
  Misère 118
Visavis 181
Waroway's Game 325
Warp and Weft 167, 394
Waterworks 309
Weave 209
Weber's Game 310
Worm 318
  Blue Nile 319
  Goldfarb's Game 319
  Snail Trail 319
  Snake Pit 319
Würmeln 311
Y 16, 41, 49, 50, 56, 62, 89, 363
  Caeth 157
  Fifty-Fifty 93
  Fortune 93
  Gem 94
  Holey 94

Master 93
Obtuse 94
Poly-Y *See* Poly-Y
Tabu 94
Why Not (Misère) 95
Yaeger's Game 299
ZeN 43, 112
Three-Player 113
Zig-Zag 16, 188
Criss-Cross 191
Four-Handed 192

Index

adjacency 19, 26, 367
  and neighbors 20, 26
anticonnection games 50
  Black Path 214
  Canoe 337
  Puddles 312
Beck, Anatole 100, 375, 377, 382
Berge, Claude 375
Berlekamp, Elwyn 29, 316
board design 38, 40
  corner effects 41
  cylinder 42, 99
  hemisphere 42, 143
  plane 43
  projective 43, 252
  recursive 247
  scale 41
  sphere 42, 140
  surface types 42
  torus 43, 112
broadsides 69
capture 72, 337, 345
  by replacement 275, 289, 291, 323
  Hamiltonian 351
  mandatory 208
  One-Capture Go 317
  pattern-based 270
  piece-surround 64, 158, 165, 205, 209, 282, 305, 315, 324, 328, 334, 350
  surround-flip 282, 338
Celtic knotwork 353
chain 20
children's games 200, 236, 299, 308, 314, 333, 361
clarity 45, 47, 51, 194, 235, 345, 347
  and depth 45, 235
  and strategy 46
classification. *See* taxonomy
cold war 54, 121, 242, 260
connection 20
  anticonnection. *See* anticonnection games
  global 72, 263
  local 72, 295
  quality 21, 331
  versus pattern 21
  virtual 58, 63, 81, 116, 128, 135, 136, 159, 184, 190, 194, 220
  vulnerable 64, 103, 152, 246
Connection-Related games
  Connection Quality 331
    Empire Building 334
    Rail Network 334
    War Games 334

Form a Pattern 336
N-in-a-Row 337
Limit the Opponent 337
Reach a Goal 335
Maze Games 335
Connective Goal games
Convergent 286
Path Making 263
Connective Play games
Path Making 295
Territorial 315
cycle 368
deadlocks 31, 53
and trivalency. *See* trivalent
defense 64
and attack 64, 79, 105, 178, 185, 347, 361
leave-two-spaces rule 222
depth 45, 233, 235, 345, 347
design. *See* board design or game design
dice games 15, 71, 188, 201, 267, 281, 292
domino
subdivision 383
tumbling 333
edge templates 82
efficiency 67, 127, 136, 152, 159, 199, 249
first-move advantage 47, 380. See also swap option
first-move equalizer. *See* swap option
flow 60, 303, 346, 356, 358, 370
Gale, David 84, 172, 377
game
game graph 28
game tree 28, 45
game design
and metarules 343
case studies 348
clarity. *See* clarity
convergence 347
correctness 344
decisiveness 346
depth. *See* depth
drama 345
evaluating a game 344
interaction 346
quality 345
gaming system
Orion 263
Gardner, Martin 16, 84, 87, 191, 381
goals 48
complementary 48
equivalent 49
multiple 50
symmetric 48
tie breakers. *See* rules, tie breakers
types 48
unequal 49
graph 367
adjacency graph 26
avoidance games 384
connected 368
disconnected 368
dual 25, 389, 394
edges 25, 367
face 367
game graph 28
nonplanar 43, 158, 176, 179, 197, 213, 218, 228, 256, 367
planar 32, 39, 42, 103, 131, 141, 355, 367
properly labeled 389
vertices 25, 367
grid
Archimedean notation 393
hexagonal 26, 38, 393
offset 395
phase spaces 170, 180, 210, 239, 260, 396

square 26, 393
three-dimensional 131, 149, 257, 321, 362
triangular 26, 393
trivalent. *See* trivalent
Hamiltonian path 351, 370
Hein, Piet 16, 81, 83
Hex and Y 16, 90, 385
hex hex board 37, 52, 344
history 16
timeline 17
Klein bottle 43
ladder 62, 68, 81, 93, 124, 137, 243, 255, 302
breaker. *See* ladder escape
escape 63, 81, 93
ludemes 37, 343, 347
map 33, 280, 313
coloring 84, 126, 344
interpretation 33, 354
matroid 371
memes 37, 348
ludemes. *See* ludemes
memetic convergence 348
metarules 38
Caeth 39, 156
either-color 39
misère 39, 44, 50, 85, 95, 118, 352
Nex 40, 164
no-pass 38, 205
Quax 40, 112
mobility 67, 349
Mobius strip 43
momentum. *See* tempo
move
cold 121, 206, 242, 260, 314, 318
double 87
forcing 56, 61, 62, 68, 83, 117, 129, 135, 152, 164, 194, 243, 260, 302
fork 58, 64, 93, 104, 128, 142, 152, 160, 196, 302
hot 52, 206
killer 61, 112, 136, 154, 194, 302
multiple 47, 93
order 93
transformer 47
multiple players 52
kingmaker effect 53
petty diplomacy 53, 114, 233
three-player games 52, 208, 238
mutators 38
Nash, John 16, 83, 191, 379
Nash equilibrium 379
neighbors 20, 26, 31, 375
no-tie property 176, 250
Hex 84, 249
Y 387, 389
overlap. *See* connection, vulnerable
path 368
pattern 21, 336
Picture Link 236
pie rule. *See* swap option
psychology
educational role 361
gender bias 359
Gestalt 356
perceiving connection 356
Pure Connection games 17, 73, 79
Absolute Path 79
Absolute Path:Cell-Based 80
Absolute Path:Edge-Based 170
Cycle Making 251
Path Majority 244
Path Race 184
Path Race:Separate Paths 185
Path Race:Shared Path 211
Reduction
Macro 134, 387
Micro 386
Partial 388

Y 158
rules
and metarules 38
convergence of rule sets 347
McCarthy's Revenge 53
Mover Loses 52
Mover Wins 52
optimizing 343
over/under 132, 138, 150, 154
Stop-Next 53, 239
tie breakers 51
sacrifice 68, 167, 199
Schensted, Craige 16, 41, 87, 90, 93, 130, 247, 298, 299, 386
Schmittberger, R. Wayne 17, 53, 67, 105, 247, 298, 318, 339, 343, 345, 347, 363
Shannon, Claude 16, 27, 93, 172, 371
Shannon game 27
Cut/Join 28, 32
Join/Join 32
on the edges 28
on the vertices 30
solution 371, 383
Shannon switching game. *See* Shannon game.
Sperners Lemma, 389
strategy 55, 58, 345
and tactics 41, 55, 265
stealing 84
symmetry 321, 382
swap option 47, 62
extended 116
inherited 157
single-move 47, 81
three-move 47, 106, 248, 252
tactics 55
and strategy 41, 55, 265
taxonomy 73
Connective Goal. *See* Connective Goal games
Connective Play. *See* Connective Play games
Pure Connection. *See* Pure Connection games
temperature. *See* move, cold or move, hot
tempo 62, 129, 152, 194
territory 22, 68, 127, 152, 206, 318, 337. *See* Connective Play games: Territorial
staking out 66, 128, 136
tessellation. *See* also grid
Indian 395
irregular 396
regular 393
semiregular 394
three-dimensional games. *See* grid, three-dimensional
three-player games. *See* multiple players
tiling. *See* grid or tessellation
trivalent 32, 35, 393
truchet tiles 268
winning conditions. *See* goals
Zermelo, Ernst 379